P9-DWJ-696

END OF DAYS

ALSO BY JAMES L. SWANSON

*Bloody Crimes: The Funeral of Abraham Lincoln and the
Chase for Jefferson Davis*

Manhunt: The 12-Day Chase for Lincoln's Killer

END OF DAYS

THE ASSASSINATION OF JOHN F. KENNEDY

JAMES L. SWANSON

WILLIAM MORROW
An Imprint of HarperCollins*Publishers*

Photograph credits: pages 5, 38, 88, 124, 126, 131, 139 courtesy of the National Archives; pages 36 (by Robert Knudson), 35, 56, 92, 159, 175, 284 (by Cecil Stoughton), 89, 91, 96, 99, 109, 179, 203, 243, 245, 248, 250 all courtesy of the John F. Kennedy Presidential Library and Museum; pages 100–101 and 141 © by Tom Dillard Collection, *The Dallas Morning News*/The Sixth Floor Museum at Dealey Plaza; pages 247, 251, 252, 254 (by Abbie Row), courtesy of the Mary Ferrell Foundation; pages 255, 256, 257, 258, 259, 260, 261 are from the author's collection.

Map on page 111 by Nick Springer/© 2013 by Springer Cartographics LLC.

Front endpaper image © by Mark Shaw/mptvimages.com; back endpaper image © by Time & Life Pictures/Carl Mydans/Getty Images.

END OF DAYS. Copyright © 2013 by James L. Swanson. All rights reserved. Printed in the United States of America. No part of this book may be used or reproduced in any manner whatsoever without written permission except in the case of brief quotations embodied in critical articles and reviews. For information address HarperCollins Publishers, 10 East 53rd Street, New York, NY 10022.

HarperCollins books may be purchased for educational, business, or sales promotional use. For information please e-mail the Special Markets Department at SPsales@harpercollins.com.

FIRST EDITION

Designed by Jamie Kerner

Library of Congress Cataloging-in-Publication Data has been applied for.

ISBN 978-0-06-208348-7

13 14 15 16 17   DIX/RRD   10 9 8 7 6 5 4 3 2 1

For my wife, Andrea.
And for my father, Lennart.

I have a rendezvous with Death . . . And to my pledged
word am true, I shall not fail that rendezvous.

—ALAN SEEGER, HARVARD GRADUATE, AMERICAN WORLD WAR I
POET KILLED AT THE BATTLE OF THE SOMME ON JULY 4, 1916,
AND AUTHOR OF ONE OF JOHN F. KENNEDY'S FAVORITE POEMS.

We are all mortal.

—JOHN F. KENNEDY, AMERICAN UNIVERSITY, JUNE 10, 1963.

If somebody wants to shoot me from a window with
a rifle, nobody can stop it, so why worry about it?

—JOHN F. KENNEDY, FORT WORTH, TEXAS,
ON THE MORNING OF NOVEMBER 22, 1963.

I should have known that he was magic all along.
I did know—but I should have guessed that it would
be too much to ask to grow old with [him] and see
our children grow up together. So now, he is a legend
when he would have preferred to be a man.

—JACQUELINE KENNEDY, NOVEMBER 17, 1964.

CONTENTS

LIST OF ILLUSTRATIONS   xi

A NOTE TO THE READER   xiii

Prologue   1

1. "Such Dangerous Toys"   3
2. "The Glow from That Fire"   23
3. "Show These Texans What Good Taste Really Is"   55
4. "A Bright Pink Suit"   84
5. "Someone Is Shooting at the President"   121
6. "They've Shot His Head Off"   144
7. "I Haven't Shot Anybody"   169
8. "We Had a Hero for a Friend"   201
9. "Lee Oswald Has Been Shot!"   230
10. "One Brief Shining Moment"   266

Epilogue: "All His Bright Light Gone from the World"   288

ACKNOWLEDGMENTS   301

BIBLIOGRAPHY   305

SOURCE NOTES   321

INDEX   383

LIST OF ILLUSTRATIONS

1. Lee Harvey Oswald, posing with his rifle   5
2. John F. Kennedy, Lyndon B. Johnson,
   and Martin Luther King Jr.   35
3. President Kennedy in Berlin   36
4. Oswald as a high school student   38
5. President Kennedy, Caroline, and John Jr.
   in the Oval Office   56
6. Oswald's blanket and rifle   88
7. Kennedy with Johnson and Governor Connally   89
8. Kennedy outside of the Hotel Texas   91
9. Kennedy speaking by the Hotel Texas   92
10. Jackie and JFK disembark Air Force One   96
11. Jackie with roses at Love Field   99
12. The Kennedys and Connally
    in the Dallas motorcade   100–101
13. The presidential motorcade proceeds through Dallas   109
14. Map of Dealey Plaza   111
15. Oswald's view of the limousine   124
16. Zapruder's camera   126
17. Side and rear views of bullet   131
18. Kennedy's second wound   139

19. The sixth-floor Texas Book Depository window 141
20. The presidential limousine at Parkland Hospital 159
21. Carrying the casket aboard Air Force One 175
22. Lyndon Johnson being sworn in 179
23. President Kennedy's casket in the East Room 203
24. The funeral procession at the United States Capitol 243
25. Jacqueline and children at the Capitol 245
26. The procession to St. Matthew's Cathedral 247
27. The caisson leaves for St. Matthew's 248
28. The coffin is carried to Arlington 250
29. Panoramic view of Arlington 251
30. Folding the flag on the coffin 252
31. The Lee mansion at Arlington Cemetery 254
32. Mourning buttons 255
33. Memorial banner #1 256
34. Memorial banner #2 257
35. Dealey Plaza desk set 258
36. The electric Eternal Flame night-light 259
37. Kennedy tribute magazine 260
38. Oswald magazine 261
39. Chief Justice Earl Warren and his commission 284

A NOTE TO THE READER

It is hard to believe it happened fifty years ago. Even in the current generation, it remains a great dividing line in American history. It was the greatest national trauma since the assassination of Abraham Lincoln at the end of the Civil War in April 1865.

Those who were alive in the fall of 1963 and were old enough then to remember it today shudder at the mere mention of the date. Find anyone you know over the age of sixty—a parent, grandparent, or friend—and speak one phrase: "November 22, 1963." At once they will tell you where they were and what they were doing when they heard the news. "I remember," they will say and then tell you their story. Many still shed tears for him.

The drama could have been scripted by Shakespeare: The elevation of a young hero scarred by war and haunted by death. This charismatic leader of his people—beside his beautiful wife, whose natural poise captured the nation—is struck down at the height of his power and fame, leaving behind a young widow to conjure his legend. It is a tale filled with irony and foreshadowing, with heroes and monsters.

This book attempts to re-create a moment when time stopped. It seeks to recapture how Americans lived through this tragedy and to resurrect the mood and emotions of those unforgettable days between President John F. Kennedy's murder and his funeral. To those of you who remember, I hope this book does justice to your memories. To those who do not and who know—or think they know—the story only through retellings in books or films, I hope this book evokes the mood, the loss, and the emotional and historical truth of the fall of 1963.

Our misguided modern-day obsession with exotic, multiple, and contradictory conspiracy theories involving tales of grassy knolls, umbrella men, magic bullets, second gunmen, Oswald imposters, doctored films, fraudulent photographs, and all-powerful government cover-ups has caused us to lose the emotional connection to the events of November 1963. We have strayed too far from the human truths of that day. A wife lost her husband. Two children lost their father. A nation lost a president.

This is not meant to be a complete history of the Kennedy assassination, if such a book could ever even be written. Nor does it travel down the rabbit hole to endorse or rebut any of those complicated webs of conspiracies. Except for some brief observations, I leave that journey to others. This book is no more than a modest attempt to tell a story, to restore the balance, to travel back in time to the days before it happened, and to return to the day when we did not know what would happen next.

John Kennedy's days ended on the sunny afternoon of a brilliant fall day in Dallas, Texas, on November 22, 1963. It was a day when the death of one man caused a nation to weep. Half a century later, Americans refuse to forget him. We mourn him still.

END OF DAYS

PROLOGUE

Historic Georgetown, in the northwest section of Washington, D.C., is one of the most prestigious neighborhoods in the nation's capital. Founded as a commercial trading hub along the banks of the Potomac River before the American Revolution, it boasts the oldest home in Washington, along with many other eighteenth-century brick town houses. Over the past two centuries, many distinguished Americans—congressmen, senators, judges, federal officials, military officers, authors, and one man who would become president—have called Georgetown home. In its long history, no public figure has been more connected to the neighborhood than John F. Kennedy, the thirty-fifth president of the United States.

He lived in several homes there beginning in 1946, during the time when he gained national recognition as a U.S. congressman and senator, married Jacqueline Bouvier, and in November 1960 was elected president. Between his election and his inauguration on January 20, 1961, the circa-1811 three-story Federal-era brick town house at 3307 N Street NW, his last home in Georgetown, served as a nerve center where Kennedy hired staff and planned for his forthcoming administration, which he called the New Frontier. Even before the election, the photographers Mark Shaw and Jacques Lowe had already made the house an iconic symbol of the Kennedy style.

For two and a half months, the lights inside the house burned late as the president-elect gathered his closest advisers around him. Journalists camped outside the home to photograph or film Kennedy. They were there whenever he opened the front door, stood on the top step, and walked down to the public sidewalk a few feet in front of the house to announce the appointment of a new person to his administration.

On January 19, the night before his inauguration, a heavy snowstorm paralyzed the nation's capital. The glow from the pair of clear glass and black metal lamps flanking John Kennedy's front door made the crystals layering the surface of the deep snow twinkle in the night. Undeterred by the weather, the Kennedys ventured out to attend the long-planned pre-inauguration parties, including the Democratic Gala, scheduled for 8:45 P.M. at the National Guard Armory.

Jacqueline Kennedy wore a shimmering white satin floor-length ball gown that mirrored the soft, thick snowy carpet that covered the capital. Photographs captured her as she walked through her front door and stepped into the night: surrounded by darkness, she shone as bright as a glimmering star. The next morning, John and Jacqueline Kennedy left their town house for the last time and embarked on a journey he would not complete, from which he would never return.

CHAPTER 1

"SUCH DANGEROUS TOYS"

He had been planning the assassination even before he bought the rifle. He had chosen his victim, scouted the location, written detailed notes in his journal, drawn maps and diagrams, even photographed the building. By the time he was ready to do it, his notebook bulged with all the information needed to carry out a successful political murder—black-and-white surveillance photos, ideas on where to hide the rifle before the attack, the best spot for his sniper's nest, estimates of the range to the target, the distance between the building and some railroad tracks nearby.

He had also planned his escape. After the shooting, he would conceal his rifle in a hiding place near the tracks and then flee on foot. He had little choice about the method of escape. He did not own a car. He did not even know how to drive. But the notebook contained all the vital information necessary to make a clean getaway on public transportation—the location of bus stops in the vicinity and, most important, the exact times that the buses would make those stops.

Yes, he planned to coordinate an assassination with a bus schedule. He would have to kill his victim, run away, hide the rifle, and arrive at the stop just before the bus approached. Timing was everything. It would be too risky to stand at a corner for several minutes, waiting for the next bus. He would be the first assassin in American history to try to escape the scene of the crime on a city bus. It was crazy. But it was crazy enough—and so unexpected—that it might work. He figured the police would never suspect or even look for an unarmed passenger riding a bus.

His target was famous but vulnerable. Aides or bodyguards could not protect the man from a clean shot with a rifle. The murder was sure to cause a national sensation and create headline news not only in Dallas, Texas, but throughout the country. The shooter had always wanted to make big news. If he could just get the target in his telescopic sight, it should work.

At first, he had thought he might use a pistol, the assassin's traditional weapon of choice for slaying an American politician. He had already bought one by mail order more than two months ago. Back on January 28, 1963, he sent an order form and a $10 down payment to Seaport Traders in Los Angeles for a $39.95 Smith & Wesson .38 special, a double-action, six-shot revolver with a five-inch barrel cut down to a stubby two and a quarter inches. The shortened barrel made the weapon easier to conceal in a pocket or a shoulder holster. He had signed the form with the false name A. J. Hidell, and he would owe the balance of the purchase price upon delivery.

But a handgun was a dangerous and even suicidal choice. You had to get close to the victim to use it, and his target was a combat veteran who had served in the military. He might fight back. He thought better of it. No, using a handgun was out of the question. Six weeks later, on March 12, 1963, he mailed a postal money order to Klein's Sporting Goods in Chicago and, under a false name, ordered an Italian, World War II military surplus 6.5mm Mannlicher-Carcano bolt-action rifle mounted with a 4-power telescopic sight.

By coincidence, the revolver and the rifle both arrived in Dallas on March 25, 1963. He picked up the rifle at his post-office box, which he had also obtained under a false name. When his wife saw the weapon, she was afraid. "What do you need a rifle for?" she asked. "What do we need that for?" Her husband told her a woman like her was incapable of understanding why a man like him needed to own a rifle.

A few days later, on the last weekend in March, he had her take photographs of him in their backyard brandishing his new rifle and wearing the pistol on his belt. "I was hanging up the diapers, and he came up to me with the rifle, and I was even a little scared, and he gave me the camera and asked me to press a certain button." She had never taken a photograph, and he had to show her how to operate

Lee Harvey Oswald poses with his rifle, pistol, and left-wing newspapers.

his cheap Imperial Reflex camera. "I asked him then," she said, "why he dressed himself up that like, with the rifle and the pistol." He was wearing all black—black shirt, pants, and boots. He had never dressed like that before. He assumed a few different, jaunty poses. "I thought he had gone crazy," she recalled.

He told her he wanted to send the photographs to a newspaper. He held two newspapers in his hand while she took the photographs—copies of the left-wing journals the *Worker* and the *Militant*. She could not understand why he would call attention to himself that way. He said that he wanted to show that he was "ready for anything." Then he told her it was not her concern—that it was "man's business." After he had developed photographic prints—for privacy he did the work himself at the printing company where he worked—he gave one to his wife and said that she should keep it for their infant daughter, June. "Of course," she thought, "June does not need photographs like that."

She hoped that he was just "playing around" and that this was just another example of her husband's immaturity. "If I had known these were such dangerous toys, . . ." she mused, after it was too late.

She had an intuitive sense of foreboding. After her husband acquired the rifle, she "knew that [he] was preparing for something." He ordered her to keep out of his private room, a little home office he had set up in their new apartment. He did not want her to find his secret notebook. He threatened to beat her if he ever caught her in there. "This is my own little nook. I've never had my own room before," he lamented. It wasn't much of a room, measuring less than four by five feet.

For a man who had shared a bed with his mother until he was ten or eleven years old, the nook, no matter how modest, represented the privacy and independence he craved. "I'll do all my work here," he boasted to his wife, "make a lab and do my photography. I'll keep my things in here. But you're not to come in here and clean. If I ever come in and find that one single thing has been touched, I'll beat you."

She knew he meant it—he was an unpredictable and violent man who had beaten her up before. He was twenty-three years old, she was twenty-one, and they had been husband and wife for only two years. But he was not the same man she had met and married in her native Russia. Now he felt free to insult, humiliate, or hit her at will. Not long before, he struck her so hard her nose bled. When he left the house, she bolted the locks. When he returned and found that he could not get in, he smashed a glass panel, reached in, and unlocked the door from inside.

"You know I am a terrible character," he had admitted to her recently. "When you see I'm in a bad mood, try not to make me mad. You know I can't hold myself in very long now."

She was furious and wanted him to know he had not beaten the pride out of her. "You weak, cowardly American. What a fool I was. I was afraid to marry a Russian because Russian men beat their wives. You! You're not worth the soles of their feet. How I wish I had woken up sooner!"

He threatened her. "I'll *make* you shut up."

She was not cowed. "Of course you can shut me up by force. But you'll never change my mind. It's better to be a drunkard than what you are."

His rage had climaxed the previous month, on February 23, 1963. In the morning, he made a special request for dinner that night. He told his wife to make him a meal of Southern beans and rice. She had never cooked that dish before and could not find a recipe. After he got home from work, he blew his top when he discovered that she had prepared the beans and rice together in the same pot. Didn't she know, he demanded, that she was supposed to cook the two ingredients in separate pots and then pour the beans on top of the rice? What difference, she asked, did it make? He was going to mix the beans and rice together on his plate anyway. He ordered her to start over and make the dinner a second time.

"I won't."

"You will."

"I won't." Then she threw the dinner she had made into the trash can.

"I'll force you to." He raised a hand to strike her.

"You have no right."

He hit her anyway. Then he turned his back and began to walk away. She retaliated by throwing a small wooden box at him. It grazed his shoulder.

He turned, charged, and threw her onto the bed. His hands encircled her throat. His eyes and face froze into a cold, murderous expression. "I won't let you out of this alive." She was sure he was going to strangle her. Then their baby began to cry.

"Go get her," he told her, as he released his grip from her neck.

"Go get her yourself."

This attack had broken her spirit. Now she felt ashamed and worthless. She went into the bathroom. She took the clothesline she used to dry her daughter's diapers. She tied the rope around her neck. He wouldn't have to kill her. She would do it for him.

He caught her in the act before she could finish. He carried her back to the bedroom. "Forgive me," he pleaded. "I didn't mean to do what I did." Then he mouthed the classic wife beater's defense. "It's *your* fault. You saw what a mood I was in. Why did you make me so mad?"

She said she had only tried to do what he had just tried to do to her. "I'm sick of it, Alka." That name, along with Alek, was one of the names she called him when they had lived in Russia. His American first name sounded very odd—Asian—to Russian ears.

"Every day we fight, and for no reason," she complained. "We fight over things so tiny, normal people wouldn't speak of them at all."

"I never thought you'd take it so hard," he replied. "Pay no attention to me now. You know I can't hold myself back." It was not his fault, he was suggesting, that he could not control his temper. "Try to understand, you're wrong sometimes, too"—as if blaming her justified the slaps and punches and choking. "Try to be quiet when you can. For God's sake, forgive me. I'll never, ever do it again. I'll try and change if you'll only help me."

"But why, Alka, why do you do it?"

"Because I love you. I can't stand it when you make me mad."

And their neighbors could not stand him when he got mad. One of them once reported him to the building manager, Mahlon Tobias, complaining, "I think he's *really* hurt her this time." On another occasion a neighbor predicted, "I think that man over there is going to kill that girl." The building's owners showed up to confront him with an ultimatum. Cut it out, or get out, they told him. He did not tell his wife about the warning. Instead, a few days later, he told her they were moving. He claimed he had found them a better apartment at 214 West Neely Street. They moved there on March 2, 1963.

Now, in their new apartment, not long after he had promised to never hit her again, he threatened to beat her if she ever intruded into his tiny private room.

SHE KNEW he had been practicing with the rifle. She had never seen him fire it, but when he went out by himself, he would sometimes say, "Well, today I will take the rifle along for practice." He had bought at least one small cardboard box of 6.5mm bullets manufactured by the Western Cartridge Company. He knew his way around a rifle. His training in the United States Marine Corps had taught him all he needed to know about that.

On one occasion, she had discovered in their apartment photographs of a strange house. What could it mean? "I asked him what kind of photographs are these?" He refused to answer.

He could never hold a job for long. On April 1, he was fired from his current one at the Jaggars-Chiles-Stovall printing company. He had hurried through his assignments, made careless mistakes, and had to redo too many jobs. He was brusque with coworkers, refusing to sit with them at lunch and often shouldering them aside in the darkroom. When his supervisor gave him his notice, all he did was stare at the floor, listen without protest, and reply, "Well, thanks." Then he just walked away.

He needed the money, but he did not beg to keep his job. He did not even seem to care. He had already gotten what he wanted—he had used the office equipment to make himself fake identification cards and had developed prints of the photographs of him posing with his rifle. It was as though his mind was distracted by something more important than holding a job. Saturday, April 6, was his last day at work. Within the next four days, he would cross the line from malcontent to madman.

On the morning of Wednesday, April 10, he put on a gray suit and white shirt, the kind of outfit a man looking for a job might wear. After waiting a week and a half to tell her, he admitted to his wife that he had been fired. "I don't know why," he whined. "I tried. I liked that work so much." He did not reveal that he had been let go for poor performance and a bad attitude. He offered a grander excuse that shifted the blame from himself to others. "But probably the FBI came and asked about me, and the boss just didn't want to keep someone the FBI was interested in. When will they leave me alone?" She accepted his explanation. She knew the government had had an ongoing interest in them since they had moved from Russia to the United States.

When he left the house that morning, his wife assumed he would spend all day looking for work. Whenever he needed a job, it was his habit to buy a newspaper, scour the want ads for leads, and walk the streets from employer to employer.

But he did not look for jobs that day. He had only one appointment to keep, and he was not due there for another ten or twelve hours. How he spent the day remains a mystery. He was in no hurry to reach his destination, the house at 4011 Turtle Creek Boulevard, in one of the better neighborhoods in town. Long ago he had decided against a daytime attack. It was too risky: Too many people, too many potential witnesses, too great a chance that the police might hunt him down. No, he would strike under the cover of darkness.

He arrived at his destination a little before nine P.M. CST. He went to the spot where he had secreted his rifle a few days earlier. It was still there—he had hidden it well. He took his position. Everything had gone as he had planned. No one saw him tote the weapon to the house. Once again, he placed the residence under surveillance, just as he had done several times before on his practice missions. He walked around to the back of the house. No one saw him crouch down. He knew that the man was in Dallas. And he was sure that at this moment his victim was somewhere inside that house on Turtle Creek Boulevard. The man he had marked for death had been in command of thousands of American soldiers, but tonight they could not protect him. He was alone.

AT THIS moment, President John F. Kennedy was in the White House, 1,200 miles away. The day before, in a 3:00 P.M. televised ceremony in the flower garden, he had conferred honorary American citizenship upon Winston Churchill. It was the first time such an honor had been given to a foreign statesman. Ancient Sir Winston, too frail to travel to the United States, watched the event on television at his home in London, via a live satellite broadcast.

After the ceremony, John and Jacqueline Kennedy hosted a reception for 250 guests in the East Room. Earlier this day, April 10, Kennedy had addressed a gathering of economics students from abroad, sent a memo to place veterans of the Peace Corps in other government jobs, and proposed the creation of a National Service Corps. Soon he would retire to bed on the second floor in the mansion's residence.

But on this night, Lee Harvey Oswald did not care where the president of the United States was. He did not want to kill John F. Kennedy.

The man Oswald wanted to kill—Major General Edwin A. Walker, United States Army—was less than forty yards away.

Oswald's fixation with Walker probably began when the October 7, 1962, issue of the Communist Party newspaper, the *Worker*, published not one but three inflammatory articles demanding that President Kennedy "arrest and prosecute General Edwin Walker . . . for murder and rebellion against the democratic authority of the Constitution of the United States."

After a distinguished military career—West Point graduate, decorated World War II and Korean War combat officer—Walker received orders from President Eisenhower in 1957 to quell civil disturbances during the desegregation of Central High School in Little Rock, Arkansas. Although he did a superb job of protecting the black students, Walker thought it was wrong to use the United States Army to flout states' rights and desegregate the South against its will. He became obsessed with the idea that Communists had infiltrated the U.S. government, and he later accused Harry Truman and Eleanor Roosevelt of being "pink"—the hot-button word for communist sympathizers. President Kennedy relieved him from command, and Walker resigned his commission. He became a political activist, giving provocative speeches throughout the nation.

On September 30, 1962, the day after Walker made virulent televised remarks opposing the admission of a black student—James Meredith—to the University of Mississippi, a campus riot left two dead and seventy-five wounded. After Walker was charged with insurrection and the violation of several federal laws, Attorney General Robert Kennedy overreached and tried to commit the general to a mental institution. The *Worker* reported that Walker "led the brick-throwing students" and warned that the racial violence at Ole Miss was the "first attempt by a conspiracy of racists and ultra-Right generals to renounce the federal constitution, to open the path to tyranny over the negro people . . . and the subjection of the entire nation to a fascist type rule."

In February 1963, Edwin Walker was in the news again when he gave a speech that seemed to advocate that the United States invade Cuba and assassinate Fidel Castro. It was too much for Lee Oswald. As soon as his guns arrived in the mail, he would show that he was "ready for anything." His justifications for the assassination came straight out of the pages of the *Worker*. Edwin Walker despised President Kennedy. In trying to kill Walker, Oswald was attempting to assassinate one of JFK's bitterest political enemies. Subsequent developments would expose the irony in that.

• • •

MOST OF the lights in the house were on. The shades were up.

Oswald looked through the windowpanes, searching for General Walker.

Then Lee spotted him, clear as day in a brightly lit room, sitting at a desk while he worked on his federal income-tax return. The deadline was in five days. Oswald raised his rifle. He leveled the barrel to the horizontal and took aim. He acquired the general in his telescopic sight. He superimposed the crosshairs over his head. He was so close. Walker was no more than 120 feet away.

Oswald steadied his shooting stance, verified the target in his scope, and squeezed the trigger. He was certain he could not miss. The man's head was dead center in the crosshairs.

The rifle fired. The supersonic bullet cracked the air.

BY AROUND nine P.M., Marina Oswald began to worry. Her husband should have returned home by now. She knew he must be up to no good. She decided to search their apartment for some clue about where might have gone. Despite his past threats, she decided to search his private office. "Then I went into his room. Somehow, I was drawn into it. . . . I was pacing around. Then I saw the note." It was in his handwriting, and it was in Russian, her native language. That meant it was for only her to understand. She still could not speak or write English well. To keep her dependent on him, Oswald had discouraged her from learning the language. They always spoke to each other in Russian.

A single key sat on top of the note. She moved the key, picked up the sheet of paper, and began reading. The note contained eleven numbered paragraphs. At first the words confused her. The first point mentioned his post office box. "This is the key," Lee disclosed, "to the mailbox which is located in the main post office . . . on Ervay Street." Yes, the key was there. He had left it for her just as he had written. But why? And why was he telling her how to locate his post office box?

The second paragraph was more mysterious: "Send the information as to what has happened to me to the Embassy and include the newspaper clippings." Marina didn't know what her husband was talking about. What was he doing tonight that a newspaper might write about? The paragraph continued. "I believe that the Embassy will come to your assistance on learning everything." Marina had no idea what those words meant.

The third through fifth points dealt with petty financial concerns: "I paid the house rent on the 2d so don't worry about it . . . recently I also paid for water and gas . . . the money from work will possibly be coming . . . to our post office box . . . go to the bank and cash the check."

The sixth through eighth points covered Lee's personal property: "You can either throw out or give my clothing . . . away. Do not keep these. However I prefer that you hold on to my personal papers . . . certain of my documents are in the small blue valise . . . the address book can be found on my table in the study."

The ninth point reminded Marina that "we have friends here," implying that a small circle of Russian-born émigrés living in Dallas would help take care of her. Then he added that "the Red Cross also will help you."

The penultimate point was about money. Lee Oswald had provided a meager sum to care for his family: "I left you as much money as I could, $60 . . . you and the baby . . . can live for another two months using $10 per week."

Yes, there was no doubt now. Lee Harvey Oswald had abandoned his wife and infant daughter.

Then Marina got to the eleventh point and her husband's final, bizarre instruction.

She could not believe it.

This was not a farewell note from an estranged husband. This was a note from a man who must have done something terrible.

"If I am alive and taken prisoner . . . the city jail is located at the end of the bridge through which we always passed on going to the city (right in the beginning of the city after crossing the bridge)."

These handwritten instructions advising Marina what to do in case Lee vanished or was killed or arrested gave no hint about the awful things he had planned to do that night—what he might be doing right now while she was reading the note. She had no clue and did not know what to do.

So she just waited for him to come home.

THE BULLET was in midflight, rocketing toward General Walker's head. He had survived two wars. He had commanded an infantry regiment in battle. Now he was about to be murdered in the tranquility of his own home. "It was right at 9 o'clock," Walker noted, "and most of the lights were on . . . and the shades were up. I was sitting down behind a desk . . . with my head over a pencil and paper working on my income tax when I heard a blast and a crack right over my head."

The bullet, traveling at 1,700 or 1,800 feet per second, had punched through the window glass and missed Walker's head by less than an inch.

Oswald had missed! But he had not fired an inaccurate shot. He did have his target in his sights. The general should be dead. But on its way into the house, the bullet nicked part of the wood window frame, which altered its flight path just enough to save the general's life.

At first, Walker did not suspect gunfire. "I thought that possibly somebody had thrown a firecracker, that it exploded right over my head through the window right behind me. Since there is a church back there, often there are children playing back there." Walker assumed that kids had tossed the fireworks through an open window missing its screen, but he was wrong. "Then I looked around and saw that the screen was not out, but was in the window, and this couldn't possibly happen, so I got up and walked around the desk and looked back where I was sitting and I saw a hole in the wall."

It did not take the general long to figure out that somebody had shot at him.

Lee Harvey Oswald could not tell if his first shot had killed Walker. The Mannlicher-Carcano had a six-round clip. Oswald could have operated the bolt, ejected the empty brass cartridge, chambered a fresh round, and fired again. But he did not attempt a second shot. Either he convinced himself he had killed the general—how could he miss at that range?—or he was too skittish to stand his ground and fire a second round. He decided to run away.

It was a wise choice. When Walker realized he had been fired upon, he did not duck for cover, cower on the floor, and hide. Instead he made a snap decision to hunt for the sniper.

"I noticed there was a hole in the wall, so I went upstairs and got a pistol and came back down and went out the back door, taking a look to see what might have happened."

Walker barged outside and headed in Oswald's direction. But Lee was already running away. Walker searched the area—"I went about halfway out to the alley"—but he found no one. If Oswald had lingered in the vicinity, hoping to fire multiple shots, Walker might have caught him in the open and shot him to death.

Instead, Lee's legs increased the distance between him and the pursuing general. Oswald fled into the night and hid his rifle in the same spot where he had concealed it prior to the assassination attempt. He didn't want to get caught with the weapon in his possession on the way home.

In a few days, when it was safe, he planned to go back and retrieve the rifle from its hiding place. For now, Oswald made his way to a bus stop and rode home. At this moment, after months of careful planning, he did not know whether he had succeeded or failed in killing Edwin Walker.

MARINA OSWALD waited for Lee to come home—ten thirty, eleven, eleven thirty. Where was he? What was he doing?

Before midnight, Lee burst into the house. He was in an excited and agitated state. His face was pale. He was sweating.

Marina spoke first. "What happened?"

"I shot Walker!"

"Did you kill him?" Marina asked.

"I don't know."

"My God," Marina said. "The police will be here any minute."

She noticed that Lee was unarmed. "What did you do with the rifle?"

He told her he had buried it.

Marina waved the note he had left her and demanded, "What is the meaning of this?" Lee did not want to talk anymore. After his confession, "He told me not to ask him any questions. He only told me that he had shot at General Walker."

Lee said that he did not know whether or not he had killed Walker. Then he revealed how he had fled the crime scene. He told Marina he had run "several kilometers" and then caught a bus home. To avoid being spotted by eyewitnesses, he decided not to stand at a bus stop too close to Walker's house.

Within a minute or two of returning to his apartment, Lee Oswald switched on the radio. He listened, hoping to hear a news flash about the attack and confirmation that the general was dead. Marina listened with him, "But there were no reports." Not one radio station broadcast a bulletin about the Walker shooting. That was odd. The assassination of an army general as well-known as Edwin A. Walker was sure to make the news.

Oswald gave up on the radio and collapsed in bed. He fell into such a deep sleep, Marina thought he looked dead.

On the morning after the shooting, Oswald bought a newspaper. The *Dallas Morning News* headlined its article, CLOSE CALL: RIFLEMAN TAKES SHOT AT WALKER. It was then that Lee learned the bad news. He had missed. He had failed to murder General Walker. The paper reported that the general had survived and that chance had saved his life: "A gunman with a high-powered rifle tried to kill former Maj. Gen. Edwin A. Walker at his home Wednesday night," police said, and "missed the controversial crusader by less than an inch."

Oswald had not missed by much. "Walker dug out fragments of the shell's jacket from his right sleeve," the story went on, "and was still shaking glass and slivers of the bullet out of his hair when reporters arrived."

The newspaper suggested that Walker's would-be assassin had fled by automobile. "It was on Monday night [April 8] that one of his assistants noticed a late-model unlicensed car parked without lights in the alley behind the Walker house . . . the car remained there for about 30 minutes while several occupants once walked up to the back door to look in, and then left." Perhaps the men had the general under surveillance.

And on the night of the assassination attempt, a fourteen-year-old neighbor boy said he saw two suspicious cars—one with a driver only and the second with several occupants—race away from the scene. But the witnesses were wrong—Oswald was alone.

When reporters asked Walker if he had any idea who shot at him, he replied, "There are plenty of people on the other side. You don't have to go overseas to earn a Purple Heart. . . . I've been saying the front was right here at home." This anticommunist crusader would have been humiliated to have been slain by a malcontent ex–U.S. Marine who had defected to the Soviet Union and was a disciple of Karl Marx. And how ironic it would have been for this arch foe of overreaching government authority to be murdered while in the act of calculating his federal income tax.

Oswald complained aloud to Marina as he read the newspaper coverage: "Americans are so spoiled! It never occurs to them that you might use your own two legs. They always think you have a car. They chased a *car*. And here *I* am sitting *here*," he boasted. "My legs had carried me a long way."

"They got the bullet," Lee informed his wife. "They said I had .30 caliber bullet when I didn't at all." He enjoyed mocking the police.

"They've got the bullet and the rifle all wrong. Can't even figure *that* out. What fools!"

He replayed the shooting in his head. "It was such an easy shot. How on earth did I miss? A single second saved him. I fired and he moved. A perfect shot if only he hadn't moved!" But the newspaper story got that detail wrong. The police had asked Walker if he had made any sudden movement. "None that I was aware of," he said. "No. Just moving with a pencil and thoroughly engrossed in my income tax." It was the wood frames of both the mesh screen and glass window that had saved Walker.

"I had it so well figured out," Oswald repeated to Marina. "I couldn't make a mistake. It was only accident that I missed."

Marina pressed her husband for more details of what he had done. "I told him that I was worried, and that we can have a lot of trouble, and I asked him, 'Where is the rifle? What did you do with it?' " Lee confirmed that he had hidden it a safe distance from Walker's home, so that gunpowder-sniffing police dogs could not locate it by its scent. Lee justified the attack, saying that Walker was "a very bad man . . . a fascist [and] . . . the leader of a fascist organization." He added that if someone had killed Hitler, it would have saved many lives.

Oswald admitted that he had been stalking Walker and planning his assassination for two months, even before he ordered the rifle. Yes, Marina thought to herself, it all made sense now. She *knew* he had been planning something. He would lock himself in his room and write in a notebook. Now Lee showed her his blueprint for killing Walker. Tucked inside it was a photograph of Walker's house. "And what," Marina asked, "do you mean to do with this book?"

"Save it as a keepsake. I'll hide it somewhere." Like many compulsive criminals, Lee Oswald desired to possess a souvenir of his crime. In the future, whenever he fondled the cherished notebook, he could relive the thrilling assassination attempt.

"Some keepsake," Marina warned. "It's *evidence*! For God's sake, Alka, destroy it."

For once, Lee did as she asked.

Marina watched him tear out each page one by one, crumple it, and light it on fire with a match. Then he dropped each crumpled, burning page into the toilet and flushed away his murder plans. Marina continued to chastise her husband. "A rifle—that's no way to prove your ideas," she insisted. "If someone doesn't like what you think, does that mean he has a right to shoot *you*? Once people start doing that, no one will dare go out of doors." She reminded Lee that when they lived in Russia, he boasted that in America there was freedom of speech, where "everyone can say as he pleases."

Lee was disappointed by his failure. "He said," Marina recalled, "only that he had taken very good aim, and that it was just chance that caused him to miss." Oswald emphasized that he was "very sorry" that he had missed. Marina was terrified. What if her obsessive husband decided to pay another call on General Walker? She begged Lee to promise that he would not make another attempt on Walker's life. Fate, she claimed, had spared him, and "he should not be shot at again." Marina threatened to save Lee's note and turn it over to the police if he tried to slay the general again or pull off any other, in her words, "crazy scheme."

Lee promised he would not. She did not quite believe him. He had promised to get rid of the Mannlicher-Carcano, but Marina observed later that "the rifle remained in the house."

Oswald was guilty of attempted murder, but Marina was afraid to report him to the authorities. If he were arrested, tried, and sentenced to prison, it would destroy their family. And what would happen to her? Marina Oswald was not an American citizen. She was afraid the government would also punish her or deport her to Russia. In the Soviet Union, the family members of criminals were often held responsible for the offenses of their relatives. Marina decided to keep Lee's secret not only to protect him but to safeguard her daughter and their life in the United States.

•   •   •

ON SATURDAY, April 13, three days after the shooting and the night before Easter, the Oswalds received unexpected visitors. It was their friend George de Mohrenschildt and his wife, Jeanne. George was an intelligent, well-educated, and exotic figure on the Dallas social scene who was acquainted with the local Russian émigré circle and had befriended Marina and Lee. Oswald enjoyed having political discussions with the fifty-two-year-old bon vivant and world traveler. People who knew them both could not understand the connection between the sophisticated European with aristocratic origins and the young, lower-class member of the Southern working poor. Lee viewed George as something of a mentor.

"Hey, Lee," George bellowed as he approached the front door, "how come you missed?"

Lee and Marina froze. They were horrified. Their eyes locked. Each suspected the other of having confessed Lee's crime to George. But George had been joking.

The Oswalds invited the de Mohrenschildts in, and both couples sat down. A nervous Lee served coffee to his guests. The assassination attempt fascinated George, and he wanted to talk about it. Lee said little, until finally he volunteered, "Oh yes, wouldn't it be fascinating to know who did it and why and how?"

The next day, on Easter Sunday, Oswald retrieved his rifle.

MARINA WAS afraid to remain in Dallas. She suggested that Lee go to New Orleans to look for work. He knew the city and had family there. Once he found a job, she promised, she and the baby would join him. But Marina had an ulterior motive: "I insisted on that because I wanted to get him further removed from Dallas and from Walker, because even though he gave me his word, I wanted to have him further away, because a rifle for him was not a very good toy—a toy that was too enticing."

Shortly after that, Oswald decided to follow Marina's advice to leave Texas and move to New Orleans. He took his rifle with him.

From New Orleans, Lee mailed a photograph to George de Mohrenschildt. It was a print of one of the photos Marina had snapped of Lee in their backyard with his rifle and pistol. On the plain white reverse of the photo were two handwritten messages. They were in Russian. One read "Hunter of Fascists—ha-ha-ha!!!" The handwriting was Marina's. Perhaps it was a humorous, mocking reference to George's joke that it was too bad Lee had missed General Walker. The second inscription, dated May 4, 1963, was in Lee's handwriting: "For my friend George from Lee Oswald."

OSWALD'S ATTEMPTED assassination of General Walker was a psychological turning point. It was the first time he had ever tried to kill a man. He had joined the Marine Corps in peacetime and had never fought in a war. Shooting a rifle at a human being was a novel thrill. It had excited him. Yes, he had missed his target, but planning and executing the sniper attack had given him immense satisfaction. He had enjoyed it. He had proven, in his own words, that he was "ready for anything."

The experience had taught him a valuable lesson. In Dallas he could shoot at a man and get away with it. Although Lee had botched his self-assigned mission, the failure of the police to catch him emboldened him and enhanced his smug attitude of superiority.

Now the taste for blood was in his mouth. He did not know it now in late April 1963, but in a little more than seven months, he would find another human target. But next time, Lee would not have to spend weeks stalking his victim. The next man who appeared in his rifle scope would come to him.

CHAPTER 2

"THE GLOW FROM THAT FIRE"

In the spring of 1963, John F. Kennedy had come a long way in the last decade. He had been what was once called, back in the day, a "confirmed bachelor." He enjoyed the social scene and had been known for his associations with a number of attractive women. But in that era, it was considered odd—especially for a politician—to be unmarried.

In May 1951, the journalist Charlie Bartlett and his wife, Martha, hosted a dinner party at their Georgetown home and invited their friend Congressman Kennedy. They also invited an attractive young woman named Jacqueline Lee Bouvier, a stylish, educated, and well-bred twenty-one-year-old debutante who had, like him, been raised to assume a place among the wealthy American elite. Jackie once teased Charlie and Martha for "shamelessly matchmaking." It was the beginning of a love story. The courtship began slowly, but Jack and Jackie fascinated each other. She liked his handsome features, wit, confidence, and devil-may-care attitude. He was drawn to her good looks and sophistication and was intrigued by her private and mysterious personality.

In September 1953, at the age of thirty-six, John Kennedy married Jacqueline Bouvier. Their wedding in Newport, Rhode Island, was the social event of the season. However, soon his lifelong health problems began to haunt him once again. He had been plagued by painful back injuries in his youth, and an incident during the Second World War had exacerbated the problem. Kennedy had narrowly escaped death in the Pacific when, on August 2, 1943, at 2:30 A.M., his boat, PT-109, was rammed and sliced in two by a Japanese warship. Two crewmen were killed, others were injured, some of them severely. One sailor was so badly burned, he could not swim. Jack clenched in his teeth a strap attached to the man's life jacket and towed him to a nearby island. He and his men were eventually rescued. The incident had tested his physical and mental endurance. Now Jack faced death again. He underwent a series of agonizing surgeries to cure the problem. He was given the last rites and almost died, but he struggled to live, and he survived.

IN JANUARY 1960, Kennedy declared himself a candidate for the Democratic nomination for president of the United States. He outmaneuvered his older and more experienced rivals, including Senator Lyndon Baines Johnson of Texas, the powerful Senate majority leader, and won the nomination.

The Republicans nominated former senator Richard Nixon of California, who had for the past seven years served as vice president of the United States under the popular president Dwight Eisenhower, the victorious supreme allied commander during the Second World War. Kennedy and Nixon shared many similarities. Both had served as navy officers in the war, both had been elected to Congress in 1946, and then to the Senate. There they worked cordially with each other, shared an interest in foreign affairs, and agreed on one of the great issues of the day—the danger posed by Communism and the Soviet Union.

But in another way, they could not have been more different. John

Kennedy's family was rich, and he had enjoyed all the privileges that money could buy—a fine Harvard education, world travel, material possessions, leisure, and his father's contacts. John Kennedy never had and never would have to work for a living a day in his life. His father wanted to free his sons from that pressure so they could pursue political careers.

Richard Nixon, by contrast, came from a poor family and grew up without privilege. Whatever he had in life—a college education, a law degree, and political office—he had to earn on his own with hard work and a keen mind. What John Kennedy was given, Richard Nixon had struggled to attain. In college, Kennedy was an indifferent student, but he developed a love of American and European history. By the time he captured the Democratic nomination, he had evolved into a mature leader who, like Nixon, was a voracious reader, a savvy politician, and a formidable debater. Both men possessed brilliant minds.

The presidential election of 1960 was one of the closest in American history. Nixon entered the contest as the favorite. He possessed a track record of significant achievements, and the majority of voters respected his years of experience as vice president. One of the top issues of the day was preventing the spread of Communism around the world and curbing the influence of America's rival superpower, the Soviet Union. And Richard Nixon had unsurpassed credentials as an anticommunist politician whose views were respected by the majority of Americans.

The most sustained effort to put Communism into practice began with the Russian Revolution of 1917 and the dictatorships of Vladimir Lenin and later Joseph Stalin. Communists claimed that their philosophy, when put into practice, would serve the common good. It proved to be a naive dream that was soon corrupted. Millions of people who resisted Communism in Russia and some thirty other countries throughout the twentieth century were killed—more than one hundred million victims in all. In pursuit of their goal, Communists established totalitarian political regimes that flouted individual rights, banned freedom of speech, eliminated free elections, set up police states, corrupted the rule of law, and imprisoned and murdered opponents.

World War II had ended with the defeat of Nazi Germany, Fascist Italy, and Imperial Japan by the Allies—the United States, Great Britain, France, and the Soviet Union. With Germany and the Axis powers crushed, the Allies emerged as the great political powers of the postwar era. Of the Allied powers, only the Soviet Union was a Communist nation and not a democracy. The end of the war resulted in a delicate balance of power—a cold war in which no shots were fired—between the democratic nations and the Soviet Union. Former British prime minister Winston Churchill warned in a famous speech that an "iron curtain" now divided Europe into the free, democratic nations in the west, and the totalitarian, Communist nations in the east. Richard Nixon owed his meteoric political career to his vigorous anticommunism at the height of the Cold War. But some Americans thought that Nixon had gone too far, and they associated him with what they believed were excesses committed at home by Senator Joseph McCarthy and the House Un-American Activities Committee during their investigations of Communists in the United States in the 1950s.

John Kennedy, too, began the race for the presidency with some disadvantages. No Catholic had ever been elected president. In that era, a prejudice that does not exist today might have prevented a person of that faith from becoming president. Kennedy argued that he would not be a "Catholic president" but merely a president who happened to be Catholic. He persuaded enough people that a person's religious beliefs should not bar him from the office.

Kennedy's other disadvantage was his lack of experience. Yes, he had served in the House and the Senate for fourteen years when he began his race for the presidency, but he was not particularly accomplished as a legislator. Nixon's supporters portrayed Kennedy as a callow young man who was only forty-three years old and who had not taken his time in Congress seriously. They said his lack of experience made him unqualified to serve as president. Indeed, fellow Democrat Lyndon Johnson referred to Kennedy as a "boy." Johnson believed Kennedy should wait his turn, until he was more mature, and not challenge him now for the nomination. Kennedy disagreed. He believed in the power of fate, and that his illnesses, injuries, and near-death experiences had marked him as a man who might be deprived of a long life. He was a man of action determined to make the most of his time, who wanted to accomplish things now.

In a series of televised debates between the two parties' presidential candidates—the first in American history—Kennedy leveled the playing field as seventy million people watched. Nixon was famous as a relentless and ruthless debater, and many expected him to vanquish Kennedy. But before the evening of the first debate, Kennedy relaxed, shaved closely, and allowed stage makeup to be applied to his face. Nixon spent the day campaigning and had aggravated a painful leg injury. He showed up at the television studio with a day's growth of beard, a five o'clock shadow. He refused makeup. People who listened to the debate on radio thought that Nixon had won. Those who watched it on television, however, thought Kennedy had won.

John Kennedy had a brilliant insight. He recognized that television would change political campaigns forever. Once, all that mattered was what a candidate said. Now it mattered just as much how he looked while he was saying it. During the first debate, John Kennedy looked relaxed, fit, and charismatic. Richard Nixon looked uncomfortable, swarthy, and nervous as he sweated under the hot lights. Kennedy also looked much younger, even though Nixon was only four years older than he. In content, the debate was almost a draw. The performances of the candidates were evenly matched. In the end, it was not necessary for John Kennedy to win the debate on the issues. It was enough that he looked as though he belonged on the same stage with Richard Nixon. He did.

When Americans went to the polls on November 8, 1960, no great issues divided the candidates. Both men advocated strong missile defense against the Soviet threat. Kennedy was as anticommunist as

Nixon; both opposed Communist expansion, including in Cuba, an island ninety miles off the coast of Florida, and both saw the Soviet Union as a dangerous rival. Neither candidate was then at the fore-front of the civil rights movement. Voters chose between the person-alities of the two men as much as they did between their stands on the issues. Kennedy presented himself as the voice of a new generation who would get the country "moving" again toward a "new frontier." Nixon argued that he, not Kennedy, had the proven leadership expe-rience to guide the nation in a dangerous world. Out of 68.3 million votes cast, John Kennedy received only about 119,450 more votes than Richard Nixon. Nixon had lost the presidency by just two tenths of 1 percent of the popular vote. It was one of the closest elections in history.

Late into the night, neither man knew who had won. Not until the morning after the election was Kennedy declared the winner. Nixon did not demand a recount and conceded victory to Kennedy. Without his father's wealth, which funded much of his campaign, and without Lyndon Johnson as his vice presidential running mate delivering the electoral votes of Texas, Kennedy would not have won.

ON JANUARY 20, 1961, John Fitzgerald Kennedy stepped forward on the East Front of the U.S. Capitol to take the oath of office as the thirty-fifth president of the United States and to deliver his inaugural address. Half of America had voted against him, but on this day, he behaved and spoke with confidence. Although he knew he had not won by a large margin at the polls, he sought to win a mandate now with his words. He summoned the American people to stand up for freedom in the shadow of the Cold War.

"Let the word go forth from this time and place, to friend and foe alike, that the torch has been passed to a new generation of Americans—born in this century, tempered by war, disciplined by a hard and bitter peace, proud of our ancient heritage and unwilling to witness or permit the slow undoing of those human rights to which this nation has always been committed, and to which we are committed today at home and around the world."

He cautioned other countries not to doubt his commitment to freedom in what he predicted would be a "long twilight struggle."

"Let every nation know, whether it wishes us well or ill, that we shall pay any price, bear any burden, meet any hardship, support any friend, oppose any foe, to assure the survival and the success of liberty."

Then he suggested that nations pursue peaceful cooperation, not military confrontation.

"Together let us explore the stars, conquer the deserts, eradicate disease, tap the ocean depths, and encourage the arts and commerce."

He reminded his audience that this would take time.

"All this will not be finished in the first one hundred days. Nor will it be finished in the first one thousand days, nor in the life of this administration, nor even, perhaps, in our lifetime on this planet. But let us begin."

Kennedy suggested that his election coincided with a special moment in history. "In the long history of the world, only a few generations have been granted the role of defending freedom in its hour of maximum danger. I do not shrink from this responsibility—I welcome it. I do not believe that any of us would exchange places with any other people or any other generation. The energy, the faith, the devotion which we bring to this endeavor will light our country and all who serve it—and the glow from that fire can truly light the world."

Perhaps the most quoted and famous line from the speech is Kennedy's call to self-sacrifice: "And so, my fellow Americans, ask not what your country can do for you; ask what you can do for your country." It was a patriotic call to the people of the United States to be civic-minded and politically active.

FOREIGN AFFAIRS and fighting the spread of Communism around the world dominated John Kennedy's first two years in office. He was a Cold Warrior who had a personal fascination with counterinsurgency warfare, covert action, and special military forces, including the Green Berets, a small, elite unit of the U.S. Army. A fan of Ian Fleming's James Bond novels, President Kennedy had an instinctive and enthusiastic appetite for secret operations. He also had an obsession with Cuba and with its leader, Fidel Castro.

A revolutionary who overthrew the Cuban government in 1959, Castro seemed at first that he might turn to America for inspiration and support. Instead he turned to the Communist Soviet Union for aid and set himself up as the repressive dictator of his nation.

During the administration of President Eisenhower, the Central Intelligence Agency had developed a secret plan to help anti-Castro Cuban exiles—living outside Cuba, trained and equipped by the United States—to invade their homeland, depose Castro, and overturn Communism. The CIA asked Kennedy to approve it.

Kennedy authorized what became known as the Bay of Pigs operation, named for the spot on the Cuban coastline where the armed exiles would land. The invasion, on April 17, 1961, was a catastrophe. The fourteen hundred freedom fighters, heavily outnumbered, found Castro's troops waiting for them. Within two days, most had been captured, wounded, or killed. The CIA and the U.S. military had persuaded President Kennedy to support the plan by arguing that its success would not require him to send American troops or air support into battle against Castro's forces. They predicted that the invasion would trigger a spontaneous uprising by the Cuban people against their leader.

That revolt never happened. The assurances by CIA officials and military generals had proven wildly optimistic—even deluded—and now they implored Kennedy to commit American forces to save the catastrophic operation. He refused. He feared that it might trigger a direct military conflict with the Soviet Union. The CIA plan had failed. It had been a humiliating disaster that would haunt his presidency. But Kennedy accepted responsibility for it and learned a valuable lesson: in the future he would be more skeptical of overconfident promises made by his military and intelligence advisers.

The Bay of Pigs episode did not stop the CIA from developing other secret plans—including one called Operation Mongoose—to overthrow or even assassinate Fidel Castro. Kennedy worried that Cuba might influence or contaminate Latin America with Communism.

Nor did the Cuban failure deter Kennedy from opposing the spread of Communism in Southeast Asia. When Kennedy took office, there were several hundred American military advisers in Vietnam. He increased their number to seventeen thousand, believing that America should make a stand against Communism there to prevent the ideology from conquering not just Vietnam but also neighboring countries.

In 1961, to prepare for the challenges ahead, President Kennedy asked Congress to increase the size and budget of the U.S. military. To promote peace and international cooperation, he also inspired thousands of young Americans to join public service by establishing the Peace Corps, an organization to help developing countries improve their public health, education, and agriculture. He wanted America to look both merciful and mighty.

Eighteen months after the Bay of Pigs, the United States and the Soviet Union almost went to war over another confrontation in Cuba. In October 1962, American spy planes detected the presence of Russian missile bases under construction there. The short distance between the island nation and the United States meant that from these sites Cuba or Russia could launch nuclear missiles at major cities and military bases in the eastern United States. The Soviets sent missiles to Cuba to deter any future invasion of the island by the United States and because, beginning in 1961, the United States had deployed in Italy and Turkey nuclear missiles that could be launched to attack the Soviet Union.

Kennedy revealed the frightening discovery in Cuba in a televised address. He warned that he would not tolerate nuclear missiles in Cuba. The volatile Russian leader, Nikita Khrushchev, had assumed, after the Bay of Pigs affair and an unimpressive personal meeting with Kennedy two months later, that the young president was weak and would do nothing when the Soviets parked their missiles in Cuba. Over thirteen days—from October 16 to 28, 1962—the United States and the Soviet Union came close to nuclear war during what became known as the Cuban Missile Crisis.

Many of Kennedy's advisers urged him to attack Cuba at once, first bombing the missile bases and then invading the island. Knowing that such a rash response might provoke war with Russia, Kennedy took his time, delaying his decision and hoping for a diplomatic solution. In the meantime, he declared a naval quarantine around Cuba and insisted that no Soviet ships carrying missiles or military supplies would be allowed to approach the island. At the last minute, to avoid a war, Khrushchev ordered his ships to turn back. But that alone did not end the crisis. Kennedy and Khrushchev negotiated a settlement: the United States agreed to remove its missiles from Turkey and promised not to invade Cuba, and in return the Russians agreed to remove their missiles from Cuba. President Kennedy had taken the United States and the Soviet Union to the brink of a nuclear war in which millions might have perished, but he had solved the dispute in a responsible manner. The most dangerous crisis of his presidency was over.

The competition between democracy and communism—between the United States and the Soviet Union—was not limited to Cuba, Southeast Asia, Eastern Europe, or even to planet Earth. Each superpower believed it could tip the balance of influence in its own favor by placing satellites in space and men on the moon. National pride was at stake: Which country would be the first to launch a rocket into space and spin a satellite in orbit around Earth? This "space race" captured Kennedy's imagination.

The Soviet Union had already beaten America into space when it launched the first satellite, Sputnik 1, in 1957. Then, on April 12,

1961, the Russians launched the first man into space. These successes shocked the American people. For reasons of prestige, and also national security, President Kennedy decided that America must catch up.

On May 25, 1961, the president addressed both the House and the Senate—a joint session of Congress: "I believe that this nation should commit itself to achieving the goal, before this decade is out, of landing a man on the moon and returning him safely to the earth. No single space project in this period will be more impressive to mankind or more important for the long-range exploration of space."

Later, in a speech on September 12, 1962, Kennedy emphasized the importance of the issue: "No nation which expects to be the leader of other nations can expect to stay behind in this race for space. . . . We choose to go to the moon in this decade and do the other things, not because they are easy, but because they are hard."

Inspired and supported by President Kennedy, the National Aeronautics and Space Administration (NASA), the agency in charge of America's space program, recruited more astronauts, designed giant rockets, and planned the Mercury, Gemini, and Apollo space programs.

Domestic issues captivated John Kennedy less than foreign affairs, although he was keen to reduce individual federal income tax rates and also corporate taxes, which he believed were too high and stifled the economy. There was one domestic issue that, above all others, he wanted to avoid: the fight for civil rights for African Americans. Kennedy was not, of course, against civil rights. He was not one of the Southerners who had disagreed with the Supreme Court's ruling in *Brown v. Board of Education*, the school desegregation case in which the Court declared it unconstitutional to ban black children from attending public schools with white children.

Nor did he want, as did many members of his own party in the South, to suppress other rights of citizenship, including voting, attending public universities, or patronizing restaurants, shops, and hotels. During the century since the Civil War and the end of slavery, African Americans had not enjoyed equal rights. Segregation and suppression were rampant. But Kennedy worried that becoming a civil rights champion was premature, and that doing so would stir up political opposition among Southern Democrats and endanger the programs and legislation that he wanted Congress to approve. Vice President Lyndon Johnson was, by contrast, a more enthusiastic advocate for civil rights; however, his authority was limited.

But a series of events made it impossible for John Kennedy to keep the civil rights movement at arm's length anymore. Back in September 1962, he was forced to send federal troops to the University of Mississippi to suppress rioting that ensued when a black man, James Meredith, was threatened with death after he tried to enroll as a student there. In May 1963, Americans—including President Kennedy— watched on television as Eugene "Bull" Connor, commissioner of public safety in Birmingham, Alabama, turned dogs and fire hoses on civil rights demonstrators. Then, Governor George Wallace refused to desegregate the University of Alabama, blocking the entrance with his own body.

John Kennedy decided he could not wait any longer. Ugly images of racist white mobs were broadcast all over the world, exposing the evil of racial discrimination in "the land of the free." This played into Communist propaganda that the United States was the land of oppression of blacks and hypocrisy, not liberty. On June 11, 1963, Kennedy gave a televised address to the nation on civil rights.

"This is not a sectional issue," he said, not wanting to single out and inflame the South. He knew that blacks also received poor treatment in the North. Indeed, Martin Luther King Jr. would say that when he led civil rights demonstrations in Chicago, the racism he encountered there was as vicious as anything he had seen in the Deep South. "This is not even a legal or legislative issue alone," Kennedy continued. "We are confronted primarily with a moral issue. It is as old as the Scriptures and as clear as the American Constitution."

In 1963, Kennedy's focus on foreign affairs gave him two of the greatest pleasures of his presidency, the first occurring in Berlin. In

President Kennedy and Vice President Johnson meet with civil rights leaders including Martin Luther King Jr. on the afternoon of the March on Washington, August 28, 1963.

1945, at the end of World War II, a treaty signed by the Allies divided all of Germany into zones of occupation. The zones controlled by the United States, Great Britain, and France became West Germany, and the zone controlled by the Soviet Union became East Germany. Berlin, the national capital, was within the Soviet zone, and the city was divided into four sectors, each occupied by a different Allied power.

In 1963, the Soviet Union still controlled East Germany. Indeed, in August 1961, during the first year of Kennedy's presidency, the Soviets began to build a concrete-and-barbed-wire wall between East and West Berlin to prevent the population of the Soviet sector from

President Kennedy addresses a crowd in Berlin, June 26, 1963.

fleeing Communism and escaping to the western zones. During the years that wall existed, Russian and Soviet-controlled East German soldiers shot to death several hundred men, women, and teenagers who tried to cross over it to freedom. On June 26, 1963, President Kennedy stood on the free side of the Berlin Wall and spoke to a throng of three hundred thousand people in the square.

"Freedom has many difficulties," he said, "and democracy is not perfect, but we have never had to put a wall up to keep our people in."

He told the massive, cheering crowd that "all free men, wherever they live, are citizens of Berlin, and therefore, as a free man, I take pride in the words *Ich bin ein Berliner.*" He had touched the German people with his empathy. And he told people around the world that if they wanted to understand the difference between Communism and freedom, "let them come to Berlin." The ecstatic crowd was the largest one that Kennedy had ever addressed. It was the high point of his worldwide popularity, and he said that he could not imagine enjoying a better day than this.

He would enjoy another foreign affairs triumph that fall.

LIKE JOHN Kennedy, Lee Harvey Oswald had always wanted to star in a historic moment. With every success that JFK had enjoyed, Oswald had matched it with failure. In the spring of 1963, John F. Kennedy was on an upward trajectory. Lee Oswald was not.

Lee Harvey Oswald was born in New Orleans, Louisiana, in October 1939, the youngest of three brothers. But it seemed as though a dark cloud had formed over him even before he entered the world. His father died two months before he was born, and during his unsettled childhood, his odd and unstable mother changed husbands, houses, jobs, and cities frequently—often turning over the care of Lee and his two brothers to orphanages or relatives.

When Lee was growing up, he lived, among other places, in New Orleans, Fort Worth, Manhattan and the Bronx in New York City, and then New Orleans and Fort Worth again. He had disciplinary problems at school, made few friends, threatened family members with knives, rebelled against any kind of authority, and missed so much school he was tracked down by truant officers and ordered to appear at court hearings.

As a teenager, Oswald became interested in the Soviet Union and the teachings of socialism, Marxism, and communism. These were strange pursuits for an American boy during the middle of the Cold War, an era in which the United States and the Soviet Union were locked in an intense ideological battle. And, of course, proclaiming oneself a communist in America at that time could trigger a government investigation.

In September 1956, Oswald dropped out of high school altogether.

And in October, after he turned seventeen, he enlisted in the United States Marine Corps. He served at bases in America and Japan, where he was court-martialed twice: once for assaulting a superior and another time for accidentally shooting himself in the arm with a pistol.

Throughout his three years in the Marine Corps, Oswald was a malcontent and constant complainer who loved to argue with his superiors to show that he was smarter than they were. He also made no secret of his interest in Communist societies. He never received a

Lee Harvey Oswald as a fifteen-year-old high school student.

better than average performance rating, but the Marine Corps managed to teach him to do one thing well—shoot a rifle with skill and reasonable accuracy.

In September 1959, under false pretenses, he was granted a dependency discharge to care for his mother, and then, in October 1959, in a series of bizarre events, he traveled to the Soviet Union, showed up in Moscow, and tried to commit suicide there when his visa expired and he was ordered to leave the country. That incident led him to the United States embassy, where he attempted to renounce his American citizenship. Soviet officials, though suspicious that he might be a spy or more likely mentally unstable, allowed Oswald to remain in the country and assigned him a job at a radio factory.

In April 1961, he married a nineteen-year-old Russian woman named Marina Prusakova. After a few years, Oswald grew dissatisfied with life in Russia, and he wanted to return to the United States. He was no longer the exotic foreigner and center of attention he had been when he had first defected. He, Marina, and their infant daughter left the Soviet Union in June 1962 and traveled to Fort Worth, Texas, where his mother and brother lived.

Between the summer of 1962 and the spring of 1963, Oswald struggled as a member of the lower class of Southern, white, working poor. After his failed attempt at killing General Walker, Oswald, at his wife's insistence, retreated to New Orleans with his tail between his legs to start over with another low-paying, entry-level job that would never allow him to fulfill his grandiose dreams.

In late May, Oswald wrote to the New York City office of an obscure organization called the Fair Play for Cuba Committee (FPCC). It was a group that lobbied for fair treatment of the island nation after its revolutionary dictator, Fidel Castro, had installed a Communist regime there. Around this time, Marina and their little girl joined Lee in New Orleans. It didn't take long for Marina to discover that Lee still possessed his rifle. He kept it in a closetlike room where he hung his clothes and stored his other belongings.

At night, Lee would often take the rifle out of the closet. "We had a screened in porch," Marina recalled, and "sometimes evenings after dark he would sit there with his rifle." Under the cover of darkness, when neighbors could not see him, Lee practiced aiming his telescopic sight. On several occasions, Marina found him on the back porch, sitting alone in the dark, fondling the rifle.

By June, Lee was distributing FPCC handbills on the streets of New Orleans. In July, he was fired from yet another job, and the U.S. Navy (which had jurisdiction over the Marine Corps) affirmed its decision to change his discharge from the Marine Corps to "undesirable" after it learned he had tried to defect to the Soviet Union. In letters to Secretary of the Navy John Connally, Oswald had argued, without success, that the service should reinstate the honorable discharge he had been given when he left the Marines, before he moved to Russia.

In August 1963, Oswald got a taste of the celebrity he had always craved. He was arrested in New Orleans after a street brawl with Cuban anti-Communists who objected to his distribution of pro-Castro literature. Oswald was jailed overnight, and Marina did not know where he was or why he did not come home that evening. It was another one of his mysterious, annoying disappearances.

The next day, he returned and explained what had happened. Marina was relieved. At least he had not tried to shoot someone. But she was scornful. She thought Lee's pro-Cuban efforts were foolish: "I would make fun of him, of his activity . . . in the sense that that it didn't help anyone really . . . I would say . . . to Lee . . . that [he] could not really do much for Cuba, that Cuba would get along well without him, if they had too." Oswald shrugged his shoulders and told Marina that she did not understand him but boasted that some people understood the importance of his work. He was about to get the recognition he longed for.

Oswald's arrest had attracted the attention of the press. He made a brief television appearance on WDSU-TV, when the station filmed him demonstrating in front of the International Trade Mart. He enjoyed watching himself on the news that night. And on August 17 and then again on August 21, Oswald participated in two New Orleans radio shows to discuss Cuba, Communism, and Marxism. His first appearance was on the WDSU-Radio program *Latin Listening Post*, hosted by William Stuckey. Oswald did not go out over the airwaves live. Stuckey taped the interview and then condensed it down to a five-minute segment for broadcast. Today it is hard to grasp the hold that Cuba had on the American mind—and on the minds of John Kennedy, Robert Kennedy, and Lee Harvey Oswald—in the early 1960s. Half a century later, the program feels dated, like a dusty, antiquated artifact from the Cold War.

Stuckey announced what he had in store for his audience.

"This is the first of a series of *Latin Listening Post* interviews of persons more or less directly concerned with the conflict between the United States and Cuba. . . . Tonight we have with us a representative of probably the most controversial organization connected with Cuba in this country. The organization is the Fair Play for Cuba Committee. The person, Lee Oswald, is secretary of the New Orleans chapter for the . . . Committee. This organization has long been on the Justice Department's black list and is a group generally considered to be the leading pro-Castro body in the nation."

Oswald was an exciting find for Bill Stuckey, a New Orleans reporter who had covered Latin American affairs in the city for several years. He had been hunting for a live specimen like Oswald for a long time. "Your columnist," Stuckey confided to listeners, "has kept a lookout for local representatives of this pro-Castro group. None appeared in public view until this week when young Lee Oswald was arrested and convicted for disturbing the peace. He was arrested for passing out pro-Castro literature to a crowd which included several violently anti-Castro Cuban refugees."

It was not hard for Stuckey to persuade Lee to appear on his show. He craved the publicity and was eager to show off his knowledge and oratorical skills.

"When we finally tracked Mr. Oswald down today and asked him to participate in *Latin Listening Post*," Stuckey explained, "he told us frankly that he would because it would help his organization to attract more members in this area."

Stuckey played up the drama by touting Lee as some sort of pro-Castro mastermind. "Knowing that Mr. Oswald must have had to demonstrate a great skill in dialectics before he was entrusted with his present post, we now proceed on the course of random questioning of [him]." Stuckey's flattery must have pleased his guest. Oswald had not won the coveted title of secretary over the other members of the New Orleans chapter because of his talent in "dialectics." He was the organization's *only* member, and he had awarded the title to himself.

During the interview, Stuckey probed Lee for details about the FPCC, questioning him about the size of the membership rolls, the specific goals of the organization, whether its members believed that Cuba was a puppet of the Soviet Union, and whether the FPCC was itself Communist controlled. For a novice, Oswald handled himself well. He said that both the number of members and their names must remain secret.

He explained that the organization's central principle was "Hands Off Cuba!"—a policy of American nonintervention in Cuban affairs—the motto printed on the handbills that he had distributed. Oswald denied Stuckey's assertion that Cuba was a Soviet colony in the Western Hemisphere. "Castro is an independent leader of an independent country," Oswald insisted.

Stuckey asked him if the FPCC would continue to support Castro if he broke off relations with the Soviet Union. "We do not support the man. We do not support the individual," Oswald explained. "We support the idea of independent revolution in the Western Hemisphere, free from American intervention. . . . If the Cuban people destroy Castro, or if he is otherwise proven to have betrayed his own revolution," Oswald added, "that will not have any bearing upon this committee."

Then Stuckey asked if the Castro regime was a Communist one. "Every country which emerges from a sort of feudal state as Cuba did, experiments," claimed Oswald, "usually in socialism, in Marxism." But he insisted that Communism had not taken over the island. "You cannot say that Castro is a Communist at this time, because he has not developed his country, his system this far. He has not had the chance to become a Communist. He is an experimenter, a person who is trying to find the best way for his country."

In any event, Oswald argued, the United States government had no right to interfere. "If he chooses a socialist or a Marxist or a Communist way of life, that is something upon which only the Cuban people can pass . . . but we cannot say . . . it is a threat to our existence and then go and try to destroy it. That would be against our principles of Democracy."

Stuckey asked Oswald to provide his definition of democracy. Lee stumbled on the answer and stalled until he could think of one. "My definition, well the definition of democracy, that's a very good one. That's a very controversial viewpoint. You know, it used to be very clear, but now it's not. You know, when our forefathers drew up the Constitution, they considered that democracy was creating an atmosphere of freedom of discussion, of argument, of finding the truth. The rights, well the classic right of having life, liberty, and the pursuit of happiness. In Latin America, they have none of those rights, none of them at all."

If the Castro regime is good for the Cuban people, Stuckey wondered, then why have fifty to sixty thousand of them fled the island? "A good question," replied Oswald. "Needless to say, there are classes of criminals there . . . people who are wanted in Cuba for crimes against humanity, and most of those people are the same people who are in New Orleans and have set themselves up in stores with blood money."

"You know," said Oswald, "it is very funny about revolutions. Revolutions require work, revolutions require sacrifice, revolutions, our own included, require a certain amount of rationing, a certain amount of calluses [sic], a certain amount of sacrifice."

Speaking of the Bay of Pigs, Oswald said, "I always felt that the Cubans were being pushed into the Soviet Bloc by American policy."

Oswald suggested that if America had adopted a different policy toward Cuba, and if the CIA had not meddled in its affairs, then "we could be on much friendlier relations" with Cuba. "If the situation had been handled differently," Oswald concluded, "we would not have the big problem of Castro's Cuba now."

Stuckey was pleased with Oswald's first appearance on radio. He was a provocative and garrulous guest, and Stuckey was eager to interview him again. But the host had no idea that Lee Oswald had withheld an explosive secret about his past.

ON AUGUST 21, Oswald was a guest on the WDSU radio show *Conversation Carte Blanche*, hosted by Bill Slatter. Joining Slatter as cohost was Bill Stuckey, who had four days earlier debuted Oswald to the public on his *Latin Listening Post* and had brought Oswald to Slatter's personal attention. Slatter introduced the program, beginning with the headliner: "Our guests tonight are Lee Harvey Oswald, who is secretary of the New Orleans chapter of the Fair Play for Cuba Committee; it's a New York–headquartered organization which is generally recognized as the principal voice of the Castro government in this country."

Slatter brought into the studio two other people to debate Oswald. "Our second guest is Ed Butler, who is the executive vice president of Information Council of the Americas, which is headquartered in New Orleans and which specializes in distributing anti-communist educational literature through Latin America. And our third guest is Carlos Bringuier, a Cuban refugee and a New Orleans delegate of the Revolutionary Student Directorate, one of the more active of the anti-Castro refugee organizations." Oswald and Bringuier eyed each other with suspicion. They had encountered each other before.

Stuckey briefed listeners on the FPCC: "The only member of the group to have revealed himself publicly so far is twenty-three-year-old Lee Harvey Oswald. He first came to public attention several days ago when he was arrested and convicted for disturbing the peace." Slatter explained: "The ruckus in which he was involved started when several local Cuban refugees, including Carlos Bringuier, discovered him distributing pro-Castro literature on a downtown street." Bringuier would have another go at Oswald tonight: "Mr. Oswald and Bringuier are with us tonight to give us opposing views on the committee and its objectives."

Stuckey boasted that Oswald was his discovery: "I believe that I was the first New Orleans reporter to interview Mr. Oswald on his activities here. Last Saturday, in addition to having him on my show, we had a very long and rambling question and answer session over various points of dogma and line of the Fair Play for Cuba Committee."

According to FPCC propaganda, Stuckey explained, "Castro's government is completely free and independent and is in no way controlled by the Soviet Union. Another cardinal point . . . is that . . . Castro seeks aid only because the U.S. Government refused."

"Mr. Oswald also gave me this rundown on his personal background. . . . He said . . . that he had lived in Forth Worth Texas before coming here to establish a Fair Play for Committee here . . ."

If Oswald was expecting another genial conversation like the one he had enjoyed with Bill Stuckey a few days ago, he was mistaken. Unbeknownst to Lee Oswald, Slatter was setting him up. In the four days since Lee's appearance on *Latin Listening Post*, Slatter and Ed Butler had both researched his life. They had found remarkable newspaper stories about things that Oswald had failed to disclose. The host of the program was about to expose him.

Now Slatter pounced: "However there were a few items apparently that that I suspect that Mr. Oswald . . . left out from this original interview, which was principally where he lived between 1959 and 1962." He told the audience that he and Butler had found old clippings that revealed that Oswald had attempted to renounce his American citizenship, had defected to the Soviet Union, and had returned to the United States in 1962.

Slatter confronted his guest: "Mr. Oswald, are these things correct?"

"That is correct, yes." He had the good sense not to try to deny the information that Slatter had collected on him.

"You did live in Russia for three years?"

"That is correct. And I think those—the fact that, uh, I did live for a time in the Soviet Union gives me excellent qualifications to repudiate the charges that Cuba and the Fair Play for Cuba Committee is communist controlled." It was not a bad recovery for a radio novice.

But it inflamed the hotheaded Bringuier, who demanded in an excited clipped tone, "I want to know exactly the name of the organization that you represent here in this city, because I have some confusion. Is it Fair Play for Cuba Committee or Fair Play for Russia Committee?"

Unruffled, Lee deflected the young Cuban's sarcasm.

Oswald had been blindsided, but he was willing to discuss his life in the Soviet Union. He was enjoying the notoriety and public attention he was receiving tonight.

"Well of course that is very provocative," Oswald replied in a calm voice, "and . . . I don't think it requires an answer." Then Bringuier recited a collection of boring statistics on the number of automobiles, telephones, and televisions per person in Cuba versus Russia, and the price difference for Cuban sugar when sold to America versus Russia. It was a long and confusing statement that Bringuier insisted Oswald explain.

Lee replied with humor: "Well, in order to give a clear and co-concise and short answer to each of those—let's say, well questions . . ." The audience could hear him chortle. To any listener, Oswald sounded relaxed and amused. He dismissed the excitable Cuban exile in one line: "This I do not think is a subject to be discussed tonight. The Fair Play for Cuba Committee, as its name implies, is concerned primarily with Cuban American relations. . . ."

Bill Slatter obliged with a pointed question on the subject: "How many people do you have on your committee here in New Orleans?"

Oswald was evasive: "I cannot reveal that as secretary of the Fair Play for Cuba Committee."

Ed Butler jumped in: "Is it a secret society?"

"No, Mr. Butler, it is not. However it is, ah, standard operating procedure for a political organization consisting of a political minority to, ah, safeguard the names and number of its members." It sounded like dialogue from a Hollywood spoof of an espionage film.

Butler dug in: "Well, the Republicans are in the minority. I don't see them hiding their membership."

Oswald parried well. "The Republicans, are not a, well [laughs], the Republicans are an established party representing a great many people. They represent no radical point of view. They do not have a violent and sometimes emotional opposition."

The conversation veered into a long discussion about whether Oswald ever tried to renounce his U.S. for Soviet citizenship. Oswald lied and denied it, at which point Stuckey contradicted him with old newspaper clippings that claimed he did. Oswald tried to get off the topic by saying "any other material you may have is superfluitous." Lee, whose vocabulary was limited, either could not remember the correct word, *superfluous*, or had forgotten how to pronounce it. So he made up a new malapropism of his own.

But Stuckey pursued him: "You apparently by your past activities have shown that you have an affinity for Russia, and perhaps communism, although I do not know that you admit to being a communist or ever have been one. Are you or have you been a communist?"

"Well, I have answered that, uh, prior to this program on another radio program."

"Are you a Marxist?"

"Yes, I am a Marxist."

"What's the difference?"

"As I said . . . very great."

Oswald launched into a long-winded and unconvincing list of countries as he tried to carve a vague distinction between Marxism and Communism. Then the host announced that it was time to break for a commercial. As soon as the show returned to the air, the interrogation continued.

Stuckey asked Oswald how he supported himself in Russia, and whether he had enjoyed a government subsidy. Lee stumbled. "Uh, well, as I, um will uh." Then he stalled. "Well with that I will answer your question directly then, since you will not rest until you get your answer." He said that he held a job.

Stuckey continued to ask Oswald about his time in Russia and pressed him again on whether he had tried to renounce his American citizenship. "Well, it's a long drawn out situation." Then he resorted to false logic. "Well, the very obvious answer is that I am back in the United States. A person who renounces his citizenship becomes by definition unable to return to the United States."

That may have been true, but it did not answer whether Lee had *attempted* to renounce his citizenship. This evasiveness would become one of Oswald's verbal trademarks.

Ed Butler demanded, "What did you do between October 31 and November 16, 1959?" and then asked Oswald if he had ever been to a particular street in Russia. The intricate question sounded obscure and meant nothing to the audience. Oswald said no, he had not been there but in the next breath he revealed that he knew the street (and pronounced its name with a proper Russian accent) and that the Soviet foreign ministry was located there. He knew that, he said, only because he had once lived in Moscow.

For a man who had lived in the Soviet Union during the Cold War, his comments about Russia were superficial and uninteresting. If this was his moment to prove that he was a sophisticated political thinker, his performance was not that impressive.

Oswald kept his cool but was getting frustrated. He objected to the whole line of questioning about his background: "Of course this whole conversation—and we don't have much time left—is getting away from the Cuban American problem." But it was fine, he added, if they wanted to grill him about Russia all night: ". . . I am quite willing to discuss myself for the remainder of this program."

Bill Slatter jumped in to rescue Oswald: "Mr. Butler, let me interrupt. I think Mr. Oswald is right. . . . We should get around to organization of which he is the head in New Orleans, Fair Play for Cuba." Slatter had omitted the last word of the organization's name. Oswald corrected him at once. "*Committee*," Oswald interjected, "Fair Play for Cuba *Committee*."

Slatter asked, "As a practical matter, what do you hope to gain through your work?"

At last, Oswald must have thought: a political question that he longed to answer. The program was almost over, but now he would have the opportunity to explain. "The principles of the Fair Play for Cuba [now Oswald failed to say the word Committee, the very lapse for which he had just chastised Slatter] consist of restoration of diplomatic, trade and tourist relations with Cuba. That is one of our main points . . . we are in a minority surely. We are honestly not interested in what Cuban exiles, or rightist . . . organizations might have to say. We are primarily interested in the attitude of the U.S. government towards Cuba."

Oswald rejected the insinuation that he was a Soviet puppet. "We are not at all Communist controlled regardless of the fact that I lived in Russia. Regardless of the fact that we have been investigated. Regardless of any of those facts, the Fair Play for Cuba Committee is an independent organization not affiliated with any other organization. Our aims and our ideals are very clear and in the best keeping with the very traditions of American democracy."

Carlos Bringuier could no longer remain silent. "Do you agree with Fidel Castro when in a speech [on July 26, 1963] he characterized President Kennedy as a ruffian and a thief?"

Oswald demurred. "I would not agree with that particular wording. However . . . does the U.S. government . . . ah, the State Department and CIA had made some serious mistakes in its relationship with Cuba . . . which put Cuba onto dogmatic Communism." No doubt he was referring to the CIA sponsored Bay of Pigs invasion in 1961 and the 1962 Cuban missile crisis ten months ago.

Then Oswald got into the question of American support for Castro's overthrow of Batista.

Ed Butler asked, "Why are people starving today?"

Oswald spoke of the unavoidable consequences of reforms and diversification of agriculture. He meandered onto topics like the production of sugar and tobacco versus sweet potatoes, lima beans, and cotton. Then time expired and the program petered out without a dramatic climax.

After the broadcast, Stuckey felt sorry for Oswald. He invited him out for a drink.

WHAT DO Oswald's little-remembered radio appearances tell us? He handled himself reasonably well when confronted by four hostile questioners on live radio. He kept his cool; he used humor to foil sarcasm and anger. He seemed to be enjoying himself. He did not come off as angry or hostile. He may have been naive or wrong, but he did not sound crazy. Yes, Oswald had made some factual mistakes and exhibited some verbal tics. Two laughable vocabulary mistakes gave clues to his lack of education: He said "superfluitous" when he meant to say "superfluous," and he said "co-concise" instead of "concise."

But Lee was proud of his Fair Play for Cuba Committee work. In a document he prepared, he describes the beginning of his involvement with the FPCC as though he had accomplished something of significance. "On May 29, 1963 I requested permission from the FPCC headquarters at 799 Brodwig New York 3, N.Y. to try to form a local branch in New Orleans. I received a cautionet [cautionary] but enthusiastic go-ahead from V.T. Lee National Director of FPCC. I then made layouts and had printed public literature for the setting up of a local FPCC. I hired person to distribute literature."

In the extract below, Oswald's overblown description of his counterintelligence operations against anti-Castro exiles sounds ludicrous. He appropriates the language of "spy talk" that in his imagination a CIA agent might use. Oswald's writings are a riot of misspelled words, grammatical mistakes, punctuation errors, and lack of capitalization. It is obvious that they are the musings of someone of limited education, even an autodidact, possibly dyslexic. They make him sound like a fool who has no idea what he is talking about:

> *I than [then] organized persons who display recetive [re-*
> *ceptive] attitudes toward Cuba to distrube [distribute]*
> *pamphlets . . . I infiltraled [infiltrated] the Cuban Student*
> *directorate and then harresed [harassed] them with infor-*
> *mation I gained including having the N.O. city atterny [at-*
> *torney] general call then in an out restraining order pending*
> *a hearing on some so-called bonds for invasion they were*
> *selling in the New Orleans area. I caused the formation of*
> *a small, active FPCC organization of members and sympa-*
> *thizers. where before there were none.*

In truth Oswald was a nonentity. He was the only member of FPCC in New Orleans, and he had failed to recruit another soul to serve Castro's cause. Oswald, as was so typical for him, was suffering from another one of his delusions of grandeur. All he did was write a letter and print some handbills. But Oswald made it sound important. "I received adive [advice], direction and literature from V.T. Lee National Director of the Fair Play for Cuba Committee of which I am a member. At my own expense I had printed 'Hands off Cuba' handbills and New Orleans membership Blanks for the F.P.C.C. local."

Oswald fancied himself a propaganda expert and claimed a hitherto unrecognized expertise. "I am experienced in Street agitation having done it in New Orleans in connection with the F.P.C.C. In Aug. 9 I was accousted [accosted] by three anti-Castro Cubans and was arrested for 'causing a disturbance' I was interrogated by intelligence section of New Orleans Police Dept. and held overnight being bailed out the next morning by relatives I subsenly [subsequently] was fined 10.$ charges against the three Cubans were droped [dropped] by the judge." Oswald was too uneducated to realize how pathetic his boast of being an expert in "street agitation" would seem to any *real* intelligence agent.

But Oswald was proud of the media coverage he had received. It allowed him to indulge the conceit that he had actually accomplished something. Finally, Oswald must have thought, people wanted to hear his opinions. He had so many of them. He was, he believed, on his way to becoming an important public spokesman for a cause. His voice would be heard.

In the Soviet Union, he had been a novelty—the American who rejected capitalism and chose socialist life in Russia. But his celebrity there had lost its sheen. Now he had a second chance at fame. Now he was another kind of novelty—the American who had chosen life in Communist Russia but rejected it and came back home, only to become a disciple of Fidel Castro and his revolution. He was in the limelight again, and he liked it. But his moment in the spotlight did not last long. The media lost interest, the invitations dried up, and it looked as though his brief brush with fame was over.

Nothing that Oswald said on the New Orleans radio programs would have given any listener cause to suspect that he had already attempted to murder one man and soon would try to kill another. In twelve weeks, he would attempt to assassinate the president of the United States. Soon Oswald would speak to a much larger audience than had heard him on a New Orleans radio station, when he would answer hostile questions in the same manner as he did on the radio show.

MARINA NOTICED that after Lee's arrest and radio appearances, he started to bring his rifle out to the porch more often: "It began to happen quite frequently after he was arrested . . . in connection with some demonstration and handing out of leaflets." Not only did he sit in the dark on the back porch peering through the telescopic sight at imaginary targets, he worked the bolt action of the Mannlicher-Carcano. "He would sit there with the rifle and open and close the bolt."

In the six months he had owned it, Marina had seen him clean the weapon four or five times. Before he tried to kill General Walker, she thought nothing of it. "I thought it was quite normal that when you have a rifle you must clean it from time to time." Now, in New Orleans, she asked him why he continued practicing with the rifle. It looked to Marina as though he was preparing again for . . . something.

His answer stunned her. He planned to go to Cuba, he said. Marina was exasperated. She warned Lee that if he did that, she would stay in New Orleans. There was no way she was going to agree to leave the United States and move to Fidel Castro's notorious revolutionary island. Lee had fantasized about killing Richard M. Nixon, JFK's republican opponent in the 1960 presidential election. Marina barricaded Lee in the bathroom until he cooled off and promised he would not try to do it. He also fantasized about hijacking a plane to Cuba. "He wanted very much," Marina said, "to go to Cuba and have the newspapers write that somebody had kidnaped an aircraft."

"For God's sakes, don't do such a thing," Marina pleaded.

Lee asked her if she would help him. Of course not, she said. "I told him that I would not touch that rifle."

He said that if she did not want to, it was fine. He could do it himself, and everything, he said, would go well. Oswald started studying airline schedules.

Despite Lee's obsession with Cuba, he never complained to Marina about President Kennedy. Oswald, a man who wanted the United States to keep its "Hands Off Cuba!" as the leaflets of the Fair Play for Cuba Committee demanded, had reasons to resent JFK for the U.S.-sponsored Bay of Pigs invasion and the Cuban Missile Crisis, but not once did he rant to Marina about the anticommunist American president.

Marina admired JFK and wanted to know more about him. "I was always interested in President Kennedy and had asked [Lee] many times to translate articles in a newspaper or a magazine for me, and he always had something good to say . . . from Lee's behavior I cannot conclude that he was against the president."

She was well aware, however, of her husband's delusions of grandeur. "He said that after 20 years he would be prime minister." How absurd, she thought. "I think that he had a sick imagination—at least at that time I already considered him to be not quite normal—not always, but at times. I always tried to point out to him that he was a man like any others who were around us. But he simply could not understand that. I tried to tell him that it would be better to direct his energies to some more practical matters."

Lee was not interested in his wife's sanguine opinion of his limitations. "At least his imagination, his fantasy, which was quite unfounded, as to the fact that he was an outstanding man. And the fact that he was very much interested, exceedingly so, in autobiographical works of outstanding statesmen of the United States and others."

Marina became convinced that Lee viewed himself in heroic terms. "I think that he compared himself to these people whose autobiographies he read. That seems strange to me, because it is necessary to have an education in order to achieve success of that kind."

A man born with what Oswald called his mean independent streak and the smug sense of superiority he exhibited in the Marine Corps could never function in Soviet society. But he could not fit in in America, either. He felt like a man without a country. Marina suspected he was incapable of being happy anywhere. She once said, "I am sure that if he had gone there (to Cuba) he would not have liked it there, either. Only on the moon, perhaps."

CHAPTER 3

"SHOW THESE TEXANS WHAT GOOD TASTE REALLY IS"

To many Americans, there was more to John F. Kennedy's administration—something intangible—than speeches, world travels, multiple crises, and anti-Communism. He had style. Kennedy possessed a glamorous effervescence that made him seem larger than life and a youthful symbol of a new era of American optimism and spirit. He was the first president born in the twentieth century. With his enchanting wife, who was just thirty-four, and two beautiful young children, the telegenic president cultivated a jaunty, athletic public image of a sailor, touch-football enthusiast, sportsman, and father. The public had nicknamed him JFK.

Jacqueline Kennedy, or Jackie, as she was known—became a star in her own right. Sometimes she even outshone JFK himself. Celebrated for her elegant, trendsetting fashion and understated beauty, she appeared on the covers of magazines looking more like an alluring movie actress than a politician's wife. She was a world traveler and lover of culture. And yet she remained a reserved, quiet person whose

President Kennedy claps his hands as his children Caroline and John Jr. dance for him in the Oval Office at the White House.

desire for privacy made her all the more fascinating to the public. A strong believer in the preservation of America's past, she undertook a much-needed historic renovation of the White House, hosted a popular and unprecedented television special on the result, and brought the president's home alive with artists, authors, and musicians.

John Kennedy's sharp wit and ability to laugh at himself enhanced his appeal. But he had a dark side. He loved history for its lessons and inspiration, but he was also drawn to its tragedy, irony, and disappointments. And below the surface of his public image of vigor, his cheer camouflaged a lifetime of physical pain and multiple illnesses, including Addison's disease. Convinced that the American people did not want to see their president as weak or sick, he battled his ailments in secret. Of all his characteristics, John Kennedy had one more important than all the rest—an ability to inspire people, through words and personal example, to attempt great things. It was the core of his mystique.

In August 1963, death whispered in Kennedy's ear again when it took the life of his newborn, two-day-old son, Patrick. It was Jackie's second failed pregnancy. But the emptiness that loss created left one saving grace. It drew Jack and Jackie Kennedy close again. They had kept it hidden from the American people, but theirs was not the picture-perfect marriage. Not until years later would any of that become public. He viewed marriage the way a royal prince might see the institution—necessary and useful but not sacrosanct.

BY LATE September, Marina was fed up with life in New Orleans and with her husband, and on the twenty-third she left New Orleans for Irving, Texas, a suburb of Dallas. She was expecting her second child next month, and her friend Ruth Paine had offered to take her in and care for her.

Ruth drove her station wagon from Irving to New Orleans to retrieve Marina and her simple household goods. Marina was pregnant, so Lee carried the boxes and loaded the car. There was also something he did not want her to see.

He had wrapped the Mannlicher-Carcano in a plaid wool blanket. Then, unbeknownst to her, he carried the rifle to the car and slipped it inside. Marina did not discover it until she got back to Texas: "After we arrived, I tried to put the bed, the child's crib together." She searched for the metallic parts. "I looked for a certain part, and I came upon something wrapped in a blanket. I thought that was part of the bed, but it turned out to be the rifle."

The rifle that Lee had used to try to murder General Walker was now back in Dallas. At the time of the Walker attack, Oswald and the Mannlicher-Carcano were barely acquainted. He had owned it for less than three weeks when he attempted to use it to assassinate the general. Now Lee and his rifle were old friends.

LEE LEFT New Orleans too, but he did not head straight to Texas or to his family. He decided to take a long and mysterious excursion between New Orleans and Fort Worth—out of the United States and into Mexico.

On September 27, four days after his departure from New Orleans, Oswald showed up in Mexico City, where he visited the Cuban embassy and applied for permission to travel to Cuba. The Cubans, scoffing at his wild tales of Fair Play for Cuba activities, gave him no special treatment and told him it would take months. Frustrated, he then went to the Russian embassy for help in getting to Cuba or returning to the Soviet Union. The Russians knew he was an odd duck and were in no hurry to allow him back into their country either. He was furious.

THAT FALL John F. Kennedy was worried about the proliferation of nuclear weapons among the superpowers. The United States and the Soviet Union possessed arsenals of several thousand nuclear weapons. Most of them were more powerful and destructive than the two atomic bombs that the United States dropped on Hiroshima and Nagasaki, Japan, in 1945 to end the Second World War. Through the 1940s, 1950s, and 1960s, the nations that possessed nuclear weapons tested their effectiveness and demonstrated their military superiority by exploding them in their own territory, either underground, in the ocean, or in the atmosphere in remote areas far removed from population centers. Nonetheless, the tests still resulted in radioactive fallout, which winds and weather systems could carry for hundreds of miles and contaminate the food supply or towns and cities.

At a historic speech at American University in June 1963, JFK laid out his quest for peaceful cooperation with the Soviet Union. He did not overlook the differences between the nations but invoked "our common interests." For in the end, he said, "our most basic common link is that we all inhabit this small planet. We all breathe the same air. We all cherish our children's future. And we are all mortal." Kennedy negotiated with the Russians to end atmospheric testing, and in October 1963, he signed a Limited Nuclear Test Ban Treaty. He considered it the most significant achievement of his presidency.

A CHASTENED and humiliated Oswald returned to the United States on October 3, 1963, and went to Dallas, where he visited Marina at Ruth Paine's house in Irving and spent the weekend of October 11–13. Paine worried that Oswald was having trouble getting a job, so she told some of her friends he needed work. On the night of October 14, Ruth told him over the phone that a girlfriend of hers said they were hiring at a place called the Texas School Book Depository on Elm Street. The next day, Oswald applied in person. The manager liked that he had served in the Marine Corps and addressed him as "sir." Lee was hired and reported for work on October 16.

Lee and Marina agreed that he would live at a boardinghouse in Dallas during the week and visit Marina at Ruth Paine's house in Irving on the weekends. Ruth, who disliked Lee and hated the way he treated his wife, hoped Marina would leave him. A neighbor of Ruth's named Wesley Buell Frazier also worked at the Book Depository, and he offered to drive Oswald from Dallas to Irving on Fridays after work and to the Book Depository on Monday mornings. Marina threw her husband a surprise birthday party on Friday, October 18, and he seemed touched. Then, on October 20, their new baby daughter, Rachel, was born. Perhaps Oswald's odd and unsettled life would finally calm down.

On October 23, Lee attended a right-wing political rally where General Walker spoke. Was this a warning sign that he was plotting another assassination attempt? Then the Federal Bureau of Investigation made a couple of visits to Marina Oswald. They told her they would like to speak to Lee. It was nothing serious, she was told, just a follow-up to chat with him since he had returned home from Russia almost seventeen months ago. The visits angered Oswald—he believed the FBI agents were harassing Marina. On November 12, he left an angry note at the FBI headquarters in Dallas for agent James Hosty, telling him to leave his family alone.

But he failed to sign the note with his name.

IT WAS the autumn of 1963, and the presidential election was just one year away. John Kennedy planned to run for a second term and hoped to win by a margin wider than he had eked out in 1960. It was essential that he again win the state of Texas. In 1963, warring factions of the liberal and conservative wings of the Democratic Party in that state were at each other's throats, fighting over which group would control politics in the state. JFK wanted the dispute to end. To increase his chances for reelection, he planned to travel to Texas in late November to campaign with Vice President Lyndon Johnson.

In 1960 Kennedy had chosen Johnson as his running mate—over the strong objection of his brother Robert Kennedy. It was a savvy move. Without Johnson and the electoral votes of Texas, JFK would not have won that election. Texas was to be a major political trip involving private meetings with Democratic leaders, public speeches, and fund-raising events. The president and first lady would visit five cities in two days—San Antonio, Houston, Fort Worth, Dallas, and Austin—and then head to Johnson's Texas ranch to rest. It was an ambitious schedule packed with activities.

Jacqueline Kennedy would accompany her husband to Texas. She had traveled with JFK to Paris, and without him on private visits to Italy, Greece, India, and Pakistan. But she had not been on a real campaign trip since the presidential election in the fall of 1960. The Texas events would come only three months after her new baby, Patrick Bouvier Kennedy, had died on August 9—two days after he was born.

• • •

WEDNESDAY, NOVEMBER 20, 1963. It was one of those legendary Kennedy parties. And although they did not know it at the time, it was also John and Jacqueline Kennedy's last night together in the presidential mansion. The occasion was a reception for the federal judiciary, including the justices of the Supreme Court of the United States. Several days earlier, the Kennedys had treated hundreds of guests to a performance by the bagpipes and drums of the Black Watch, the legendary British military unit. Kennedy receptions and dinners were already the stuff of legend. One of the most memorable was a dinner that the president hosted for Nobel Prize recipients. He quipped that never had such talent and brilliance been assembled at the White House— since Thomas Jefferson had dined there alone.

The Kennedys looked forward to more social events the next week. On Monday, November 25, they would host their next dinner party—it was to be a state dinner for the new chancellor of West Germany, Ludwig Erhard. The invitations were already in the mail. They loved to entertain—not just at official events but also intimate, private dinner parties and small dances. Often these evenings concluded with guests being invited upstairs to the private family quarters. It was at a small private dinner party in Georgetown where they had met twelve years earlier.

Earlier in 1963 they had celebrated the tenth anniversary of their marriage. Monday, November 25, would also be their son John's third birthday. Jackie would organize a little party for him before the adults arrived for dinner.

Then, for Thanksgiving that Thursday, November 28, they would travel to Hyannis Port, Massachusetts, and spend the holiday at a Kennedy family home there. Already on JFK's calendar for December 6 was a special event over which he was eager to preside—the first ever presentations of the new Presidential Medal of Freedom to a group of distinguished Americans. He had created the award and then vetted the names of potential recipients. The ceremony would epitomize the Kennedy style and its celebration of achievements in literature, culture, and the arts.

Just last month, when he spoke at Amherst College in a tribute to Robert Frost, JFK had paid homage to the arts. It was similar to President John Adams's famous observation that he studied politics and war so that his grandchildren could study poetry and music. True, Jackie was the more passionate cultural devotee in the family (JFK might prefer a private screening of the Hollywood spectacular *Spartacus* over a sublime solo performance by Pablo Casals), but the president's enthusiasm for American achievement in all fields was genuine. He was entranced by excellence. When he nominated former football star and Justice Department lawyer Byron White to be a Supreme Court justice, Kennedy said that one of White's qualifications was that he had excelled in everything he had ever attempted.

After the Medal of Freedom awards, the Kennedys were looking forward to Christmas in Palm Beach, Florida, at the family compound there. The president and Jackie had just signed the first batch of their Christmas cards for mailing later.

John Kennedy loved his job. He savored every moment of it—even the times of crisis and testing. He was never one of those presidents who complained about the debilitating stress and awesome burdens of the office. He knew how lucky he was—not just to be president of the United States, but to be alive at all. In addition to his own narrow escapes, many of those close to him, even some of his own siblings, had died young. He and Jackie had lost their stillborn daughter, who Jackie would have named Arabella, and their son Patrick, who had lived less than two days. Just as JFK had described Robert Frost in Frost's own words in that October speech at Amherst, he too was "one acquainted with the night."

But in November 1963, the Kennedys looked forward to the future. Jackie had emerged from her mourning for Patrick. JFK had started to think about the presidential campaign of 1964. His prospects looked decent. To improve them, he had scheduled the three-day political swing through Texas in the week before Thanksgiving, where he would visit the five cities and make numerous speeches, as many as three in a single day.

Jackie had even agreed to join him on the trip. She had never liked campaigning and had not done it since the election of 1960. But she was eager to make this journey. In the fall of 1963, as unlikely as it might sound to a jaded modern reader with fifty years' worth of cynicism and hindsight in the wake of all the subsequent revelations about JFK's private life, John and Jacqueline Kennedy might have been more in love with each other this November than they had been since the year they married.

Once they returned from Texas, they could begin again.

The president had received warnings that he might not be entirely welcome in Texas. In Dallas, various political enemies criticized him as either too liberal on civil rights or too soft on Communism. Weeks earlier, the president's opponents had protested a visit to the city by Adlai Stevenson, Kennedy's ambassador to the United Nations. In the auditorium where he spoke, hecklers disrupted his speech. Stevenson did not help his cause when he insulted one of them with a prissy tongue-twister: he might as well have said it to the whole state. "Surely, my dear friend, I don't have to come here from Illinois to teach Texas manners, do I? . . . I believe in the forgiveness of sin and the redemption of ignorance."

It was exactly the aspect of Stevenson's prickly, egghead personality that made JFK dislike him. When, outside, the ambassador asked an angry woman how he might help her, she hit him in the head with a cardboard picket sign. Jackie may have been a fan of Adlai—they enjoyed private dinners in New York City—but the two-time Democratic presidential nominee and two-time loser rubbed Jack the wrong way. Thus Stevenson was never accepted into JFK's inner circle.

Several people had warned JFK to stay out of Dallas. The previous month, on October 3, Senator J. William Fulbright of Arkansas suggested during a meeting with President Kennedy that he not go there. "Dallas is a very dangerous place. *I* wouldn't go there. Don't *you* go." The hostile reception that Adlai Stevenson received in Dallas on October 24 seemed to confirm Fulbright's counsel.

And on November 4, Byron Skelton, Democratic national committeeman from Texas, sent a letter to Attorney General Robert Kennedy that warned him about the hostile political climate in the city.

"I am worried," he confessed, "about President Kennedy's proposed trip to Dallas." He mentioned a prominent figure in the city who had complained that JFK "is a liability to the free world." The insult troubled Skelton. "A man who would make this kind of statement is capable of doing harm to the President." Skelton suggested that Kennedy bypass Dallas.

Skelton was so worried, he also wrote a warning to an aide of Lyndon Johnson, and the following week, when Skelton was in Washington, he alerted two other men at the national committee. One woman from Dallas sent a letter to presidential press secretary Pierre Salinger. "Don't let the president come down here. I'm worried about him. I think something terrible will happen to him." Others expressed misgivings about the visit.

But no warnings would intimidate John Kennedy enough to stay out of Dallas. First, he was a fatalist. Second, he dismissed such warnings as exaggerations. And third—and this was the most important reason he refused to drop Dallas from his itinerary—JFK believed that no president of the United States should ever be afraid to visit an American city.

Any man who showed such fear was not, in John Kennedy's opinion, up to the job. He was going to Dallas.

Before the Texas trip, JFK was less interested in threats than he was in what Jackie planned to wear to Texas. In general he was happy to leave such matters to her personal taste—except when he saw the bills. Often she tried to conceal the cost of her purchases. On one occasion, she insisted that reports of her extravagance were so exaggerated they could not be true. "A newspaper reported that I spent $30,000 a year buying Paris clothes and that women hate me for it," she moaned. "I couldn't spend that much unless I wore sable underwear." But aside from occasional spats about the expense involved, JFK did not micromanage Jackie's wardrobe. Of course he wanted her to look good, and she always did.

But he was concerned how she would look in Texas. "There are going to be all these rich Republican women at that lunch [in Dallas on November 22]," he told her, "wearing mink coats and diamond bracelets, and you've got to look as marvelous as any of them." But don't, he cautioned, go over the top. "Be simple—show these Texans what good taste really is."

He need not have worried—Jackie rarely overdid it. In the family quarters of the White House, she began to show him the outfits she had chosen for the trip, holding them up in front of her body one after the other. Three of the most attractive outfits included a black velvet dress, a matching white wool suit, and a matching pink suit.

AT THE White House on the morning of Thursday, November 21, John and Jacqueline Kennedy said good-bye to their daughter, Caroline. On the lawn, three helicopters, their rotor blades spinning, prepared to leave for Andrews Air Force Base, where their jet was waiting. Their son, John, who would celebrate his third birthday in a few days, loved flying in the helicopter; as a special treat, they agreed to take him along for the ride to Andrews. The president sat with his boy and a few Secret Service agents in the first copter. Staff members climbed aboard the other two. But the president's helicopter could not take off. Jackie was nowhere in sight. Then, dressed in a white two-piece suit, she left the White House and walked to her husband's aircraft. She was the last person to board.

Awaiting the party at Andrews was Air Force One, the sleek new jet that had become the symbol of the modern presidency. The plane had gone into service on October 21, 1962, making John Kennedy the first and only president to have flown in it so far. It was a beautiful aircraft decorated with two tones of bright blue paint and a big red, white and blue American flag painted on the tail. Raymond Loewy, the famous industrial designer, had given the aircraft his imprimatur, and JFK himself had approved the design.

Kennedy considered the aircraft one of the great perks of the presidency and a big step up from his own private plane—the *Caroline*, on which he had flown all over the country during his pursuit of the presidency. Kennedy's blue and white *Caroline* was a twin-engine Convair 240 that seated sixteen; it had been purchased by the president's father, Joe Kennedy, in 1959. The plane had a galley and a bedroom, and Kennedy and his staff used it to great effect during the campaign. But it was a puddle jumper compared to the majestic presidential jet.

At Andrews, John Kennedy Jr. wanted to get on the plane and fly to Texas with his parents. The president told his son he could not come on the trip, and explained that he would see him in a few days. His parents would be home before his birthday. The Kennedys said good-bye to their son and boarded Air Force One for the flight to San Antonio. The plane took off, bearing the president and first lady west to Texas and two busy days of back-to-back events. John Jr., accompanied by Secret Service agent Bob Foster, returned to the White House by helicopter.

John Kennedy had left some unfinished paperwork behind on his desk in the Oval Office, including an autographed photograph he intended as a gift for a supporter. After inscribing the photo, he had neglected to sign it. It was of no consequence. He could add his name once he returned from Texas.

THE DAY the Kennedys left the White House for Texas, a man waiting twelve hundred miles away in Dallas was eager for the president to arrive. He was not an important politician who wanted to discuss business with President Kennedy. He was not a supporter who hoped to shake his hand, nor one who had purchased a ticket to the November 22 breakfast to be held for several hundred people in nearby Fort Worth, or for the big lunch scheduled in Dallas that afternoon.

Nor was he a political opponent of John Kennedy's who planned to protest his policies with a homemade, hand-lettered cardboard sign. No, this man who awaited John Kennedy in Texas had something else in mind. He wanted to kill the president.

But the man's timing was strange, because these feelings had come on all of a sudden. Just two days earlier, on the morning of Tuesday, November 19, 1963, when Lee Harvey Oswald awoke in Dallas, Texas, he did not know that within the next three days, he would decide to murder the president of the United States. If a fortune-teller had prophesied this future, the twenty-four-year-old married father of two children might not have believed it.

Indeed, among Oswald's corrosive obsessions—and there were many—John F. Kennedy was not one. There is no evidence that Oswald hated the president. Much evidence suggests he did not think about him much at all. He had no long-standing fixation with Kennedy. He had not made him the primary subject of his everyday conversations. He had not been stalking the president or, as far as can be told, fantasizing about killing him. Among Lee Harvey Oswald's list of long-simmering resentments, frustrations, and grievances, the Kennedy presidency was not one of them.

This is not to say that Lee did not harbor violent fantasies. From Moscow, he had written to his brother Robert on November 26, 1959, that he would like to see the government of the United States overthrown: "In the event of war I would kill *any* American who put a uniform on in defense of the American government—any American." But he had never spoken that way about John Kennedy.

ON THE nineteenth, the *Dallas Morning News* had published the details of the route President Kennedy's motorcade would follow when, in three days, Air Force One would lift off from Fort Worth and land at Love Field, Dallas. From there, the presidential limousine—a big, custom-built Lincoln Continental convertible—would take JFK on a long, circuitous motorcade through downtown Dallas to a lunch for more than 2,000 people at the Trade Mart, a huge convention center and wholesale merchandise mart.

The motorcade was unnecessary. There were shorter, quicker routes from Love Field to the Trade Mart. But a parade would allow tens of thousands of Dallas citizens who would not otherwise glimpse JFK to assemble on the sidewalks and streets to see the president in person.

In addition, many people working in office buildings along the route could open windows overlooking the street to enjoy a good, un-obstructed view of the president. After the limousine drove through downtown Dallas, it would turn right from Main Street to Houston Street, proceed one block, turn left on Elm Street, and finally, as the crowds thinned in an area known as Dealey Plaza, pick up speed, vanish under an underpass, and follow the Stemmons Freeway for a short trip to the Trade Mart lunch.

Anyone familiar with the streets of Dallas would know that after the president's car turned left onto Elm, it would pass directly below a seven-story office building and warehouse known as the Texas School Book Depository. Everybody knew the Depository. It had a big, yellow clock atop the roof—put up by the Hertz rental car company—that made it a local landmark.

Lee Harvey Oswald was familiar with Dealey Plaza. Since mid-October, before the president's Texas trip had been finalized, he had held a job at the Depository as a low-level order filler who moved cardboard boxes of school textbooks around the building and pulled boxes to fulfill customer orders. But he didn't know that John F. Kennedy would be driving right past the place he worked in three days because he probably failed to read the newspaper on the day the story ran. Too cheap to buy a daily paper, Oswald was in the habit of reading stale, day-old newspapers left behind in the lunchroom by coworkers. Thus, it is possible that it was not until the morning of Wednesday, November 20, two days before President Kennedy was scheduled to arrive in Dallas, that Oswald learned for the first time that the president of the United States would drive past his work-place.

Oswald must have realized the implications of what he had just read: someone with the mind to do it could open a window on one of the upper stories of the Book Depository, wait for the president's mo-torcade to drive by, and shoot Kennedy as he passed. The distance between an open window on, say, the fifth, sixth, or seventh floors and Elm Street was too great to fire a pistol at a stationary target below, let alone at a moving car. A pistol's short barrel could not guarantee sufficient accuracy at that range. But Oswald would have known from his military training that someone would need to use a rifle to hit someone from such a distance.

It had never been attempted before: no American president had ever been assassinated from long distance by a rifle. Three of them— Abraham Lincoln, James Garfield, and William McKinley—had all been murdered at close range—no more than several feet—by lone gunmen firing pistols in 1865, 1881, and 1901. But sometimes pistols were not enough to get the job done.

In 1912, former president Theodore Roosevelt had been shot in the chest with a revolver during his campaign for reelection as a third-party candidate, but he survived the wound. On February 15, 1933, an assassin in Miami, Florida, fired a pistol at a convertible car occupied by president-elect Franklin Roosevelt. The gunman missed his target but wounded the mayor of Chicago, Anton Cermak, who was standing next to Roosevelt. Cermak died the next month.

And on November 1, 1950, two Puerto Rican nationalists who wanted complete independence from the United States tried to assassinate President Harry Truman by fighting their way with pistols into Blair House, the government guesthouse where he was living during White House renovations. The terrorists shot three policemen, wounding one fatally. One of the assassins was killed, and the other was captured. The gunmen never got into the president's residence.

Four years later, on March 1, 1954, while Congress was in session, a gang of four other Puerto Rican nationalists sitting in the visitor's gallery of the House of Representatives opened fire with semiautomatic pistols wounding, but not killing, five congressmen. To this day, bullet holes from this attack scar the furniture in the House chamber. No, a pistol was not a foolproof weapon for an assassination.

• • •

SO ULTIMATELY John Kennedy and Lee Harvey Oswald were brought together by a staggering coincidence. It is likely that Oswald would never have thought of killing Kennedy at all if the publicized motorcade route had not taken JFK to the doorstep of Oswald's place of employment. It was a once-in-a-lifetime opportunity—the president was coming to him!

Earlier in the year, on March 15, less than a month before he tried to assassinate General Walker, Oswald wrote to his brother Robert: "It's always better to take advantage of your chances as they come along."

Oswald must have thought about it. He possessed the necessary skill and equipment. He had learned to shoot in the U.S. Marine Corps, and he owned a rifle. He could do it. Yes, he could. But would he? And why?

Sometime between the morning of November 19 and the afternoon of November 21, Oswald decided to assassinate President Kennedy. No one knows exactly when he made that decision. It could have been as early as the morning of Tuesday, November 19, but only if he broke his habit and read the morning paper the same day it came out. If he followed his usual custom, then he would not have read Tuesday's paper until the following day, the morning of November 20. Once Oswald read the day-old paper, perhaps he also consulted Wednesday's *Dallas Times Herald*, the afternoon paper, which confirmed the motorcade route. Then, on Thursday, November 21, to make sure that the public knew where to go to see the president, the morning paper published a map of the route that Kennedy's limousine would follow.

WHEN AIR Force One landed at San Antonio International Airport, Jackie Kennedy was first to disembark, ahead of her husband. He wanted the crowd to see her first. JFK loved the excitement she caused. Waiting to receive them was Vice President Lyndon Johnson, along with various dignitaries. Johnson had flown from his ranch to San Antonio on his own private plane, arriving about an hour before the president. He killed the time by having lunch with his wife and getting a haircut. A reception committee presented Jackie with a bouquet of yellow roses.

Also waiting was the presidential limousine, flown in the night before from Washington. It was a Lincoln convertible that had been custom-built by the Ford Motor Company. A second car had also been flown to San Antonio—a 1955 open-top Cadillac that followed close behind the president's car in motorcades and that the Secret Service had nicknamed the Queen Mary.

One hundred twenty-five thousand people lined the motorcade route, including a large number of high-school students carrying flags and friendly signs. At one high school, each student waved an American flag. In honor of the president's visit, merchants had put up their Christmas decorations one week early. A handmade sign tried to entice Jackie to slip into a swimsuit: JACKIE, COME WATERSKI IN TEXAS. Other signs welcomed the Kennedys in Spanish.

Outside the airport, JFK had spotted one disturbing sign. It did not protest his visit—but warned him of racial injustice: MR. PRESIDENT, YOU ARE NOW IN A SEGREGATED CITY.

Along the whole route, crowds cheered the motorcade. They were loud and would not stop screaming. It was a fantastic welcome that put the president in a great mood.

The motorcade took him to Brooks Medical Center, a United States Air Force facility, where twenty thousand people waited for him to give an outdoor speech on the "New Frontier." When the Kennedys got out of their limousine, the crowd went wild. People clamored to present gifts. One wild woman broke from the crowd and reached Jackie. "Mrs. Kennedy, Mrs. Kennedy, please touch my hand!"

As in da Vinci's painting of God touching his fingertip to the form of man to give him life, Jackie touched the woman. In ecstasy the Jackie disciple cried out, "O, my God! She really did touch me. She really did."

John Kennedy gave an invigorating speech on the challenges of the future.

"We . . . stand on the edge of a great new era filled with both crises and opportunity. . . . It is an era which calls for action, and for the best efforts of all those who would test the unknown and the uncertain in every phase of human endeavor. It is a time for pathfinders and pioneers."

The president turned to one of his favorite topics.

"This nation has tossed its cap over the wall of space—and we have no choice but to follow it . . . with the help and support of all Americans, we will climb this wall with safety and with speed—and we shall explore the wonders on the other side."

From Brooks Medical Center, the Kennedys rode in a half-hour motorcade to Kelly Field, a military airfield adjacent to San Antonio International, where Air Force One had been moved. Five thousand people waiting for them there cheered their arrival. They boarded the plane at 3:48 P.M. for the forty-five-minute flight to Houston.

EVEN BEFORE John Kennedy had touched down in San Antonio at 1:30 P.M. on the afternoon of November 21, Oswald had already decided to kill him. A deviation from Oswald's normal behavior offers an intriguing clue. On the morning of November 21, he ate breakfast at the Dobbs House restaurant. He was not in the habit of eating breakfast out, and he couldn't afford to do it. Did breaking his routine by treating himself to a special breakfast signal that something was different and that he had decided by the morning of the twenty-first to assassinate the president?

Depending on the exact timing of his decision, Oswald had about twenty to fifty hours to make—and carry out—his scheme. He was a trained and experienced rifleman who would have known that a successful assassination required careful advance planning.

He could not just poke his rifle out of a random Book Depository window on the spur of the moment on November 22 and start shooting, hoping to hit his target. No, a proper sniper attack combined angles, timing, stealth, concealment, and patience. And an escape route. Oswald could not leave these details until the last minute.

Where, Lee Oswald might have asked himself, was the best location in the Texas School Book Depository from which to shoot someone driving by on Elm Street?

One of the upper floors, high over the street, would place him above the sight lines of parade watchers and the passengers in the multicar motorcade. That way he could position himself to shoot down at the president, from an angle that would make it difficult for witnesses to spot him. Oswald chose the sixth floor. It was a floor he knew well. He had spent a lot of time up there wandering around with his clipboard and searching for boxes of books to fill customer orders.

Additional circumstances made the sixth floor an even better choice. A portion of the wood flooring was being refurbished, so workers had moved many heavy cardboard cartons full of books to the south side of the floor, near the row of windows looking down upon Elm Street. There were twice as many boxes on that side of the building as normal. Stacks of them obstructed a clear sight line across the room and would shield anyone who wanted to remain hidden from anyone else on that floor. It would be as easy as a child stacking building blocks to move some of those cartons and arrange them into a wall on November 22. Oswald could set up his position at the window of his choice. He could even shift a couple of boxes to the edge of the windowsill and rest his rifle on the top one to steady his aim.

The large number of boxes on the floor would also make it easy for Oswald to stash his weapon. On November 22, he would need to hide the rifle from other Book Depository workers for four hours, between the time he brought it to work by around eight A.M. until around noon, when he would retrieve it. Then he would take his position behind the boxes and wait for his prey.

After he shot the president, what would he do with the rifle? Oswald planned to leave the weapon behind at the Book Depository. There was almost no chance he could descend five flights of stairs (or take the elevator) without encountering coworkers or any policemen or Secret Service agents who might storm the building after they heard the shots. It wouldn't look good if he was clutching the rifle in his hands. Even if he didn't pass anyone while escaping the building, he could hardly expect to stroll unnoticed down Elm Street with a rifle slung over his shoulder. He could try to disassemble it on the sixth floor and stuff it into the same brown paper bag he planned to use to carry it to the Book Depository the morning of the twenty-second, but that would cost too much time, at least one minute and maybe two. Every second would be precious to Oswald's escape. He would have to abandon his weapon on the sixth floor.

No criminal wants to leave a murder weapon behind at the scene of the crime. A firearm is an incriminating piece of evidence. It can bear fingerprints that identify a gunman. Firearms possess serial numbers that can be traced. The inside grooves of a rifle barrel leave unique marks on a bullet so that a spent round can be identified later as having been fired from that particular weapon. A brass cartridge case ejected from a rifle after the bullet has been fired can bear telltale signs that match it to the weapon from which it came.

It was risky for Oswald to leave his rifle at the Book Depository. Yes, he had purchased it by mail under a false name using a postal money order, not a personal check, and he had directed it to be shipped not to his home address, but to a post-office box, which he had rented using a false name. But even taking those steps did not guarantee secrecy. There was still a chance that law-enforcement officials could trace the weapon to him, but that, he figured, should take a while.

The would-be assassin had now chosen his floor and his method of disposing of his weapon. Now he had to pick his window. Fourteen large, tall double-hung windows ran along the south wall facing Elm Street. President Kennedy would drive within view of all of them. Oswald had a choice of any of them from which to aim his rifle. He selected one, the window at the far southeast corner. At some point he might have rehearsed his plans, performing a walk-through of the assassination.

Perhaps he walked the length of the wall, peering down to Elm Street through each window, assessing the suitability of its angle of view. At the last window on the left, Oswald must have noticed its two advantages over all other windows on the floor. First, it looked straight down Houston Street, the route that Kennedy's limousine would follow to Elm Street on November 22. As Oswald watched the traffic come up Houston Street, he had to have seen what an easy target a car on that road would be.

The president would follow that identical route. In other words, for one block, the president would drive directly toward the Depository, up Houston Street, and toward that window. From Kennedy's point of view, the window was the one on the far right end of the sixth floor. From Oswald's point of view, he would have an unobstructed, head-on look at the president's car as it drove closer and closer toward the Book Depository.

Second, right below the Depository, the president's car would have to slow almost to a stop to make the hairpin, tight-angled left turn onto Elm Street. Then, from almost the moment the car made the turn and then continued along the length of the Book Depository, anyone standing in that window would have, for at least ten or fifteen seconds, a perfect view of the back of the presidential limousine. That was plenty of time to get off two to four well-aimed shots. No other window on the sixth floor offered an earlier look at the president's approach up Houston Street or a longer look at the back of the car once it turned onto Elm. This was where Oswald would build his sniper's nest.

BY THE afternoon of Thursday, November 21, Oswald was willing but not yet equipped to carry out an assassination the next day. Oswald was spending weekdays at a Dallas rooming house at 1026 North Beckley Street, while Marina lived in Mrs. Ruth Paine's house in Irving. He kept his .38-caliber snub-nosed revolver and its leather belt holster (designed for concealment) in his room. Police-style revolvers, unlike Oswald's, had longer barrels, which made them more accurate but harder to hide from view. But he needed his rifle. It was at Mrs. Paine's house, in her garage, lying flat on the floor, wrapped in a blanket. But it was a Thursday, not a Friday, and he was not supposed to drop in at Ruth Paine's unannounced. He had never gone to Irving on a weekday. Tonight, he would have to make an exception to that rule. Lee, who did not own a car or have a driver's license, asked Buell Wesley Frazier for a ride to Irving. Buell had been giving Oswald a lift to Irving every Friday.

Between eight and ten o'clock on Thursday morning, Oswald approached Frazier at a first-floor work table, handling orders.

"Could I ride home with you this afternoon?"

"Sure. You know, like I told you. You can go home with me anytime you want to, like I say anytime you want to go see your wife that is all right with me."

Frazier's house, at the corner of Westbrook and Fifth, stood about half a block east from Ruth Paine's, both on the north side of Fifth. Frazier realized it wasn't Friday, Lee's customary day to visit Marina for the weekend.

"Why are you going home today?"

"I am going home to get some curtain rods. You know, put in an apartment."

"Very well."

Frazier asked Oswald if he would also like a ride home after work tomorrow afternoon too, on Friday afternoon after work, but Lee said no, he did not need a ride on November 22. He would not be going home to Marina that day. For the rest of Thursday, they did not talk any more about the ride. They got off work at 4:40 P.M., and Buell drove Lee to Irving. It usually took until 5:20 P.M. to 5:35 P.M. to get there, depending on traffic and the length of stops at train crossings. During the trip, neither man mentioned President Kennedy's visit to Dallas.

Oswald's morning conversation with Frazier offers a clue that Lee had decided by no later than ten A.M. on Thursday, November 21, to assassinate President Kennedy. And the fact that he did not bring his revolver to work that morning is persuasive evidence that he had not decided by the night of Wednesday, November 20, to commit the murder. If he had, he might have brought the pistol to work the next morning and hidden it overnight in the Book Depository. So Thursday morning was Oswald's last chance to carry the revolver from his rooming house and sneak it into the Depository. He could not go to his Dallas rooming house on Thursday evening to get it—that night he needed to drive with Wesley Frazier to Irving to get the rifle from Mrs. Paine's garage.

Lee also wanted to see the babies and Marina one last time. There was something he wanted to ask her.

ABOUT TEN minutes before Lee Oswald's workday at the Depository ended, Air Force One touched down in Houston at about four thirty P.M. There was another enthusiastic reception committee, another three-dozen gorgeous yellow roses for Jackie, another eager, straining crowd held back by a railing. The Kennedys worked the line, reaching over the top and shaking hands as they walked.

Several women grabbed both of Jackie's hands simultaneously, making it seem as if they might lift her off the ground and pull her over the railing. She called out to Congressman Albert Thomas, whom JFK would honor that night at a huge dinner. "Don't leave me!" Jackie pleaded. "Don't get too far away." Thomas and some of the other dignitaries rescued her and escorted her toward the car.

In Houston, she and the president would not ride in their Lincoln limousine. It had been flown ahead to Dallas for the big, forty-five-minute motorcade that would happen there tomorrow. About 175,000 people had turned out for the Houston motorcade. When the Kennedys arrived at the Rice Hotel, they retired to their suite.

WHEN LEE Oswald showed up at Ruth Paine's home in Irving, Marina was surprised, and not too happy, to see her quarrelsome, violent husband. She showed her anger by refusing to speak to him. "He was upset over the fact that I would not answer him. He tried to start a conversation several times, but I wouldn't answer."

Then Mrs. Paine came home after shopping for groceries. She was not pleased to find him at her home on a day he was not supposed to be there. Marina was exasperated because she had discovered that Lee was living under an alias at the Beckley Street rooming house. Recently, when she had telephoned there to speak to him, she was told that no one by the name of Lee Oswald lived there. Marina thought all this mystery and secrecy was too much. It reminded her of his behavior when he had attempted to assassinate General Walker. She had demanded that her husband explain why he was using a false name again.

Oswald's visit was a pretext to go into the garage and retrieve his Mannlicher-Carcano. He planned to spend Thursday night in Irving and then on Friday morning bring it to work. Perhaps the sight of his attractive young wife, his two-year-old daughter, and his newborn baby girl softened his murderous heart. Maybe he would not kill President Kennedy the next day after all. He spoke kind words uncharacteristic of a killer.

Lee told Marina he loved her and asked her to move back to Dallas and live with him there. "He suggested that we rent an apartment [there]. He said that he was tired of living alone." She said no. He said he would rent a nice apartment for them and the children where they could begin their lives anew. She refused. Then she tested him. "I told him to buy me a new washing machine."

It was hard for her to keep up with the laundry for two small children. When he agreed to do it, she rebuffed him and said she didn't want it after all. Marina told him it would be better if he spent the money on something else, for himself.

If Oswald was not reconsidering killing Kennedy, he would have had no reason to find a better apartment or purchase a washing machine.

Like a lot of married couples that day, the Oswalds discussed President Kennedy's forthcoming trip to Dallas. Marina wanted to see JFK and Jackie. "I asked Lee whether he knew where the president would speak, and I told him that I would very much like to hear him and see him. I asked how this could be done."

Lee did not tell Marina that she could go to Love Field to watch the president's plane land. He did not tell her that she could catch the motorcade as it crawled through downtown Dallas. And he did not reveal what he already knew—what he was already counting on—that the president's motorcade would pass through Dealey Plaza, turn onto Elm Street, and drive right past the Texas School Book Depository, where he had worked for the past six weeks, and that she could meet her husband outside and see the president there. No, to Marina he pleaded ignorance. "He said he didn't know how to do that," she recalled, "and didn't enlarge any further on that subject."

Marina did not know that Lee knew exactly where to find President Kennedy. Nor could she know that her treatment of him tonight would help him decide to follow through with his plans to see the president tomorrow.

"I was angry," Marina said about that night. "He was not angry—he was upset. I was angry." Lee told her that he was lonely because he had not been to visit her the previous weekend. Ruth Paine had planned a family event, and she did not want him there. "He said that he wanted to make his peace with me." Marina saw that "he tried very hard to please me. He spent quite a lot of time putting away diapers and played with the children on the street." But Marina would not give in. Perhaps, she wondered later, whether she had been too hard on him.

WHILE THE Oswalds were bickering about their family life, John and Jacqueline Kennedy rested in Houston's Rice Hotel and dressed for the evening. The president put on a fresh shirt and suit. He liked to look good, and he was in the habit of changing clothes several times a day, even when home at the White House. His personal valet, George Thomas, was with him on the Texas trip and took care of all of his wardrobe needs. Jackie dressed in a black velvet suit. Her dark hair and the dark fabric framed her pale face and seemed to make her skin glow. Then she fastened on a pearl necklace and diamond earrings.

Now, dressed for the big dinner they would attend that night, the president of the United States and his wife sat down together in their hotel suite and dined alone. It was impossible for any president to eat a meal in peace at a political dinner. So many people wanted to talk to him that he never had a moment even to raise a fork or spoon to his mouth, let alone chew the food.

So the Kennedys ate their own dinner before the public one. They did not know it, but this was to be their last private dinner together. So this was John Kennedy's last gaze across the private dinner table at a beautiful young woman, the mother of his two children, dressed in understated but gorgeous elegance, practicing in her head the speech she would deliver in Spanish that night in the hotel ballroom, to the League of United Latin American Citizens (LULAC).

John Kennedy could not have known it, but that night another man having *his* last dinner with *his* wife was still planning to bring his rifle to work tomorrow.

After the First Couple finished their meal, Lyndon Johnson visited their suite for an animated private talk about the political rivalries and bruised egos that existed among Texas elected officials and were causing some ill will on this trip. Jackie went to another room to practice her speech, but she could hear loud voices on the other side of the door. She thought the two men were arguing, but they were not.

When Johnson left, Jackie came out to see her husband. Jack told her that he and the vice president had not had a fight. Oh no, Jack assured her. Then, apropos of nothing, Jackie said she could not stand Texas governor John Connally.

"Why did you say that?"

"I [couldn't] stand him all day. He's just one of those men—oh, I don't know. I just can't bear him sitting there saying all these great things about himself. And he seems to be needling *you* all day."

The president gave his wife a brilliant piece of advice that revealed what an astute political psychologist he was. "You mustn't say you dislike him, Jackie. If you say it, you'll begin thinking it, and it will prejudice how you act toward him the next day."

Kennedy explained. "He's been cozying up to a lot of Texas businessmen who weren't for him before. What he was really saying in the car was that he's going to run ahead of me in Texas [in the next election]. Well, that's all right. Let him. But for heaven's sake don't get a thing on him, because that's what I came down here to heal. I'm trying to start by getting two people in the same car. If they start hating, nobody will ride with anybody."

The Kennedys went downstairs for the LULAC reception. After the president's remarks, he said to the crowd, "In order that my words will be even clearer to you, I am going to ask my wife to say a few words to you also." Jackie gave her little talk in Spanish, the crowd loved her, and when she finished they yelled "Olé!" A fugitive piece of lost but recently discovered amateur film footage taken that night with a home movie camera, without sound, shows how excited the audience was.

From the Rice Hotel ballroom, the Kennedys drove to the Houston Coliseum for the huge dinner honoring Congressman Thomas. Tonight the president made no attempt at soaring oratory. Instead he gave more of a bread-and-butter talk designed to flatter local interests and leaders, especially the honoree. But he slipped in an eloquent passage from the Bible. Indeed, it was so fine that it seemed out of place in what was otherwise a pedestrian effort. It sounded prophetic: "Your old men shall dream dreams, your young men shall see visions. Where there is no vision, the people perish."

The Kennedys left the Coliseum at a little after nine thirty P.M. and drove to the airport.

At 11:07 P.M., Air Force One landed at Carswell Air Force Base, outside Fort Worth.

Air Force One had closed the distance in geography and time that separated John F. Kennedy and Lee Harvey Oswald.

LEE WATCHED television for a while and then went to bed before Marina and turned off the light. But when Marina joined him, she sensed that he was still awake, only pretending to be asleep. And when she, in a gesture of intimacy in the darkness of the night, extended her leg to touch his, he kicked it away.

And now, as the night lengthened on November 21, 1963, he had failed in love. If Marina had told him this evening that she loved him, that he was a good man, that she and their daughters would live together again—then tomorrow Lee might look for a new apartment for his family instead of carrying a long package of "curtain rods" to work. But tonight, without her approval, he was helpless, alone, and drifting toward oblivion. It was Oswald's habit to copy sentimental Russian-language poetry into journals. Perhaps it was to please Marina. Some evidence suggests that she read some of them and annotated a few of the pages in her own hand. Or maybe Lee just wanted to practice his Russian comprehension and handwriting. From the hundreds of lines he copied into his commonplace books, one phrase leaps out. It reads as though he was describing himself on the night of November 21, 1963. "Not everyone can understand his own life. . . . Life is boring, empty and uninteresting."

Tomorrow he would change that.

This was the last night he would spend with his wife.

NEAR MIDNIGHT on Thursday, November 21, John Kennedy prepared to turn out the lights in his bedroom in Fort Worth's Hotel Texas. He was tired. In one day, he had flown from Washington, D.C., to San Antonio, Texas; had driven in a motorcade; had given a speech to twenty thousand people; had toured a hospital; had flown from San Antonio to Houston; had spoken at a reception; had spoken at a major political dinner; and had flown from Houston to Forth Worth, where Air Force One had landed at 11:07 P.M. In Washington time, it was already past midnight. Then he had been driven to the Hotel Texas in Fort Worth, where he found himself now. It was a grueling schedule, typical of a presidential campaign.

Tomorrow, on November 22, the president faced another tough schedule, which included informal remarks on the street outside his hotel; a breakfast and talk at the hotel; the flight to Dallas; the long motorcade from Love Field to the Trade Mart for a big political lunch and speech; then back to the airport for the flight to Austin, another motorcade, and another speech at a major Democratic dinner that evening.

Kennedy was in full campaign mode. In late summer and fall of 1960, he had spent weeks just like this on the campaign trail. This trip to Texas was a dress rehearsal for next year, and it had proven what an asset Jackie was going to be during his second presidential campaign in the fall of 1964.

He wanted her to know that.

"You were great today," he told Jackie. "How do you feel?"

"Oh, gosh," she said, "I'm exhausted."

He told her she did not have to get up early. "Don't get up with me. I've got to speak in the square downstairs before breakfast, but stay in bed. Just be at that breakfast at nine-fifteen."

They did not know that this was the last night they would spend together.

SOMETIME THAT night, after John and Jackie Kennedy had gone to sleep—no one knows exactly when—Lee Oswald slipped into Mrs. Paine's garage, turned on the light, and lifted up the blanket and its deadly contents. The Mannlicher-Carcano was still there. Tonight, in addition to his rifle, Lee also needed ammunition. He had depleted his supply during target practice. All he had left in his possession were four bullets. These would have to be enough. He disassembled the weapon, removing the barrel from the stock, and slipped the shorter pieces into a bag he had made at the Book Depository with brown paper and tape.

Later that night, probably not until long after midnight, Lee Harvey Oswald drifted off to sleep.

CHAPTER 4

⸺ ◦≡◦ ⸺

"A BRIGHT PINK SUIT"

The Kennedys were staying in suite 850 of Fort Worth's Hotel Texas. The president had just been handed a copy of the November 22 edition of the *Dallas Morning News*, and he was irate. It contained a full-page ad about his visit. At first glance, its boldface headline, WELCOME MR. KENNEDY TO DALLAS, seemed to be a friendly greeting. But the ad was bordered in black, the way a newspaper might announce a tragedy or a death. As the president read further, he saw a long list of complaints attacking him and his administration. The ad demanded that Kennedy answer twelve defamatory questions that accused him of being soft on Communism and unpatriotic. "Why is Latin America turning either anti-American or Communistic, or both, despite increased U.S. foreign aid, State Department policy, and your own Ivy-Tower pronouncements?" it began. Then the ad claimed that "Thousands of Cubans" had been imprisoned, starved, and persecuted as a result of Kennedy's policies. The ad even accused the president of selling food—"wheat and corn"—to communist enemies engaged in killing American soldiers in South Viet Nam. Attorney General Robert Kennedy did not escape attack. "Why have you ordered or permitted your brother Bobby . . . to go soft on Communists, fellow-travelers, and ultra-leftists in America, while permitting him to persecute loyal Americans who criticize you, your administration, and your leadership?"

"We demand answers," the ad concluded, "and we want them now."

The charges incensed JFK. Was this the kind of welcome he should expect when he flew from Fort Worth to Dallas later this morning?

There was more. Unbeknownst to Kennedy, last night in Dallas someone had printed several thousand leaflets headlined WANTED FOR TREASON. The handbill resembled an Old West–style reward poster with front and side mug shots of the "criminal" on the loose—in this case the thirty-fifth president of the United States. To ease tensions, the Dallas police chief had already gone on television to ask his fellow citizens to receive the president with respect.

Kennedy warned his wife, "We're heading into nut country today. But Jackie," he added, "if somebody wants to shoot me from a window with a rifle, nobody can stop it, so why worry about it?"

Kenneth O'Donnell—JFK's longtime aide and a trusted member of the "Irish Mafia" trio that had guided his political career for years, and who was with the president in Texas that morning—agreed with that sentiment. "The President took a fatalistic attitude about the possibility of being assassinated by a fanatic, regarding such a danger as being part of his job, and often talked about how easy it would be for somebody to shoot him with a rifle from a high building."

Then JFK said an eerie thing to his wife. He reminded her of their harried, late-night arrival at the hotel. There were a lot of people pressing near them, and some of them had gotten too close for JFK's comfort. "You know," he told Jackie, "last night would have been a hell of a night to assassinate a president. There was the rain, and the night, and we were all getting jostled. Suppose a man had a pistol in a briefcase and melted away into the crowd?"

The president had already survived one assassination attempt. In 1960, when he was president-elect, a madman had plotted to blow him up with a bomb in Palm Beach, Florida. But that plot had been thwarted.

Perhaps JFK recalled a letter he wrote in 1959, the year before the election. Kennedy replied to a man who wrote to him about the "thought-provoking . . . historical curiosity" that since 1840 every man who entered the White House in a year ending in zero had not lived to leave the White House alive. JFK replied that "the future will necessarily answer" what his fate will be if he should have "the privilege of occupying the White House."

John Kennedy was a fatalist who lived with a sense of detachment and ironic humor. He had an intuition that he might not live a long life. One of his favorite poems was one written by a fellow Harvard graduate, Alan Seeger, who had been killed in the First World War. "I have a rendezvous with death," Seeger had prophesied. Kennedy often asked Jackie to read the poem aloud to him.

But he also believed he was lucky and that luck had taken him all the way to the White House, where, at age forty-three, he was the youngest man ever elected to the presidency. On November 22, he was forty-six years old. He had been president of the United States for two years, ten months, and two days.

As John Kennedy spoke in the safety of his Fort Worth hotel suite about rifles and assassins, a man who wanted to kill him was already waiting for him in Dallas. That man had a rifle, and he was in a tall building.

ON THE morning of November 22, Lee Harvey Oswald woke up early and was out of bed before Marina. He told her not to get up. He went to the kitchen and made himself a cup of coffee. She did not see what he did next. He placed $170 in cash, almost all the money he had in the world, on the top of the dresser for Marina. He had less than $20 in his wallet. And in a porcelain cup she had brought with her all the way from Russia, Lee Oswald left his wedding ring.

He told Marina he would not be back tonight. Sleeping on the idea had not changed his mind. He left the house, went into the garage, and emerged with his package.

Oswald walked the half block over to Buell Wesley Frazier's house. Buell's sister, Linnie Mae Randall, lived there too with her husband and three daughters. Last night, Linnie remembered, her brother told her "that Lee had rode home with him to get some curtain rods from Mrs. Paine to fix up his apartment."

Friday morning, after Linnie made breakfast and while she was packing Buell's lunch, she saw Oswald cross Westbrook Street, then cross her driveway, and walk toward the carport. She noticed that Oswald had something in his right hand. "He was carrying a package in a sort of heavy brown bag, heavier than a grocery bag it looked to me." Oswald held the bag in a vertical position. "It almost touched the ground as he carried it."

Linnie watched Oswald go to Buell's car, open the back door, and lay the package down. She could not see if he had laid it on the floor or the backseat. Then Oswald approached the house and peered through the kitchen window while Buell was sitting down eating his breakfast.

Frazier's mother saw him and asked, "Who is that?"

"That is Lee," her son replied.

It was the first time Oswald had ever walked to Frazier's house and peered through the window. Usually Frazier just picked him up on the street while Oswald was walking to his house.

Frazier rose from the table and said, "Well, it is time to go."

Oswald waited for him a few feet outside the back door. They walked together toward the car and got in the front seats. When Frazier turned his head, he saw the long paper bag. "I noticed there was a package laying in the backseat, I didn't pay too much attention and I said: 'What's the package, Lee?'"

And he said, "Curtain rods."

The blanket that Oswald used to hide the rifle in Mrs. Paine's garage, the brown paper bag he used to carry the rifle on his way to work, and the disassembled Mannlicher-Carcano.

Frazier replied, "Oh yes, you told me you were going to bring some today."

Frazier didn't think any more about it. "I asked him did he have fun playing with them babies, and he chuckled and said he did." Buell noticed Lee did not bring his usual small paper sack containing his lunch. "Right when I got in the car I asked him where was his lunch, and he said hc was going to buy his lunch that day."

They did not discuss President Kennedy's visit to Dallas.

Frazier had no way of knowing it, but Oswald did not need cur-

President Kennedy, Governor Connally (second from right), and Vice President Johnson (far right) outside the Hotel Texas.

tain rods. His small room at the boardinghouse was already fully furnished with blinds, curtains, and curtain rods.

The men got into the car and drove to Dallas. Unbeknownst to Buell Wesley Frazier, he was chauffeuring an assassin—and his murder weapon—to the scene of a crime that had not yet happened.

Oswald and Frazier arrived at the outdoor parking lot located several hundred feet behind the Texas School Book Depository at around 7:52 A.M. They were early, so Frazier stayed in the car for a minute: "I was letting my engine run and getting to charge up my battery."

Oswald jumped out of the car, reached into the backseat for his package of "curtain rods," and began walking fast toward the Depository, getting as far ahead of Frazier as he could without breaking into a run. It was the first time Lee had ever walked ahead of Buell. They usually walked to the Depository together, but not this morning: "Eventually he kept getting a little further ahead of me," Frazier noticed.

Frazier also noticed that Lee carried the package in an unusual manner, vertically, with one end tucked under his right armpit and the other end in his cupped hand. The billowing sleeve of Oswald's jacket almost concealed the package, so that it was almost invisible to anyone—including Frazier—walking behind Oswald. By the time Lee neared the Depository, he was well ahead of Buell: "[He was] I would say, roughly 50 feet in front of me but I didn't try to catch up with him because I knew I had plenty of time so I just took my time walking up there."

Oswald opened the back door of the building and slipped inside. Then he went to the sixth floor, using either the staircase at the rear of the building or one of the two freight elevators. He probably took the stairs to avoid riding in an elevator with any curious coworkers who might ask questions about his package. Once he was sure no one was watching him, he hid the bag containing the still-unassembled rifle between stacks of book boxes on the sixth floor. No one knows for sure when he assembled it. It is possible he assembled the rifle that morning before hiding it, in case something later prevented him from doing so during the critical moments before Kennedy's motorcade arrived. With a screwdriver it would have taken between two and two and a half minutes to assemble the weapon. But Oswald did not need a tool to remount the barrel to the stock—a thin dime would turn the screws and transform the rifle into firing condition.

With his rifle concealed from view, Oswald picked up his clipboard and began filling orders for books, just as he did on any other normal day at work. At least for the next three hours until the presidential motorcade approached Elm Street, he would earn his pay.

JOHN AND Jacqueline Kennedy were still in their suite at the Hotel Texas in Fort Worth when Lee Oswald walked into the Book Depository with his rifle. The president dressed first. He was scheduled to go downstairs soon to deliver brief remarks to a crowd that had assembled in a parking lot across the street. He looked outside. It was a gloomy, rainy day. The Secret Service worried about the weather in

Dallas. The president's limousine had already been sent there ahead of Kennedy, so it would be in position when he landed at Love Field. The car was a convertible, and the agents wondered if they should install the plastic bubble top to protect the president from the rain. John Kennedy preferred to ride in an open car so the people watching the motorcade could get a better look at him. It created an intimacy with the crowd. Jackie did not like the open car—the wind would play havoc with her stylish hair.

While in Fort Worth, Kennedy would travel in a rented convertible. It was unarmored. If a madman jumped from the curb and fired shots, the bullets could penetrate the metal doors of the car. During Kennedy's trip to Berlin, two women had broken through the security cordon and ran right up to his car in a motorcade. They turned out to be overenthusiastic but harmless fans.

The president went downstairs without Jackie at 8:45 A.M. to speak to the cheering crowd standing outside in the rain. "There are no faint hearts in Fort Worth," he complimented them, "and I appreciate your

President Kennedy greeting the crowd outside the Hotel Texas in Fort Worth on the morning of November 22, 1963.

President Kennedy speaks to the crowd outside the Hotel Texas in Fort Worth. Note the open windows in the background.

being here this morning." Some people began chanting "Where's Jackie? Where's Jackie?" JFK could tell they were disappointed at not seeing her, and he used wry humor to soften the blow and explain her absence: "Mrs. Kennedy is organizing herself. It takes longer, but, of course, she looks better than we do when she does it. But we appreciate your welcome . . . here in this rain."

Back in their hotel room, as Jackie got dressed, she was able to hear his voice over the loudspeakers.

At 9:00 A.M., the president returned to the hotel to speak at a public breakfast in the ballroom. But Jackie's seat at the head table was empty. The disappointed guests wanted to see her. Kennedy told one of his Secret Service agents to call her room and tell her to get down to the breakfast as soon as possible.

"Where's Mrs. Kennedy," he said. "Call Mr. Hill. I want her to come down to the breakfast."

She had forgotten. When Secret Service agent Clint Hill escorted her downstairs in an elevator, she thought he was taking her to the car. "Aren't we leaving?" she asked. "No," Hill replied, "you are going to a breakfast."

On the podium, JFK stalled for time. Then he joked with the crowd: "Two years ago, I introduced myself in Paris by saying that I was the man who had accompanied Mrs. Kennedy to Paris. I am getting somewhat that same sensation as I travel around Texas. Nobody wonders what Lyndon and I wear." Then a commotion began at the back of the room. Jackie had arrived, and the crowd shouted its approval as she marched to the podium.

Wall Street Journal reporter Al Otten took cynical amusement in Jackie's service as a political prop. "When are you going to have her come out of a cake?" he wisecracked to Dave Powers, JFK's Irish longtime top political operative.

Powers was not amused. "She's not that kind of Bunny."

She did not need a cake. Jackie did not disappoint. She was wearing a bright pink, nubby wool jacket (Jackie called the color raspberry) faced with dark blue lapels, with a matching pink skirt and a pink pillbox hat. She wore short, bright white cotton gloves, a fashionable accessory for women at that time. It was one of the most flattering outfits in her wardrobe, and she had worn it on several prior occasions. The colorful suit seemed almost to glow and created a striking contrast against her rich black hair and pale white skin. She looked radiant.

After the Chamber of Commerce breakfast, the Kennedys went upstairs to their suite to rest before leaving Fort Worth. Jackie offered to do more campaigning. "I'll go anywhere with you this year."

"How about California in the next two weeks?"

"I'll be there."

"Did you hear *that*?" the president asked his aides and laughed. They had an hour until their departure.

President Kennedy's motorcade left the Hotel Texas for Fort Worth's Carswell Air Force Base. It would be a short flight—just thirteen minutes—to Dallas. It was such a short trip that they could have driven. But that would have eliminated the ceremonial glamour of a presidential arrival of Air Force One at Love Field. The landing of the beautiful jet, the emergence of Jack and Jackie from the rear door, the welcoming committee, the crowds, the photographers, and the television cameras—all enhanced the excitement and appeal.

The president boarded the plane at 11:23 A.M. Air Force One took off from Carswell at 11:25 A.M. (CST) and touched down at Love Field in Dallas at 11:38 A.M. It was an hour later in the nation's capital, and most officials in Washington were at lunch. Several members of the cabinet were out of the country, aboard a plane flying to Japan for trade talks with government officials.

At Love Field, reporters and television cameras prepared to cover the president's arrival. While JFK's jet taxied off the runway and slowed to a stop, one journalist, Bob Walker of WFAA-TV (Dallas channel 8), the ABC affiliate, broadcast a live report: "Security precautions at this luncheon they are going to attend range from the distance from the president's car door to the Trade Mart entrance, and to how many doors and windows there are in the building, and even to the method of choosing the steak that the president will eat."

But the reporter failed to mention how many windows—more than twenty thousand—the motorcade would drive past on the way to the Trade Mart, before the president would have the opportunity to be protected there from chewing a piece of poisoned steak. The Secret Service planned to select John Kennedy's plate at random from among the two thousand meals that the kitchen prepared. To kill the president, an assassin would have to outdo the legendary poisoner Lucrezia Borgia. He would have to poison *all* of the steaks served that day.

The reporter glanced at the crowd at Love Field: "Handkerchiefs are being waved, the placards are being held high, and hundreds of tiny American flags are now being waved toward the presidential jet."

The plane parked near a reception committee of dignitaries standing on the tarmac. The reporter, making no effort to conceal his obvious excitement, continued his broadcast: "And here is the presidential jet, U.S. Air Force Number 1 . . . the doors fly open and the loading ladders are being wheeled to the plane. . . . This is a split second timed operation for the Secret Service . . . nothing is left to chance, every possible precaution has been taken."

But it was not true.

None of the people at Love Field had been searched, not even the suspicious ones displaying unfriendly signs. Anyone in that crowd could have been carrying a concealed pistol. All that separated the spectators from the president was a hip-high chain-link fence.

At the Texas School Book Depository, the lunch hour would begin soon. Most of the employees would start their noontime descent from the upper floors to congregate in the lunchroom on the second floor or the "domino room" on the first floor, or to go outside and watch President Kennedy drive by on Elm Street. The biggest crowds waiting to see the president—tens of thousands of people—had already gathered downtown, choking the sidewalks several bodies deep before Kennedy had even landed. Hundreds of people had perched in windows and now looked down on the route. The Secret Service and the Dallas police did not have the manpower—or the desire—to search every building, to scrutinize every window, or to analyze every face. It was impossible. "Every possible precaution" had *not* been taken. On the contrary—almost none had.

But here, in Dealey Plaza, where the motorcade would cover the last leg of its ten-mile route before it picked up speed and took the highway to the Trade Mart, the crowds were much thinner. It was easier for the scattered spectators to stake out a spot right at the edge of the curb and stand just a few feet away from where the president's limousine would soon pass.

Indeed, people standing on the right side of the car would be closest to Kennedy, in some cases less than six feet from him. Things were much calmer and quieter in Dealey Plaza, in the vicinity of the Texas School Book Depository, than they were downtown.

• • •

AT LOVE Field, the president looked out one of the plane windows and saw the enthusiastic crowd. He told Ken O'Donnell, "This trip is turning out to be terrific. Here we are in Dallas, and it looks like everything in Texas is going to be fine for us."

Jack and Jackie stood by the rear door of Air Force One, waiting for it to open. Dave Powers told them, "You two look like Mr. and Mrs. America." The president and First Lady exited Air Force One and descended the portable stairs that airport workers had rolled out to the plane. Jackie came down first.

"There is Mrs. Kennedy," exclaimed the reporter, "stepping off the plane, wearing a bright pink suit . . . and a matching pink hat."

Mike Quinn, a reporter for the *Dallas Morning News*, was taken

Jackie and JFK disembark Air Force One in Dallas.

by Jackie's outfit: "[I]t was beautiful, and the color seemed to reflect the sun. As she stepped out ahead of the president the crowd seemed awestruck, then started applauding and . . . squealing."

Waiting for them on the ground were a number of local dignitaries and politicians, including Vice President Lyndon B. Johnson, who had flown to Dallas ahead of the Kennedys. Protocol dictated that LBJ, a Texan, would welcome the president at each stop within the state.

The wife of the mayor presented Jackie with a big bouquet of flowers. The yellow rose was the official state flower, but so many of them had been ordered for the various events the Kennedys were attending in Texas that local florists had run out of them. Instead, the mayor's wife gave Jackie roses of another kind. "Mrs. Kennedy has been presented her bouquet of brilliant red roses," the reporter told his listeners, "and they make a lovely contrast to the bright pink suit she's wearing." He was right. The bloodred crimson petals looked striking against the pink fabric. The combination of colors seemed to intensify the hue of each, rendering them almost fluorescent.

While the Kennedys greeted the members of the airport reception committee, the journalist encouraged listeners to throng the motorcade route. "The weather couldn't be better, we have a brilliant sun. . . . Now those of you who are waiting along the parade route, just to be sure that you find yourself in the proper location, let's give it to you again."

Lee Harvey Oswald was not listening. He already knew *he* was in the proper location. He had not brought a portable, battery-powered transistor radio to his window perch. He did not need one to tell him when the president was coming. The crowds would do that for him. When the spectators one block away at the far end of Houston Street saw the motorcade coming up Main Street, they would begin to wave and cheer. Oswald would know the president was nearing Dealey Plaza.

The radio announcer continued: "The party is now ready to depart Love Field. It will go Mockingbird Lane to Lemmon Avenue, then travel south on Lemmon to Turtle Creek, Cedar Springs, through the downtown area to Main. West on Main to Houston, through the triple underpass to Stemmons Freeway, then on to the Trade Mart."

But JFK's car did not leave Love Field just then. The reporter said there was some kind of delay, but he could not discover its cause. Then he saw what was happening. "The president is not *in* his limousine. This is the moment where the Secret Service has its point of tension. As we have talked with many of these Secret Service men in the past few days—they say, when the president stops moving, that's when we are concerned because that is when the possibility of trouble comes to the forefront."

JOHN KENNEDY saw that a few thousand people had turned up behind a chain-link fence at Love Field to greet him. Some carried flags and signs, most of which were friendly. Someone brought a Confederate battle flag, a possible sign of protest against the president. Kennedy, who was treating the whole Texas trip like a campaign stop in preparation for the November 1964 election, made an impulsive decision to work the fence line. Jackie, cradling the red roses, followed him.

At that moment, a news photographer named Art Rickerby, who had gotten ahead of the Kennedys, turned around and took a series of color photographs of the couple as they walked side by side toward his lens. The president's trim blue-gray suit and blue tie and Jackie's bright pink suit and red roses saturated the camera's color film. Behind them was the big red, white, and blue American flag painted on the tail of Air Force One. The decorative blue highlights painted on the fuselage of the plane matched the color of the bright blue Texas sky. When the people in the crowd realized the president of the United States was coming over to visit them, they went wild.

"And here comes the president now," radio listeners were told. "In fact he's not in his limousine. He's departed the limousine. He is walking."

Kennedy strolled right up to the crowd and plunged his hands

Jackie, having just been presented a bouquet of red roses at Love Field in Dallas.

and arms over the hip-high fence. In response, hundreds of hands grabbed his. "He is reaching across the fence, shaking hands, shaking hands, with many of the people who have come to see him. He is closely accompanied by Dallas police officers and of course the Secret Service."

As Jackie walked the line, women begged to touch her and shrieked with delight if they succeeded. Jackie, who could be timid in crowds, smiled and shook hands with people too. Her big, genuine smile suggested that she was enjoying herself. Charles Roberts, White House correspondent for *Newsweek* magazine, asked Jackie how she liked campaigning. "It's wonderful, it's wonderful," she said, and seemed to mean it. The limousine crawled along behind the president.

The Kennedys and Texas governor Connally smile at the crowds as the motorcade makes its way from the airport. Agent Clint Hill, in sunglasses, is visible behind Mrs. Kennedy. Nellie Connally is on her husband's left.

"Security is tense at this time but is going beautifully," gushed the radio reporter.

AFTER KENNEDY shook a last hand, he and Jackie turned to their car. The bubble top was off. They would ride through Dallas in an open car. Earlier that morning, the agents gambled that the gray skies and rain would disappear by noon. Their bet paid off. It was a bright, gorgeous, and sunny fall day.

The president sat in the right rear passenger seat. Jackie took the seat to his left. In front of the Kennedys, sitting at a slightly lower level in folding jump seats that also faced forward, were the governor of Texas, John Connally, and his wife, Nellie. In the front seats sat the driver, Secret Service agent Bill Greer, and the head of the Secret Service detail, Roy Kellerman. Two other Secret Service agents trotted alongside the rear fenders. Behind the president's car, the Queen Mary convertible was filled with Secret Service agents.

Following behind them were several other vehicles, whose passengers included Vice President Lyndon Johnson and his wife, Lady Bird, several Texas politicians, members of the press, and White House staff, including the president's personal secretary, Evelyn Lincoln. As the motorcade got under way, preceded by a pilot car that drove ahead to spot any trouble, and followed by police motorcycles and a white car carrying the Dallas chief of police, Jesse Curry, a television news camera filmed the Kennedys' departure from Love Field. Everything was in order. Then the camera caught something odd.

It appeared that Secret Service agent Don Lawton, trotting behind the presidential limousine, was about to reach for the handholds on the back of the car and step up onto the footholds attached to the rear bumper. In that position, he could stand behind the president as the car moved at a slow speed through downtown Dallas. It would be an extra precaution that placed an agent closer to the president. If a spectator rushed the car, Lawton could leap off and intercept him. And if the agent stood erect on the bumper, he could stand between the president and anyone who might try to take a shot at him from behind.

Then a Secret Service agent in the trail car, Emory Roberts, stood up and shouted to Lawton. He motioned him to back off from the president's limousine. Lawton appeared to object to the order by shrugging his shoulders three times and waving his arms in the air, as if to say, "what are you *doing*?" But then he walked away from the car.

The television camera captured it all. But this brief vignette was not what it appeared to be. It was just a joke between Lawton and Roberts. Lawton was not even assigned to ride in the motorcade through Dallas. He knew that JFK did not want any agents riding on the limousine, and Lawton thought it would be funny if he pretended to defy the president's order. It was too bad that it was just a joke. The camera continued to track the limousine until it vanished from sight.

The president and First Lady were on the way to downtown Dallas. The route was ten miles long, and it would take about forty-five minutes to drive it. They were scheduled to arrive at the Trade Mart lunch by 12:30 P.M. Kennedy's impromptu visit with the crowd at the airport had put him a little behind schedule. If the motorcade experienced no more delays, they should arrive at their lunch by around 12:35 P.M.

THE PRESIDENT'S speech was ready. He would not deliver informal, off-the-cuff remarks as he did that morning in Fort Worth when he stepped outside into the rain to speak to the crowd. No, this was an important stop on the trip. Like the breakfast speech he gave inside the Hotel Texas, his words for the lunch at the annual meeting of the Dallas Citizens Council—his third talk of the day—had been prepared in advance.

He would speak about American exceptionalism—the nation's role in the world, national defense, foreign aid, and the strength of American science, industry, education, and the free-enterprise system. He planned to criticize naysayers: "There will always be dissident voices heard in the land, expressing opposition without alternatives, finding fault but never favor, perceiving gloom on every side and seeking influence without responsibility. Those voices are inevitable." There was more. "But today," he warned, "other voices are heard in the land—voices preaching doctrines wholly unsuited to reality, wholly unsuited to the sixties, doctrines which apparently assume that words will suffice without weapons . . . and that peace is a sign of weakness."

The text of the speech echoed many of the themes from his Inaugural Address: "Our adversaries have not abandoned their ambitions, our dangers have not diminished, our vigilance cannot be relaxed. But now we have the military, the scientific, and the economic strength to do whatever must be done for the preservation and promotion of freedom." Kennedy promised that America's strength "will never be used in pursuit of ambitions—it will always be used in pursuit of peace." The speech ended with a full-throated appeal for peace through strength.

"We in this country, in this generation, are—by destiny rather than by choice—the watchmen on the walls of world freedom. We ask, therefore, that we may be worthy of our power and responsibility, that we may exercise our strength with wisdom and restraint, and that we may achieve in our time and for all time the ancient vision of 'peace on earth, good will toward men.' That must always be our goal, and the righteousness of our cause must always underlie our strength. For it was written long ago: 'except the Lord keep the city, the watchman waketh but in vain.' "

THE PRESIDENT was scheduled to deliver another speech that night in Austin. Its text focused more on domestic issues: "For this country is moving and it must not stop. It cannot stop. For this is a time for courage and a time for challenge. Neither conformity nor complacency will do. Neither the fanatics nor the faint-hearted are needed. . . .

So let us not be petty when our cause is so great. Let us not quarrel amongst ourselves when our Nation's future is at stake. Let us stand together with renewed confidence in our cause—united in our heritage of the past and our hopes for the future—and determined that this land we love shall lead all mankind into new frontiers of peace and abundance."

At least these were the words he planned to say today. He was in the habit of revising his typewritten speeches with his undecipherable handwriting, often at the last minute. Later in the day, before he gave them, he might look them over once more at the Trade Mart, and on the trip to Austin.

The cover of the glossy dinner program promised John Kennedy an unforgettable "Texas Welcome." Inside, the brochure printed Governor Connally's eerie introductory remarks: "This is a day long to be remembered in Texas."

BACK IN Dealey Plaza, the electric digital clock on the big, yellow Hertz-Rent-a-Car sign atop the Depository roof flashed the time. It was past noon. Buell Wesley Frazier had skipped lunch, and he was already standing near the top step of the Elm Street entrance to the Depository.

Frazier was eager to see John Kennedy: "He was supposed to be coming by during our lunch hour so you don't get very many chances to see the President of the United States and being an old Texas boy, and [he] never having been down to Texas very much, I went out there to see him just like everybody else."

While most of Lee Oswald's coworkers came down from the upper floors for the lunch hour, he proceeded to the sixth floor. A few asked him if he was having lunch or planning to watch the president with them. He lingered behind and gave vague answers. According to what the newspapers said, Kennedy's car should be in front of the Depository by about 12:15 or 12:20 P.M.

Lee Harvey Oswald was now alone on the sixth floor. He retrieved his rifle, either in pieces from inside the brown paper bag, or already fully assembled. He walked to the southeast corner of the building and settled into his position at the window, behind stacks of book cartons. If anybody came up to the sixth floor, they would not see him now. If the lower window pane was not already open, he slid it up now and secured it in position. He inhaled the fresh, crisp fall air. The breeze hit his face. It was about sixty-five degrees.

If he had looked down Houston Street, he would have heard no sounds nor seen any signs that the motorcade was near. No police motorcycles were in sight yet, and people on the street were not fidgeting or craning their necks the way excited crowds do when a president approaches.

Perhaps Lee adjusted some of the boxes to ensure that no one who came up to this floor could see him as he hid behind them. He still had a few minutes. He checked his rifle. It was ready to fire. One round was already in the chamber. Three more rounds waited in the clip, if Oswald needed to use them.

If he heeded the warnings drilled into him by his Marine Corps training, he would have switched the safety mechanism of his rifle to the on position. That device would prevent the weapon from firing a round accidentally if the shooter dropped and jarred it or snagged the trigger. As long as the safety was engaged, squeezing the trigger would fail to release the firing pin, and the rifle could not fire the chambered round. Once the president came within sight, Oswald could, by disabling the safety with the flick of a thumb, render his weapon lethal in an instant.

He waited in silence. The only sounds were the occasional voices that floated up from the street below before they faded into nothingness.

MINUTE BY minute, President Kennedy's motorcade drew closer to the Book Depository. The journey through downtown Dallas was more of a parade than a motorcade. On the nation's streets and highways, the president of the United States traveled at two speeds: fast and efficient to save time and minimize danger, and slow and leisurely to show himself to the crowds and wave to people as he drove by.

On November 22, John Kennedy did not want anyone in Dallas to miss seeing him because his car was traveling too fast.

Accordingly, the Secret Service had slowed the limousine to parade pace of ten to fifteen miles per hour. This made the agents nervous. When the car was topless, as it was today, the agents preferred to ride the running boards—retractable metal shelves protruding from the side of the car. They would also stand on the two steps at the rear of the limousine. Metal bars mounted to the body of the car provided handholds. Having agents stand on the Lincoln gave the president at least some protection. They could try to intercept any objects thrown at the president.

Once, on another trip, someone tossed a bouquet of flowers into the car. It was harmless, but what if had concealed a hand grenade? And the agents' bodies could block gunfire. This was especially true of the two men assigned to stand on the back steps behind the trunk. The presence of these agents would make it almost impossible for a sniper to shoot the president from behind. Had a Secret Service man been standing on the right rear step when John Kennedy drove through Dealey Plaza, and had Lee Harvey Oswald chosen to shoot the president from behind, the agent's body would have obscured Oswald's sight line. The assassination attempt would have failed.

But President Kennedy did not like having his bodyguards ride on the car because he thought it made him look less approachable to the people. On a number of occasions, he complained when agents did so. He often told his "Ivy League charlatans"—his affectionate nickname for them—to get off his car. Today in Dallas, all the men in the security detail knew they risked irritating the president if they rode on the outside of the car during the trip from Love Field, through downtown, through Dealey Plaza, and to the Trade Mart.

The Secret Service hated that open car. But the president and the people loved it.

Aboard Air Force One at Love Field, Dave Powers had given the

Kennedys advice on how to behave during the journey: "Mr. President, remember when you're riding in the motorcade downtown to look and wave only at the people on the right side of the street. Jackie, be sure to look only at the left side, and not to the right. If the both of you ever looked at the same voter at the same time, it would be too much for him!"

The ride through Dallas was a triumph. There were no ugly incidents, and people cheered extra loudly whenever the president drew near. Jackie put on her sunglasses. "The sun was so strong in our faces," she recalled. But her husband told her to take them off—the people wanted to see her face, he said.

Ken O'Donnell was impressed. "We could see no sign of hostility, not even cool unfriendliness, and the throngs of people jamming the streets and hanging out of windows were all smiling, waving and shouting excitedly. It was by far the greatest and the most emotionally happy crowd that we had ever seen in Texas."

Kennedy was happy too. O'Donnell noticed that "the President seemed thrilled and fascinated by the crowd's noisy excitement. I knew he had expected nothing like this wild welcome."

People waved and clapped. Many snapped photographs or took home movies, which in that era did not record sound. At one point along the route, JFK turned around and looked over his right shoulder in the direction of someone making a home movie. The president smiled, and with a playful hand gesture he beckoned the cameraman to follow along. "Come on!" the president seemed to tease as his car drove past the bystander.

The motorcade made two unscheduled stops. Children held up a homemade sign that asked the president to stop and greet them. When Kennedy spotted the sign, he shouted to his driver to step on the brakes. In the middle of a motorcade surrounded by thousands of onlookers, the gleaming Lincoln limousine halted. The Secret Service agents went on high alert, fearing the president was about to get out of the car and stand up. Kennedy was famous for plunging into friendly crowds and leaving nervous bodyguards behind in his wake. If he did that now,

The motorcade proceeds through downtown Dallas.

then thousands of excited people would rush forward to try to shake his hand. At Love Field, at least there was a fence. On this street, no barrier separated the president from the throng, just feet away.

It was dangerous to stop a motorcade. In 1958, when Vice President Richard Nixon visited Venezuela, an anti-American mob intercepted his car, smashed the windows, almost tipped it over, and came close to murdering him and his wife. Kennedy had the good sense not to leave the car. In Dallas, the children were allowed to approach him.

Then the limousine drove on. At a couple of points, Secret Service agent Clint Hill hopped off the trail car and stepped onto the back of the presidential limousine, assuming a crouching position behind

Jackie. From that location he could leap off the car to intercept any pedestrian who rushed the Lincoln, and if he stood up straight, his body would shield her and absorb any bullet fired at her from behind.

But Hill did not stand up, perhaps to avoid an admonition from the president or the head of the Secret Service detail. After a short time, agent Hill stepped off the car and jumped back onto the running board of the trail car. Farther along the route, JFK spotted some Catholic nuns standing curbside. He ordered his driver to pull over for the sisters. A fellow Catholic, the president wanted to pay them his respects. Kennedy remained in the car and soon bade farewell to the delighted nuns.

LEE HARVEY Oswald was not able to see any of this. He never indicated later what was in his mind while, for the second time in his life, he prepared to kill a man. His sniper attack in April on General Walker was an absurd failure. That night he was certain he had his target fixed in the center of his telescopic sight, but his first shot struck the framing of the window screen and that deflected the bullet's trajectory.

The morning after the Walker shooting, Oswald wondered how he possibly could have missed. He believed his aim was perfect. It was not his fault, he told himself and Marina. Fate had thwarted his plans. He had no time to fire a second shot.

Lee had hidden his rifle and run away into the night. Now, alone in the quiet of the sixth floor, did he think about the ridiculous, failed attempt on General Walker's life? Did he think of Marina and his two little girls one last time? Did he reconsider? Did he ask himself what on earth he was doing at this window with a rifle in his hands? Oswald left behind no journal, diary, or manifesto, no last-minute letter of explanation or justification to his wife or to the country.

Oswald waited by the window. Whenever his eyes searched Houston Street for the first signs of the motorcade, he had to be careful to not hold the rifle high in his hands. Someone on the ground might spot a man with a rifle and warn the authorities.

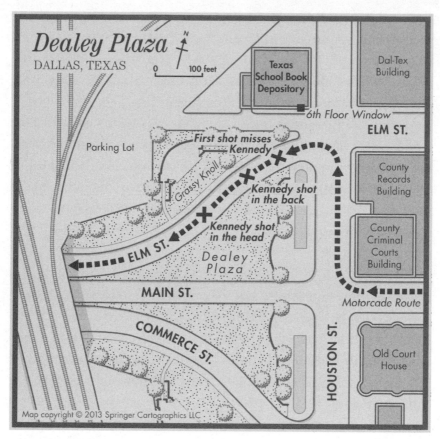

Map of Dealey Plaza.

Finally, the police motorcycles, their red lights flashing, trailed by the lead car carrying the Dallas chief of police, reached the corner of Main and Houston and then turned right onto Houston. The Texas School Book Depository stood one block ahead. Oswald could see the motorcycles first. Then the police chief's white car. Then he could see what he had been waiting for—the big, gleaming, midnight-blue limousine carrying the president of the United States. (Today, presidential motorcades travel with as many as three hardtop limousines—two decoys, plus one occupied by the president—to confuse potential assassins. In John Kennedy's era, there was only one.)

From the moment Oswald saw that car turn onto Houston Street, he knew that one of its six occupants had to be John F. Kennedy. Even from a distance, if Oswald fixed his gaze on the car, he must have noticed what appeared to be an unusually bright pinpoint of color toward the right rear of the vehicle. It was Jacqueline Kennedy, her pink suit and pillbox hat glowing like a signal beacon. Indeed, he probably saw her first, before he spotted the president.

Inside the limousine, Nellie Connally was delighted with how well the day had gone. The motorcade had passed through the cheering crowds in downtown Dallas without incident. As the car drove closer to the Book Depository, she turned around to congratulate JFK: "You can't say that Dallas doesn't love you today, Mr. President." Kennedy looked at her and smiled.

FROM THE moment Oswald saw the presidential limousine, he knew the odds that he would be able to successfully assassinate President Kennedy had just tipped in his favor. Luck was with him that day. First, Kennedy had come to Dallas. Oswald would have never stalked him on a presidential trip to another city. But now the president had come to him.

It was pure chance five weeks earlier, on October 16, before the details of Kennedy's Dallas trip had even been planned, that Oswald would start a job at a building that would turn out to be on the motorcade route. Even President Kennedy's itinerary proved lucky for Oswald. At any other time of day, the Book Depository employees working on the upper floors might have discovered Oswald hiding in his sniper's nest. But JFK would drive by during lunch hour, when the employees would vacate the upper floors and go down to eat or leave the building to watch the president pass by. By a little after noon, Oswald could expect to have the entire sixth floor to himself. But even this advantage offered no guarantee of success.

As the limousine got closer, Oswald could see that the bubble top was off. This plastic top was not exactly bulletproof, but it would have been protective because a bullet would have needed to penetrate it at a perfect angle in order to get through without being deflected. But this wasn't something Oswald would have to worry about anymore.

He could also see that no Secret Service agents were standing on the back of the Lincoln. That meant Oswald would have a clear line of sight to John Kennedy's back once the limousine turned left onto Elm Street and drove past the Texas School Book Depository. Nothing would block his view of the president.

At this moment Oswald was also in imminent danger. Three other men who worked at the Book Depository—James "Junior" Jarman, Harold Norman, and Bonnie Ray Williams—were heading to one of the upper floors to get a bird's-eye view of the motorcade.

Williams remembered that they had planned to watch the parade from the sixth floor. He had gone downstairs to retrieve his lunch—a chicken sandwich, a bag of Fritos, and a soda. Then he went back up to the sixth floor, sat by a window along the Elm Street side of the building, and ate his meal while he waited for his friends to come back up.

Visibility on that floor was limited: "I couldn't see too much of the sixth floor," Williams said, "because the books . . . were stacked so high. I could see only in the path I was standing . . . I could not possibly see anything to the east side of the building. So far as seeing to the east and behind me, I could only see down the aisle behind me and the aisle to the west of me."

What Williams could not see, concealed just a few dozen feet away from him, behind stacks of book cartons, was Lee Oswald hiding with his rifle. The men were almost within spitting distance of each other. Lee didn't make a sound, and Williams never realized that someone was lurking near the corner window.

Williams finished his lunch and, impatient for his tardy friends to join him, he took an elevator down to the fifth floor to search for them. If he did not find them there, he planned to go to the first floor and watch the motorcade from street level, in front of the Depository. But there were Norman and Jarman standing along the windows facing the Elm Street side. Hank Norman was right below Oswald's window, Junior Jarman was two or three windows over, and Williams took a position between them.

They had staked out the windows closest to the far southeast corner of the building, knowing it would give them a great view straight up Houston Street as the president's car drove toward the Depository and then turned left onto Elm right under their window.

Their perch there gave them a commanding, unparalleled view of Dealey Plaza, better than anyone else waiting for Kennedy that day. Only Oswald—on the sixth floor, right above their heads—could see the president better. Oswald was lucky that the men did not ride the elevator up one more floor, where they might have discovered him hiding with his rifle in his little fort of book boxes.

A man on the street had noticed Jarman, Norman, and Williams looking out the fifth-floor windows. Arnold Rowland was an eighteen-year-old newlywed high-school graduate, taking classes in preparation to attend college and working part time as a pizza maker at Pizza Inn. Rowland and his wife, who was still in high school, finished their classes early and went downtown: "I had to go to work at 4, so we were going downtown to do some shopping. We went early so we could see the President's motorcade," Rowland recalled.

They arrived about 11:45 A.M. and spent the next fifteen minutes walking five or six blocks to find a good vantage point. Frustrated, they tried out several locations but could not find one that suited them. They noticed a lot of policemen and commented on the security precautions being taken. By about 12:15 P.M., they had settled on a spot near the Book Depository. Arnold's watch read 12:14, but when he looked up at the Hertz clock atop the Depository, he observed that his timepiece was running one minute behind. He reset it to quarter past.

More than most people in Dealey Plaza that afternoon, the Rowlands took an unusual and keen interest in presidential security. "It was a very important person who was coming, and we were aware of the policemen around everywhere, and especially in positions where they would be able to watch crowds," Arnold noted. "We talked momentarily of the incidents with Mr. Stevenson, and the one before that with Mr. Johnson, and this being in mind we were more or less security conscious."

Then Rowland noticed several people in the windows of the Book Depository: "My wife and I were both looking and making remarks that the people were hanging out the windows . . . the majority of them were colored people, [and] some of them were hanging out the windows to their waist."

Mrs. Rowland began watching a "colored boy" in one window. Her husband continued to scan the Depository facade. "At that time," he continued, "I noticed on the sixth floor of the building that there was a man [standing] back from the window, not hanging out the window." It was then, Rowland explained, that "I saw the man with the rifle."

Rowland stared at the open window. "He was standing there and holding a rifle. This appeared to me to be a fairly high-powered rifle because of the scope and the relative proportion of the scope to the rifle, you can tell what type of rifle it is. You can tell it isn't a .22."

The man stood still, grasping the rifle with both hands, and holding it in front of his body at a forty-five-degree angle, with his right hand near his waist and his left hand near his left shoulder.

The man was light complexioned with dark hair, and "he was rather slender in proportion to his size." Rowland guessed he might weigh 140 to 150 pounds. The man looked to be "either a light Latin or a Caucasian." His hair was "either well-combed or close cut." He was young: "I remember telling my wife that he appeared in his early thirties." He was wearing, Rowland observed, "a light shirt, a very light colored shirt . . . this was open at the collar . . . it was unbuttoned about halfway, and then he had a regular T-shirt . . . under this . . . he had on dark slacks."

Rowland thought that the man was standing three to five feet back from the window to avoid being seen but still close enough for the sun to shine on him.

Rowland asked his wife if she wanted to see a Secret Service agent.

"Where?" she asked. He pointed at the Book Depository.

"In that building there." But he was mistaken. No Secret Service agents had been deployed to the Book Depository.

At that moment, the Rowlands noticed a man in Dealey Plaza who was suffering an epileptic fit. They watched policemen come to his aid and call an ambulance. Then Arnold told his wife to look at the building again, and at the open sixth-floor window at the southeast corner of the Book Depository. But the man was gone. "He is not there now," he told his wife.

She asked what he looked like. Her husband gave her a brief description of the man, including his clothing—"open collared shirt, light-colored shirt, and he had a rifle." Mrs. Rowland said she wished she could have seen him, speculating that he had gone to another part of the building to watch the crowd.

The young couple continued discussing the rough treatment that Adlai Stevenson had received in Dallas a few weeks ago. "This was fresh in our mind," Arnold recalled.

Rowland could not take his eyes off the window. "I [looked at it] constantly . . . I looked back every few seconds, 30 seconds, maybe twice a minute . . . trying to find him so I could point him out to my wife."

But Rowland never saw anything else in the window.

Another man standing in Dealey Plaza had noticed the epileptic man near the Book Depository too. He was Howard Leslie Brennan, a forty-five-year-old steamfitter employed by the Wallace and Beard construction company. That day he was working on a fabrication for pipe at the Republic Bank Building. At noon he ate lunch at a cafeteria on the corner of Main and Record Streets. When he finished, he glanced at a clock—it read 12:18 P.M.

"So I thought," he recalled, "I still had a few minutes, [and] that I might see the parade and the President."

He walked over to the Southwest corner of Houston and Elm, across the street from the Book Depository. It was about 12:22 P.M.

A couple minutes later, about twenty yards away from the corner, he noticed the man having an epileptic fit.

Then Brennan walked over to a retaining wall of a little park and jumped up on the top ledge. He sat down on the top of the wall. It was between 12:22 P.M. and 12:24 P.M. Soon another bystander making a home movie of the motorcade captured an image of Brennan sitting atop the wall, wearing his dark gray construction worker's hard hat and gray khaki work clothes. From here Brennan could survey Dealey Plaza.

"I was more or less observing the crowd and the people in different building windows, including the fire escape across from the [Book Depository] on the east side of the [Depository], and also the [Depository] building windows."

Then Brennan scanned the upper floors.

"In particular, I saw this one man on the sixth floor [who] left the window . . . a couple of times . . . at one time he came to the window and he sat sideways on the window sill. That was previous to President Kennedy getting there. And I could see practically his whole body, from his hips up."

Brennan did not see anyone else at any of the other sixth-floor windows.

He did see men in the windows on the fifth floor. "There were people on the next floor down, which is the fifth floor, colored guys." Brennan had just spotted Junior Jarman, Bonnie Ray Williams, and Harold Norman. Brennan got a good look at them. Later, he was able to recognize the faces of two of the three men.

Just then the motorcade approached Dealey Plaza, and Brennan turned his eyes to the parade. "I watched it . . . as it came on to Houston" and headed to Elm and the Book Depository.

THE CARS following the president copied the turn onto Houston. Oswald could see a whole line of them now—the Secret Service car bearing eight agents—four in the car and four standing on the side running boards—plus longtime JFK aides Dave Powers and Ken O'Donnell, then the car carrying Vice President Johnson, then the other vehicles filled with the reporters, White House staff members, and others.

All of the passengers in Kennedy's car could see the Texas School Book Depository now, looming only one block ahead. And Oswald had a clear view of the president, who was now within range and getting closer with every passing second. Soon Oswald could raise his rifle, place the crosshairs of his scope on Kennedy's forehead, and squeeze the trigger. One well-aimed shot through the head would be sure to kill him.

Arnold Rowland was eager to see the motorcade too. "As the motorcade came along, there was quite a bit of excitement. I didn't look back [at the Book Depository] from then. I was very interested in trying to see the President myself. I had seen him twice before, but I was interested in seeing him again."

Rowland did not realize he had seen something much more important than John Kennedy. A few minutes before the president was about to drive past the Book Depository, Rowland had spotted the man waiting inside to assassinate him.

Oswald was in jeopardy now. He and his rifle had been spotted. What if Rowland found a policeman and told him of what he had seen. Or what if he shouted a warning to the crowd? "There's a man with a rifle in the window of that building!" Police officers might have run into the street to stop the motorcade in its tracks. It was a historic but fleeting opportunity to save the life of the president of the United States.

"We thought momentarily that maybe we should tell someone," Rowland admitted. "But then the thought came to us that it was a security agent. We had seen in the movies before where they have security men up in windows and places like that with rifles to watch the crowds, and we brushed it aside as that . . . and thought nothing else about it."

Rowland remembered films he had seen about the failed attempts on the lives of Theodore and Franklin Roosevelt: "Both of these had Secret Service men up in windows or on top of buildings with rifles, and this is why . . . it didn't alarm me."

Mrs. Rowland was more interested in looking at Jackie Kennedy than the mystery man in the sixth-floor window. "My wife," Mr. Rowland remembered, "remarked on Jackie's clothing." The pink suit was having its desired effect. "We made a few remarks on [her suit] and how she looked, her appearance in general. . . . Everyone was rushing, pressing the cars, trying to get closer. There were quite few people . . . trying to run alongside the car."

The couple was still discussing Jackie's outfit as President Kennedy's car drove away from them. "My wife likes clothes," Mr. Rowland explained.

IN DEALEY Plaza, another man waited to get President Kennedy in his sights. He was Abraham Zapruder, a dress manufacturer whose office was nearby. He owned a portable color 8mm Bell & Howell movie camera, a popular compact recorder that served as the unofficial memory maker of the 1960s.

Zapruder walked over to Dealey Plaza with one of his female employees, after another one had encouraged him to bring his camera and film President Kennedy. Zapruder selected an optimal vantage point along Elm Street in the middle of the plaza, on the same side of the street as the Book Depository.

To get a better view above the heads of people gathering to see the president, Zapruder stood on top of a low concrete pedestal. He asked his employee to hold his legs and steady him once the president's car came into view. It was the perfect spot. From here, Zapruder would enjoy a panoramic vista of the limousine from the time it turned onto Elm Street until it disappeared below the Stemmons Freeway underpass and out of sight.

When the police motorcycles leading the motorcade turned onto Elm Street and came within sight, Zapruder held down the RUN button of his camera and started shooting. But the president's car had not yet made the turn. To save film, he stopped after a few seconds to wait for John Kennedy.

LEE OSWALD waited too. He was still waiting to get the president in his sights.

CHAPTER 5

"SOMEONE IS SHOOTING AT THE PRESIDENT"

If Oswald shot now, as Kennedy drove toward him, he would have very little room for error. To hit a passenger in a vehicle moving at so close a distance, Oswald would have to lower the barrel of his rifle in a dipping, continuous motion to keep his target sighted. His aim would have to be perfect. If he shot too low, he risked hitting the windshield or the metal horizontal crossbar above the middle of the car used to attach the top. If his aim was a little better, he might hit Governor John Connally, who was sitting in a jump seat in front of the president.

Only if Oswald's aim and timing were perfect would he hit Kennedy. And he might get off only one shot. To fire a frontal shot, the assassin would have to step forward and poke his rifle through the open window. Even before he aimed and fired, he would be visible, in plain sight, to anyone in Chief Curry's car, in the presidential limousine, or in the Secret Service trail car. Kennedy's driver might take evasive action, swerving or accelerating to disrupt Oswald's aim. Agents might open fire on him, forcing him to duck for cover. The assassination would most likely fail.

And if Oswald got off a shot, the sound of gunfire would draw eyes to the upper floors of the Book Depository. If the police or Secret Service were quick to pinpoint Oswald's location, he risked capture or death before he could even run down the stairs from the sixth floor to the first. No, the risk was too great. Oswald held his fire as he watched the president's limousine drive straight at him, until it reached the corner of Houston and Elm and slowed to make its hairpin left turn onto Elm.

A radio reporter in Dealey Plaza described the scene: "The president's car is now turning onto Elm Street, and it will be only a matter of minutes before he arrives at the Trade Mart."

In the Secret Service car behind the presidential limousine, Dave Powers looked at his watch and spoke to Ken O'Donnell: "It's just twelve thirty. That's the time we're due to be at the Trade Mart."

"Fine," said O'Donnell. "It's only five minutes from here, so we're only running five minutes behind schedule."

By now, the president was passing right below Oswald. If he leaned out of the window now, aimed down, and fired as the president's car slowed to almost a stop to make the turn, he could hardly miss. But that would expose his position to the rest of the motorcade and to the people in the street. It must have been difficult for the impulsive Oswald to restrain himself. Kennedy was so close that if Oswald had wanted to, he could have thrown a hand grenade out of the window and it could have landed in the president's lap.

Six floors below Oswald, standing on the front steps of the Depository, Buell Wesley Frazier—the man who had driven Oswald and his rifle to work that morning—got a great view of the president too.

Oswald readied to take full advantage of his well-chosen sniper's nest. Now the shiny Lincoln entered Abraham Zapruder's field of vision, and the amateur cameraman started filming again.

Jackie Kennedy shifted her eyes away from the crowd and looked ahead a few blocks to the spot where Elm Street merged into the triple underpass. "We saw this tunnel ahead," Jackie said. "I thought it would be cool in the tunnel . . . the sun wouldn't get into your eyes."

Amos Lee Euins, a fifteen-year-old high-school student, was in Dealey Plaza standing at the corner of Houston and Elm directly across the street from the Depository. Late that morning, about eleven thirty A.M., he was excused from school to watch the motorcade: "The teachers called us and told us the ones who wanted to go downtown to see the President could come down to the office and get an excuse and they could go."

Euins's mother drove him from Franklin D. Roosevelt High School to a spot near the motorcade route. She had to go to work, so she left him there alone.

The boy did not even look for a street sign that would tell him what road he was on: "I was just trying to keep an eye on the President." Euins never forgot what he saw next: "I was standing here on the corner. And then the President came around the corner right here . . . and I was waving, because there wasn't hardly no one on the corner right there but me. . . . He looked . . . and waved back at me."

Then Euins looked at the Book Depository. He saw something protruding from a window on one of the upper floors.

It was a long, thin, horizontal object.

At that moment, the boy continued, "I had seen a pipe, you know, up there in the window, I thought it was a pipe, some kind of pipe . . . right as [President Kennedy] turned the corner."

As Oswald peered out the sixth-floor window and looked to his right, he was facing the president's back. A single tree partially obstructed his view. Oswald positioned his body in a shooting stance—he was either standing or kneeling before he took his first shot—and thrust the barrel of his rifle through the open window. He might have rested it on a cardboard box of schoolbooks to stabilize his aim. He pointed down and aimed for the back of John Kennedy's head—the only sure kill shot.

By now several eyewitnesses on the ground had seen the barrel protruding from the window. Amos Euins still thought it was a pipe.

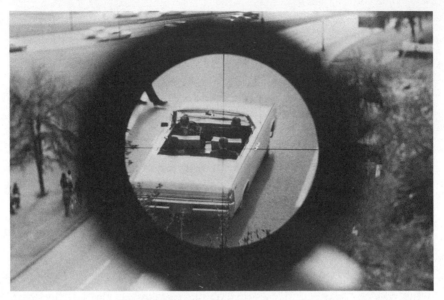

A re-creation of Oswald's view of the presidential limousine.

Another person spotted a man at the window with a sneer on his face. It was too late for either of them to stop Oswald. The president was now in the crosshairs of Oswald's scope.

This was it.

Oswald had not lost his nerve.

He was really going to do it.

He squeezed the trigger. *Never jerk the trigger back with a quick pull,* he learned in the Marine Corps. Squeeze the trigger slowly, apply gradual, increasing pressure. When that pressure reached a few pounds, his trigger released the firing pin. It struck the rear of the bullet in the chamber, instantaneously igniting the gunpowder in the brass cartridge case.

The rifle spit a 6.5mm conical, jacketed lead bullet traveling at 1,700 or 1,800 feet per second at the president of the United States. John Kennedy was close—less than one hundred feet away.

The Book Depository clock read 12:30 P.M.

Abraham Zapruder's finger maintained constant pressure on his camera's RUN switch, and he continued filming.

Through his telescopic sight, Oswald expected to see the evidence of his deed. But the president's body displayed no reaction to being hit by the bullet. He did not recoil from the impact, slump in his seat, or even twitch.

Oswald had missed!

Not only did he fail to shoot Kennedy, he was not even close. He did not hit Jackie, who was sitting a few feet to the president's left, nor did he hit Governor Connally, who was sitting in front of Kennedy. The shot was so off the mark it had even failed to hit the car, which, at this close range, was a huge target in Oswald's scope.

But those in the motorcade certainly heard the gunfire. In the car, John Kennedy heard the shot. At that moment he stopped waving to the crowd and lowered his right arm. Jackie heard it too. She had been looking to her left at the people standing along the curb. At the sound of the gunfire, she spun her head to the right and looked to her husband's side of the car.

"They were gunning the motorcycles," she remembered. "There were these little backfires. [Then] there was one noise like that. I thought it was a backfire."

Witnesses heard different things. Some people traveling in the motorcade—and many bystanders along Elm Street—thought it sounded like a firecracker. What kind of jerk would play a joke like that, they wondered. Jack Bell, a reporter for the Associated Press wire service riding in the press-pool car, thought to himself, *My God, these Texans don't ever know to quit . . . they are shooting off firecrackers and cherry bombs.*

Others thought it sounded like a car or a motorcycle backfiring. To the pigeons atop the roof of the Depository, it was a sound of danger that caused them to flee to the skies. To others—some of the policeman, Secret Service agents, ex-military men, or hunters in Dealey Plaza that day—it sounded like something else.

It was a rifle shot.

"After the President had passed my position," steamfitter Howard Brennan observed, "I really couldn't say how many feet or how far, a short distance I would say," there was a sound. "I heard this crack that I positively thought was a backfire." Brennan thought it was a motorcycle.

"Then something, just right after this explosion, made me think that it was a firecracker being thrown from the Texas Book Store. I glanced up."

The man he had seen earlier in the sixth floor window was back.

He was holding a rifle.

Abraham Zapruder heard it too. It caused him to jostle his camera involuntarily, almost imperceptibly. Through his viewfinder, he, like Oswald, observed no signs of distress to the president or the other passengers. He continued to film the car, keeping it in his sights as it moved slowly in front of him from left to right.

Arnold Rowland also heard the first report. "This I passed off as a backfire, so did practically everyone in the area because gobs of people, when I say gobs, I mean almost everyone in the vicinity, started laughing that couldn't see the motorcade . . . a lot of people laughed."

James Worrell, a high-school senior, had decided to cut class to see the president. At around eight A.M., he had hitchhiked a ride to Love Field to watch John Kennedy's arrival. He got there early, about nine A.M., "and just messed around until the President come in." But

Abraham Zapruder's Bell and Howell camera.

Worrell did not have a good vantage point—"I didn't get to see him good at all."

So he caught a bus from the airport to downtown Dallas with the hope of intercepting the motorcade as it drove through town. "I just, I don't know, happened to pick that place at the Depository, and I stood at the corner of Elm and Houston." When President Kennedy's motorcade drove by, Worrell was standing with his back to the building, about four or five feet from the wall, facing Elm Street.

When the president's car had moved "oh, at least another 50, 75 feet on past me," Worrell heard the sound too. "I heard the first shot, [and] it was too loud to be a firecracker . . . there was quite a big boom . . . just out of nowhere, I looked up . . . just straight up."

The student tilted his head up at a ninety-degree angle and looked straight back over his head, toward the sky. That's when he saw it, protruding from a window above him. "I saw about 6 inches of the gun, the rifle. It had . . . a regular long barrel but it had a long stock and you could only see maybe 4 inches of the barrel."

Buell Wesley Frazier heard the noise too. "Just right after he went by, he hadn't hardly got by, I heard a sound, and if you have ever been around motorcycles, you know how they backfire, and so I thought one of them motorcycles backfired."

It did not take Frazier long to realize that it was something else.

Amos Euins had a better view from across Elm Street: "I was standing [there], and as the motorcade turned the corner, I was facing, looking dead at the building. And so I seen this pipe thing sticking out the window. I wasn't paying too much attention to it. Then when the first shot was fired, I started looking around, thinking it was a backfire. Everybody else started looking around."

From inside the Book Depository, the three men on the fifth floor knew exactly where the shot had come from—right above their heads, one floor up. Bonnie Ray Williams remembered, "After the President's car had passed my window, the last thing I remember seeing him do was—it seemed to me he had a habit of pushing his hair back. The last thing I saw him do was he pushed his hand up. . . . I assumed he was brushing his hair back. And then . . . [there] was a loud shot—first I thought they were saluting the President . . . even maybe a motorcycle backfire."

Williams was not sure where the first shot—if it was a shot—came from. "I really did not pay attention to it, because I did not know what was happening."

Lee Harvey Oswald had botched the first shot. The same man who had failed to assassinate General Walker had just failed in his attempt to assassinate the president of the United States. But how had he missed? Lee was too good a rifleman and the limousine too close for him to have missed it completely. Even if, in his excitement, he had rushed the first shot, he should have hit *someone* or *something* in the car. Chance had played a role in saving General Walker, and it had just saved President Kennedy from the assassin's first bullet.

At the moment Oswald had trained his scope on JFK, the presidential limousine had driven under an oak tree on the Book Depository side of Elm Street. For a few seconds, the tree's branches had acted as a semitransparent screen that stood between Oswald and Kennedy. Looking between the branches, Oswald could still see the president, so he had fired that first shot. But before it could find its target, the bullet had probably nicked a tree branch, which deflected its trajectory and probably stripped it of its outer metal jacket. Instead, the bullet struck a concrete curb beyond the limousine on the far side of Elm Street. It is also possible that the bullet had glanced off the horizontal beam of a traffic signal light. The impact showered fragments into the air that hit a bystander in the face and made a small cut in his cheek.

ANYTHING MIGHT happen next. If Kennedy's driver hit the gas and accelerated the car from its present speed of 12 to 15 miles per hour to just 25 or 30 miles per hour, he could carry the president away from further danger. It might have been enough to just swerve the car violently from right to left. But the driver did not react. The leader of John Kennedy's Secret Service detail, sitting in the front passenger seat, could have ordered the driver to race out of Dealey Plaza. But he did not. The president himself—a decorated World War II navy veteran who had heard gunshots before—could have shouted orders to get him out of there. But he did not.

What would Oswald do now? He could run away, just as he'd done that night at General Walker's house. If he stepped back from the window right now and withdrew his rifle, he could hide it between the boxes and return to work and pretend that nothing had happened. If no one on the ground had seen him, he might have reasoned, and if witnesses convinced themselves that the sound *had* been nothing more than a firecracker or engine backfire, then police might not even investigate the Book Depository. He might escape. Lee Harvey Oswald could go home that night and scold himself for yet another failed attempt to become part of history.

But Oswald did not have a faint heart in Dallas that afternoon. He did not release his grip on the rifle. At the limousine's present speed and direction, the president would be within range for the next ten to twelve seconds. The clock had started ticking with Oswald's first shot. Now he prepared to fire a second one.

With his right hand, Oswald grasped the bolt, raised it, pulled it back, and ejected the empty brass cartridge casing, which popped into the air and made a hollow ping when it landed on the wood floor. The ejection of the cartridge caused a spring in the clip to push another round into position in front of the bolt. Oswald slammed the bolt forward, chambered the round, and turned the bolt handle down. He had practiced this movement countless times and had distilled it into one quick, fluid motion.

On the floor below him, Junior Jarman, Harold Norman, and Bonnie Ray Williams had heard the first explosion a few feet above their heads. They heard the brass casing hit the floor. The loud shot sent a vibration through the floorboards and produced a shock wave that rattled the windows.

It was easy for Oswald to track the president in his scope. The car was traveling at a slow speed in almost a straight line down Elm Street away from the sixth-floor window. Thus Oswald did not have to swing his rifle from left to right to track a target moving horizontally. The car's speed and direction created for Oswald the advantage of an optical illusion in which the car seemed to be almost a stationary object that was slowly getting smaller. All he needed to do was make a small vertical adjustment. He raised the barrel of his rifle a few degrees.

James Worrell watched the rifle barrel: "I looked to see where he was aiming." It was pointing up Elm Street, tracking the president's car.

Almost 2.7 seconds had elapsed since Oswald had fired the first shot. No one in the presidential entourage had reacted to it yet. For the second time, Oswald took aim at the back of John Kennedy's head. Abraham Zapruder continued to film the motorcade. For a few moments, Kennedy disappeared from Zapruder's viewfinder—a large Stemmons Freeway roadside sign temporarily blocked his view of the limousine and President Kennedy. For the second time, Oswald squeezed the trigger. Three seconds. The president was 190 feet—a little more than 60 yards—away.

3.4 seconds.

The rifle fired.

To Bonnie Ray Williams: "The second shot, it sounded like it was right in the building. . . . And it . . . even shook the building, the side we were on. Cement fell on my head."

Through wide cracks between the floorboards above their heads, loose debris drizzled down on them. Williams saw it come down: "Cement, gravel, dirt, or something, from the old building, because it shook the windows and everything. Harold [Norman] was sitting next to me, and he said it came from right over our head."

"No bull shit!" Williams replied, "and we jumped up."

Then Hank Norman confirmed it came from directly over their heads. "I can even hear the shell being ejected from the gun and hitting the floor."

Junior Jarman moved toward Williams and Norman and said, "Man, somebody is shooting at the President."

"No bull shit," Williams said a second time.

WITH HIS second shot, Oswald missed the president's head, but the bullet struck Kennedy in the upper back, to the right of his spine, bored a tunnel through his body at a downward angle—because Oswald was shooting from high above the car—and exited through his lower throat.

John Kennedy's tissue absorbed some of the bullet's energy, and it exited his body at a slower speed—1,500 to 1,600 feet per second—than it had entered it. Upon exit, the bullet, still deadly, tumbled and struck Governor Connally in the back, exited his chest at 900 feet per second, hit his wrist, and then, its speed reduced to 400 feet per second, lodged in his thigh. Oswald's bullet had traveled through the bodies of two men and had inflicted serious wounds to both. These multiple impacts had flattened and deformed the bullet's shape, but it was still in one piece.

Although they had just been shot, President Kennedy and Governor Connally were still alive and conscious . . . and might survive.

For the second time, the sound of gunfire had echoed in Dealey Plaza. If the president had heard this shot, his auditory senses did not register it until after he felt the bullet's impact. And if he had heard it,

Side view of the bullet that struck President Kennedy and Governor Connally. Rear view of the same bullet, showing how it was deformed from impact.

he would have known it was a rifle and he was in the process of being assassinated. No one could doubt the sound now.

Tom Dillard, chief photographer of the *Dallas Morning News*, was riding in one of the motorcade's press cars, and he warned the other passengers: "It's heavy rifle fire."

Rowland heard the shot too. "After the second report, I knew what it was. . . . I knew that it was a gun firing." And it was not a pistol. "It gave the report of a rifle." "It appeared to me he was standing up and resting against the left window sill, with the gun shouldered to his right shoulder, holding the gun with his left hand and taking positive aim . . . as I calculate a couple of seconds."

Howard Brennan noticed the direction the barrel was pointing. It was "at somewhat 30 degrees downward and west by south." In other words, down Elm Street, away from the Depository, in the direction of the railroad underpass.

The rifle was pointed at President Kennedy's limousine.

Now Brennan lost sight of the presidential limousine. It had gone too far down Elm Street for his eyes to follow. It did not matter. "I knew what he was firing at."

WESLEY FRAZIER knew it too. "It wasn't just a few seconds that . . . I heard . . . the same type of . . . sounds, and by that time people was running everywhere, and falling down and screaming. . . . I knew something was wrong . . . somebody was shooting . . . I figured somebody was shooting at President Kennedy." Frazier froze. "People were running and hollering so I just stood still. I have always been taught that when something like that happened, . . . it is always best to stand still because if you run that makes you look guilty sure enough."

But Frazier had already done something that would soon make him a subject of suspicion. Five hours earlier, as he drove to work that morning, the rifle that was shooting at the president was lying flat in the backseat of his car. And the man who would fire it had been sitting in the front seat next to him. Frazier would have a lot to explain that afternoon.

SECRET SERVICE agents standing on the running board of the trail car turned their heads and looked over their right shoulders at the Book Depository. The shots sounded as though they had come from behind them. A moment after the second shot, a news photographer, James Altgens, snapped a picture of the front of Kennedy's limousine and of the Secret Service trail car, freezing the agents in position with their heads turned back to the Depository. The photograph shows something else.

The impact of the bullet propelled President Kennedy's elbows and forearms up, parallel to the ground, and pushed his clenched fists, thumbs in, toward his throat. He could not move his arms from this position—they were locked in place by nerve damage. In the photo, a white-gloved hand—Jackie Kennedy's—touched the president's left forearm.

Dave Powers, riding in the Secret Service car just a few feet behind the president's, turned to Ken O'Donnell. "Kenny, I think the President's been shot."

"What makes you think that?"

"Look at him. He was over on the right, with his arm stretched out. Now he's slumped over toward Jackie, holding his throat."

Abraham Zapruder did not capture on film the moment the president was shot. Kennedy was uninjured when he disappeared behind the Stemmons Freeway sign. When he emerged from behind the sign, he had already been hit and had raised his arms. Zapruder kept filming.

Jackie Kennedy knew something was wrong. Governor Connally, an experienced hunter, already knew. He shouted "No, no, no!" and "They're going to kill us all!"

A Secret Service agent riding in the trail car saw the bullet hit the president and tear a hole in the back of his suit coat. Jackie rotated her body toward her husband. She held his left forearm in her hands. Puzzled, she looked at Governor Connally, who was twisting strangely in his seat. Then she faced her husband again. The startled, wide-eyed expression on his face frightened her. She leaned in closer. Their eyes were just inches apart. Then she seemed to ask, "What's wrong, Jack?" He did not answer. He could not speak.

Lee Harvey Oswald operated the bolt of his rifle and ejected a second brass cartridge onto the floor. He chambered a third round. He elevated the barrel a few degrees. He took his time. This shot might be his last. Yes, he still had a fourth and final bullet in the clip. But he would probably not have the opportunity to fire it before Kennedy's car drove out of range, especially if it sped up to interfere with his aim. More time elapsed since the first shot. Seven seconds now.

Secret Service agent Clint Hill, riding the left-front running board of the trail car, knew what was wrong. He leaped off the car and sprinted for the presidential limousine. Fellow agents rooted for him to close the distance before the sniper could fire a third shot. Hill ran as fast as he could. The thirty-one-year-old agent had been with Mrs. Kennedy from the beginning and had protected her on trips all over the world. He had to help her now. She was sitting so close to the president that another rifle shot might blow her head off. If only Hill could get between the Kennedys and the sniper before he fired another shot.

Hill, who was not wearing a bulletproof vest, could block the bullet with his own body and save the president. It was a sacrifice he would be happy to make. He was an excellent athlete and closed on the president's car fast. Just a few more yards, and he could grab for the big handle on the trunk and yank himself up on the rear foothold. If he could do that he would be a human shield—the assassin would have to shoot through him to get the president. Only Jackie's agent could save John Kennedy now—the president's own Secret Service agent had not yet left the trail car. Seven seconds.

Oswald's second shot was not necessarily fatal. It had not struck Kennedy's head, spine, or heart. His other vital organs—lungs, liver, kidneys, spleen—were undamaged. The chief danger to the president was shock and loss of blood. But it was a survivable wound. Many soldiers in World War II and the Korean War had survived worse injuries. If the driver gunned the engine and increased the car's speed to 80 miles per hour, the president could be in the emergency room of nearby Parkland Hospital in less than ten minutes.

Eight seconds. Oswald began to squeeze the trigger. The Newman family, two young parents and their two small children, were standing just a few feet away from the president. Abraham Zapruder kept filming. Nellie Connally held her husband in her arms. As he lost consciousness, he believed he was dying. Bystander Mary Moorman raised her Polaroid camera and prepared to click the shutter. Clint Hill knew he would make it in just a couple more steps. The president remained upright in the backseat. He had not spoken a word. The driver did not race away to escape the gunfire. Instead, the limousine seemed to slow down. Jackie Kennedy gazed into her husband's uncomprehending eyes. They had been married ten years that fall.

Howard Brennan could see most of the rifle now. "I calculate 70 to 85 per cent of the gun." And he could see most of the man. "I could see practically his whole body, from his hips up." He looked to be "a man in his early thirties, fair complexion, slender but neat . . . possibly 5-foot 10 . . . from 160 to 170 pounds." He was a white man, and he wore "light colored clothes, more of a khaki color."

Amos Euins watched too. After he heard the second shot, he did not take his eyes off the Depository. "I was still looking at the building. . . . I was looking where the barrel was sticking out." After the first two shots, the teenager was afraid he might get shot too. For protection, he ducked behind a water fountain. "I got behind this little fountain," Euins said, but he kept the sixth floor window in sight. "Then he shot again."

"I could see his hand, and I could see his other hand on the trigger, and one hand was on the barrel thing." Euins saw that the barrel was tracking JFK's car as it moved farther away from the Depository. As the distance grew, the gunman thrust more of the rifle out of the window. "After the President's car had come down the street further . . . he kind of stuck it out more." When Euins first saw the barrel, only the tip protruded from the window. By the time Oswald was ready to fire for the third time, it looked to Euins like three feet of rifle was visible. "It was enough to get the stock and receiving house and the trigger housing to stick out the window."

There was nothing Euins or Brennan could do. There was no time to shout a warning. There was no time to find a policeman. Or to stop the man in the window.

Brennan could only watch as the man "fired his last shot."

8.4 seconds.

The rifle fired.

The president was less than one hundred yards away. The Marine Corps had taught Oswald to shoot at targets with fixed, iron sights— without the aid of a telescopic sight—at distances of two hundred, four hundred, and even six hundred yards.

The accuracy of Oswald's third shot would determine if the president of the United States lived or died. For Oswald, everything depended on this last bullet. The first two bullets had already guaranteed his infamy. The third would determine whether he was remembered as a failure who missed his chance or as a man of history who changed the future.

AT THE third shot, Bob Jackson, a photographer riding in the same press car as Tom Dillard, glanced at the upper floors of the Depository. "There's a rifle," Jackson said, "in that open window!"

Dillard's instant conclusion was that "they've killed him."

Euins watched Oswald fire the third shot: "I had seen a bald spot on this man's head. I was looking at the bald spot. I could see his hand, you know the rifle laying across his hand. And I could see his hand sticking out on the trigger part. And after he got through, he just pulled it back in the window." But the boy could not see Oswald's face. He did not get a very good look at him. "The cardboard boxes near the window were throwing a reflection, shaded."

Bob Jackson saw two men on the fifth floor "leaning out and looking up." Then he saw Oswald's gun. "[M]y eyes went up to the next window, and I could see the rifle on the ledge and I could see it being drawn in. I could not see who was holding it."

Jackson's camera was out of film. That caused him to miss what would have been one of the most dramatic and important pictures in American history—the assassin's rifle shooting at the president, and possibly an image of Lee Harvey Oswald firing it. In the days ahead, Jackson would have a chance to redeem himself.

Tom Dillard brought his camera up in an instant and snapped a photograph. But it was too late. All he got was a picture of the empty sixth-floor window and of the Depository employees still in the windows one floor below. Oswald had stepped back just in time to avoid being photographed.

Malcolm Couch, a cameraman for WFAA-TV, saw it too: "I looked up and saw about a foot of the rifle going back in the window."

Brennan kept his eyes on the man in that window. "He drew the gun back from the window as though he was drawing it back to his side and maybe paused for another second as though to assure hisself [sic] that he hit his mark, and then he disappeared."

WHITE HOUSE reporter Merriman Smith, riding in the front seat of the press-pool car, said out loud that "those were gunshots." To Bob Clark from ABC News, who was riding in the same car, the three shots sounded "equally loud and equally clear" and were "clearly fired from almost over our head." Clark was sure that someone "was firing from almost directly above us." Pierce Allman from WFAA-TV and radio heard it too. "There were three shots. They were very distinct."

Mrs. Robert A. Reid, a clerical supervisor who had worked at the Texas School Book Depository for seven years, was standing on Elm Street in front of the building: "I was . . . watching for the car as the President came by. I looked at him and was very anxious to see Mrs. Kennedy, I looked at her and I was going to see how she was dressed, and she was dressed very attractive, and she put up her hand to her hat and was holding it on, the wind was blowing a little bit and then [they] went right on by me."

A few seconds later, Reid heard the three shots. She turned to Mr. Campbell, who also worked at the Depository, and said, "Oh, my goodness, I am afraid those came from our building." It seemed to her that the shots "came just so directly over my head." Then she looked straight up. She saw "three colored boys up there, and I only recognized one because I didn't know the rest of them so well." It was James "Junior" Jarman. Then she said, "Oh, I hope they don't think any of our boys have done this."

THIS TIME Lee Harvey Oswald did not miss.

The third bullet hit John Kennedy in the back of the head and sliced through his thatch of thick, reddish brown hair. It tore a neat hole through his scalp and punched a round hole through the back of his skull. The velocity, the pressure, and the physics of death did the rest. The right rear side of the president's skull blew out—exploded really—tearing open his scalp, and spewing skull fragments, blood, and brains several feet into the air, where they hung for a few seconds, suspended in a pink cloud. It splattered the motorcycle windshield and face of a police officer riding close to the left side of the car.

To Clint Hill, the impact resembled the squashing sound of a melon being smashed on cement.

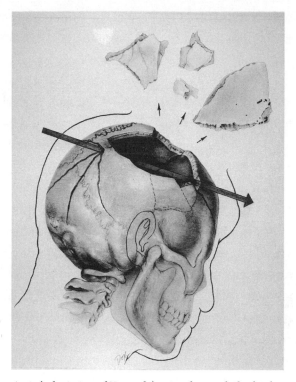

Artist's depiction of Kennedy's second wound, the fatal head shot.

The president of the United States never heard this shot. He lost consciousness the moment the bullet hit him. His wife's mouth opened wide in shock as his limp body began to tip toward her. "Oh no!" she exclaimed. She was so close to her husband when Oswald shot him in the head that her hair, face, white gloves, and pink suit were all stained with gore. "I could see a piece of his skull coming off," she said. "It was flesh-colored, not white . . . I can see this . . . piece detaching itself from his head. Then he slumped in my lap; his blood and brains were in my lap."

Powers and O'Donnell watched it happen from just fifteen or twenty feet behind Kennedy's car. "While we both stared at the President, the third shot took the side of his head off. We saw pieces of bone and brain tissue and bits of his reddish hair flying through the air."

O'Donnell was horrified. "He's dead," he said to Powers. It would be impossible for a man to suffer a wound like that and live.

OSWALD, HIS eye glued to his rifle scope, must have seen the spray of red mist. Then he lowered the stock from his right cheek, operated the bolt, ejected the third cartridge case, and chambered the fourth round. He was ready to shoot again. If he wanted to fire a last round, he had to be quick. He could not take as much time as he had between the second and third shots. He would have to aim this shot by instinct, and maybe even use the iron sights and not the scope. Locating the car in the telescopic sight might take a fraction of a second too long. And Kennedy was slumping over, about to disappear into the backseat. Jackie and Clint Hill presented finer targets, and by now Oswald had a better chance of hitting them than the president. Would he fire his fourth and last round?

He paused, and then he stepped away from the window. His work was done. Now it was time to escape.

The disappearance of the shooter from the window did not convince Brennan that no more shots would be fired in Dealey Plaza. What if there were other gunmen? He did not want to get shot. He jumped off the retaining wall and took cover. "It occurred to me that there might be more than one person, that it was a plot which could mean several people." He feared that "there were going to be bullets flying from every direction."

THIS WAS not a suicide mission for Oswald, despite his ritualistic behavior that morning at Ruth Paine's house. Leaving behind his wedding ring and almost all of his cash symbolized a shedding of worldly things, a fatalism that suggests that Oswald believed he was about to undertake a dangerous—even suicidal—task from which he might never return. At the Book Depository, once he shot the president, he could have surrendered himself to the police as a political prisoner. Once upon a time, he had fancied the romanticism of that. "If I am taken prisoner . . ." he had written last April in that frightening note of instructions to Marina before he went off to assassinate General Walker. But this morning, although he had left his ring and cash, he left no note, letter, or political manifesto on the bedroom dresser.

No, Lee Oswald was not ready to give up. His heart raced. Adrenaline pumped through his system. His survival instincts kicked in and whisked him away from the sixth floor window. He had done it. He had summoned all the will and the discipline buried inside him to shoot John F. Kennedy. And he had done what no man had ever done before—he had become the first presidential assassin to ever commit the act from long distance, with a rifle. But this was no time for self-congratulation. He had to get out of that building.

Photograph taken moments after Oswald fired the third shot and disappeared from the sixth-floor window.

• • •

JACKIE KENNEDY panicked. She saw that something had fallen onto the dark blue, mirrored finish of the limousine trunk. It was a piece of her husband's skull. She rose from the backseat, turned around, kneeled on the blue upholstery, and reached for the top of the trunk. She was fully exposed to the assassin now. If Oswald had lingered in the window and desired to fire a fourth shot, he could have used his last bullet to kill Jackie. Still kneeling on the backseat, she stretched her body toward the skull fragment. The president, she might have convinced herself, would need that piece of his head when the doctors fixed him at the hospital.

Clint Hill could not believe what he was seeing. If the car accelerated now, Jackie might be thrown onto the trunk, slide off it, fall to the pavement, and get run over by the Secret Service car. He lunged forward and caught the left handle mounted on the back of the limousine. He'd made it! But at that moment, the car lurched forward, and Hill lost his footing. Unless he released his grip, in another second the car would drag him down to the pavement. The "Queen Mary," traveling just a few yards behind SS-100-X, would run him over and kill him.

But he would not abandon Mrs. Kennedy. He mustered all the strength he had and pulled hard against the handle, yanked his body up, and caught a tenuous foothold on the step. Then he launched his body across the trunk, gathered Jackie in his arms and pushed her back into the car. He used his body to cover her and the president.

Later, Jackie had no memory of this terrifying scene. Clint Hill never forgot it: the president was lying faceup in her lap, and she shouted, "My God, they have shot his head off!"

Hill looked at the president, appalled at what he saw: "The right rear portion of his head was missing."

As the car sped away, other agents saw Hill pound his fist on the trunk and shake his head in despair. Then he signaled them with a thumbs-down. The president's car accelerated and, along with the trailing Secret Service car, reached the shadows of the triple underpass and disappeared from the view of the dumbfounded witnesses in Dealey Plaza and the rest of the motorcade.

Abraham Zapruder was so shaken by what he had just seen, he lost the ability to speak coherently.

In Vice President Johnson's car, Secret Service agent Rufus Youngblood yelled "Get down," climbed from the front seat into the back, pushed LBJ to the floor, and covered him with his body to protect him from gunfire. The car raced after the others. Inside, Lyndon Johnson had no idea whether President Kennedy was dead or alive.

The radio reporter in Dealey Plaza described what he saw: "It app . . . It appears as though something has happened in the motorcade route. Something, I repeat, has happened in the motorcade route. There's numerous people running up the hill alongside Elm Street. Several police officers are running up the hill. . . . Stand by. Just a moment please . . . Parkland Hospital—there has been a shooting— Parkland Hospital has been advised to stand by for a severe gunshot wound. I repeat, a shooting in the motorcade . . . the president's car is now going past me. The limousine is now traveling at a very high rate of speed . . . it appears that someone in the limousine might have been hit by the gunfire."

CHAPTER 6

"THEY'VE SHOT HIS HEAD OFF"

Lee Oswald abandoned his three spent brass cartridges where they fell and, still gripping his rifle, ran the 96-foot diagonal length of the sixth floor, heading for the back stairs. He shoved the weapon—horizontally, scope up—into a narrow space between two stacks of boxes. Now, as he raced down the stairs, he possessed no evidence that would link him to the shooting. Oswald descended the floors—fifth, fourth, third—and encountered no one coming up. When he reached the second floor, he must have heard someone below him ascending the stairs because he ducked into the lunchroom.

A Dallas policeman, the first one to enter the Depository after the shooting, found the building manager, Roy Truly, and together, they hurried up the stairs. When the policeman reached the second floor, he spotted Oswald through a window in the lunchroom door, ordered him to stop, and asked the manager if Oswald worked there. When Truly answered yes, the policeman let Oswald go and continued racing to the roof.

Oswald was lucky. It was a spectacular error for the officer to assume that the assassin was not an employee at the Depository.

MRS. REID had hurried back to the front door of the Depository after the third shot. "It was just a mass of confusion," she said. "I saw people beginning to fall, and the thought that went through my mind, my goodness, [was that] I must get out of this line of shots. They may fire some more." So she sought shelter: "I ran into the building . . . I ran up to our office."

She passed through the lobby, went up the stairs, and, within two minutes of the last shot, Mrs. Reid entered the front door to her second-floor office. That office had two doors—the one she had just entered and another in the back.

Oswald was still in the building. He continued his descent. He had just had a close call with Officer Baker and Mr. Truly. More policemen would rush up the stairs soon. He needed to get out of the building fast. Within a minute or two, the police and Secret Service might seal off all the exits.

As soon as Mrs. Reid walked several feet past her desk, a man entered the office through the back door. She and the man faced each other. It was Lee Harvey Oswald.

"Oh, the President has been shot" she told Lee, "but maybe they didn't hit him."

Oswald mumbled something unintelligible to her.

She did not regard her coworker as suspicious. "I didn't pay any attention to what he said because I had no thoughts . . . of him having any connection with it all because he was very calm. He had gotten a coke and was holding it in his hands, and I guess . . . I thought it was a little strange that one of the warehouse boys would be up in the office at the time, not that he had done anything wrong."

Oswald strode past her, walked out the front door of the office, and descended the stairs to the first-floor lobby. He was headed for the first-floor exit onto Elm Street.

. . .

OSWALD COULD have chosen to escape via the back door, the one through which he and Buell Wesley Frazier—and the rifle—had come in to work that morning. A man with something to hide might have preferred to sneak out the back way and run. But evidently Oswald shared Frazier's philosophy—if you run, people will just think you are guilty. To avoid suspicion, Oswald decided to walk out right through the front door.

At that moment, a man grabbed Oswald—the assassin thought the man was a Secret Service agent—but he only asked Oswald where he could find the nearest telephone. It was just a reporter, Robert MacNeil. Oswald gave him directions.

Then, as policemen and citizens converged near the front door of the Texas School Book Depository, and as Dealey Plaza devolved into the chaos of sirens, police radios, shouting, and people running in every direction—Lee Harvey Oswald strolled away from the scene of his crime. He had shot the president, and he had, at least for now, escaped. Seven months after his failed attempt to slay General Walker, his second attempt to shoot a man had proven more successful.

In the press-pool car, Merriman Smith of United Press International, a news service that distributed reports to media outlets all over the world, grabbed for the radiotelephone mounted near him in the front seat. Another reporter sitting behind him, Jack Bell from the rival Associated Press news service, grabbed for the phone too. The two newshounds struggled over their only link to the outside world. Neither would give it up. Each wanted the credit for transmitting the first news of the shooting to the world. The men almost came to blows. Smith won the struggle and wrenched the phone away.

Just four minutes after Oswald's first shot, at 12:34 P.M., Smith communicated a brief message to his employer: "THREE SHOTS WERE FIRED TODAY AT PRESIDENT KENNEDY'S MOTORCADE IN DOWNTOWN DALLAS."

Bonnie Ray Williams remembered that after the gunshots, he and his two coworkers "got kind of excited . . . we all decided we would run down to the west side of the building." He said they saw policemen and others, "running, scared, running—there are some tracks on the west side of the building, railroad tracks. They were running towards that way. And we . . . know the shots practically came from over our head. But since everybody was running . . . to the west side of the building, towards the railroad tracks, we assumed maybe somebody was down there. And so we ran all that way, the way that the people was running, and we was looking out the window."

Hank Norman said, "Man, I know it came from [above our heads]. It even shook the building."

Then Norman said to Williams, "You got something on your head."

James Jarman spoke. "Yes, man, don't you brush it out."

Then Jarman added, "Maybe we better get the hell out of here."

Williams concurred. "And so we just ran down to the fourth floor and came on down." Why didn't they run up to the sixth floor? "I really don't know. We just never did think about it . . . going up to the sixth floor. Maybe it was just because we were frightened."

Jarman reminded Hank, "That shot probably did come from upstairs, up over us," and Hank said, "I know it did, because I could hear the action of the bolt, and I could hear the cartridges drop on the floor."

If they had quickly run across the fifth floor and bounded up the staircase, they might have encountered Oswald on his way down. Even if Oswald had already descended below the fifth floor, they would have been close enough to him on the staircase to have heard him running down it. Again, luck was with Oswald this day.

HOWARD BRENNAN watched as law enforcement officers started to run west past the Book Depository. They were going the wrong way!

"They were directing their search," he observed, "toward the west side of the building and down Houston Street," toward the underpass and the railroad tracks.

He decided to do something. "I knew I had to get to someone quick to tell them where the man was." He walked fast or ran across the street and found a police officer standing near the corner of the Book Depository. "I asked him to get me someone in charge, a Secret Service man or an FBI. That it appeared to me that they were searching in the wrong direction for the man that did the shooting."

Brennan said that the gunman was definitely in the Depository.

"Just a minute," the cop said. Then he took Brennan to a car parked in front of the building. Inside was agent Forrest Sorrels. Brennan told Sorrels what he had seen. As Sorrels discussed the information with some other law enforcement officers, two of the three men that Brennan had spotted looking out the fifth-floor windows emerged from the front door of Book Depository and ran down the front steps, close to where Brennan was now standing. "That's them!" exclaimed Brennan. "Those were the two colored boys that was on the fifth floor . . . on the next floor underneath the man that fired the gun." They got no farther and were brought in for questioning.

THE ROUTE to Parkland Hospital would take President Kennedy right past the Trade Mart, where a thousand people awaited his arrival. Ron Jenkins, a reporter for Mobile Unit 6 at station KBOX, was in position outside the Trade Mart. He saw that the limousine was approaching too fast. It should be slowing down to pull up to the building.

"The presidential car is coming up now," he reported. "We know it is the president's car. We can see Mrs. Kennedy's pink suit. There is a Secret Service man spread eagle on the top of the car."

As Jenkins watched the motorcade speed past him, the wail of multiple sirens almost drowned out his voice. "Something is wrong here," he warned, "something is terribly wrong."

• • •

BACK IN Dealey Plaza, Arnold Rowland and his wife already knew something was terribly wrong. He regretted his silence about the man in the window.

"I never dreamed of anything such as that," that the man he saw in the window could be an assassin. "Perhaps if I had been older," the eighteen-year-old ruminated, "and had more experience in life it might have made a difference. It very well could have."

Many things might have made a difference. If Lee Harvey Oswald had succeeded in assassinating General Walker seven months ago in April, perhaps his thirst for blood would have been sated. And if Marina Oswald had reported that attempted murder to the police, perhaps on November 22 her husband would have been languishing in jail instead of lurking in a sixth-floor window. At least the police would have confiscated his rifle.

If Lee Oswald had obtained permission that fall to return to Russia or enter Cuba, perhaps he would not have been in Dallas this day. If Ruth Paine and Linnie Mae Randle had not helped Lee get a job at the Texas School Book Depository, perhaps he would not have been working in a tall building along President Kennedy's motorcade route. If Marina had disposed of her husband's rifle when she had discovered it in Ruth Paine's station wagon after her trip back home from New Orleans to Texas, perhaps Lee would not have bought another one. If on the night of Thursday, November 21, Marina had agreed that she and the girls would move to a new apartment in Dallas with Lee, perhaps that gesture of love would have turned his heart away from murder. If Marina had just moved the rifle from its hiding place in Ruth Paine's garage, perhaps Lee would not have found it in time to bring it to work on the morning of Friday, November 22.

There were other ifs. If on that morning the weather forecast for cloudy skies and precipitation over Dallas had proven true and the Secret Service had installed the rainproof plastic bubble top on the president's limousine, perhaps it would have deflected any bullets fired at the car. If John Kennedy had not been so stubborn about his habit of riding through cities in a convertible, perhaps Oswald would have never tried to kill him. If Kennedy's agents, the "Ivy League Charlatans," had disobeyed their boss and had stood on the back of the open car during the motorcade, perhaps Lee Harvey Oswald might never have had a clear shot at the back of JFK's head. If Oswald had never enlisted in the United States Marine Corps, perhaps he would have never learned the marksmanship skills he needed to shoot JFK. If Junior Jarman had peeked behind the boxes in the southeast corner when he went up to the sixth floor to eat his lunch, perhaps he would have surprised his coworker hiding in his sniper's nest. If Bonnie Ray Williams and Harold Norman had gone up to the sixth floor to watch the parade with Jarman, perhaps they would have stopped Oswald from ever firing a second shot—or a third one. If Arnold Rowland had told someone that there was a man with a rifle in the window, perhaps the assassination might have been prevented. And if Special Agent William Greer had swerved or accelerated the limousine after the second shot, perhaps the president would have survived his wound and Oswald would have missed the third shot, just as he had the first.

Ten minutes after the shooting, at 12:40 P.M. (CST), at the New York headquarters of CBS News, the television network interrupted its regular programming and displayed its company logo on television screens across the nation. CBS did not possess the technology to interrupt scheduled, pretaped programming and get a reporter's face on the air immediately. The television camera needed twenty minutes to warm up, and it was not even in the vicinity of the news desk. At first, CBS could broadcast only a voice. Walter Cronkite, one of the most important journalists on television, spoke in an urgent, clipped tone: "Here is a bulletin from CBS News. In Dallas, Texas, three shots were fired at President Kennedy's motorcade in downtown Dallas. The first reports say that President Kennedy has been seriously wounded by this shooting."

Marina Oswald, knowing that the president was in Dallas, had been watching television all morning. The black-and-white images flickering on the screen were a poor substitute for the full-color, live, and in-person sights she would have enjoyed if Lee had granted her wish to see the president's motorcade. She had not even gotten dressed yet.

When Ruth Paine left the house earlier in the morning to take her children to a doctor's appointment, she had left the television on for Marina. When Ruth returned home not long after one P.M., she joined Marina in front of the TV. She told Marina, who could hardly speak English, that the announcer on TV had just said the president had been shot. Marina was stunned.

"It was hard for me to say anything," Marina said. "We both turned pale. I went to my room and cried."

Ruth told Marina there was more news: "By the way," Ruth said, "they fired from the building in which Lee is working."

Right away her instincts made her suspect him. It could not be. But if the shots had been fired from the School Book Depository, how could it *not* have been him? There was only one way to know.

"I then went to the garage to see whether the rifle was there, and I saw that the blanket was still there, and I said, 'Thank God.' "

She did not try to pick up the blanket. "I didn't unroll the blanket. It was in its usual position, and it appeared to have something inside."

Seconds after Cronkite announced the news, another printed bulletin was thrust into his hands. Listeners could hear him shuffling papers as he spoke. "More details just arrived. These details about the same as previously. President Kennedy shot today just as his motorcade left downtown Dallas. Mrs. Kennedy jumped up and grabbed Mr. Kennedy. She called, 'Oh no!' The motorcade sped on. United Press says that the wounds for President Kennedy perhaps could be fatal."

The nation waited in suspense. Would the president live or die? Everyone knew the story of PT-109 and John Kennedy's close brush with death during the Second World War. Maybe he would beat the odds again. Millions of people began to hope and pray that Kennedy's wounds were not fatal. Only a handful of people—Clint Hill and some of the Secret Service detail and Merriman Smith and a few other journalists—knew the awful truth. But Smith was reluctant to report it without official confirmation from the White House staff.

Once the CBS network was able to broadcast both image and sound, Cronkite appeared on camera for the first time and continued to read news bulletins. It was the beginning of four unprecedented days of television coverage on all the national networks—CBS, NBC, and ABC. Within one hour of the shooting, more than three quarters of the American people knew what had happened.

AS LEE Oswald walked down Elm Street, the driver of President Kennedy's car raced to Parkland Hospital, and Dallas police officers radioed ahead to advise the emergency-room staff to stand by to receive the victim of a severe gunshot wound. On the hands of a clock, it was a brief ride. In the minds of the five unwounded passengers, it seemed to go on forever.

The two Secret Service agents in the front seat could do nothing to help the president other than get him to the hospital as fast as possible, reaching speeds of more than 80 miles per hour.

In the jump seats behind the agents, Nellie Connally tended to her stricken husband. His wounds were painful, he was losing blood, and he lapsed in and out of consciousness. As Nellie held him in her arms, she promised him that everything would be all right.

Behind the Connallys, the backseat was a tangle of intertwined arms and legs and bodies. After Clint Hill had saved Jackie from falling to the street and pushed her back inside the car, she grabbed the president and held him close. Hill sprawled on top of Jackie and the president to shield them from any further gunfire. In his awkward, spread-eagled position over Kennedy and his wife, it was hard to hang on as the car reached top speed. The wind blew the sunglasses off Hill's face.

As the car sped on, Jackie cried out, "They've shot his head off! I have his brains in my hands."

Then she was heard speaking to her husband: "Jack, Jack, can you hear me? I love you, Jack. Jack, Jack, what have they done to you?" He did not answer.

Under Hill's human shield, the Kennedys rode to Parkland in contorted, sideways positions, lying partly on the backseat and partly on the floor. Blood pooled in the footwells of the floor.

Jackie tried to close Kennedy's gaping wound with her hands. "I tried to hold the top of his head down, maybe I could keep it in . . . but I knew he was dead."

She remembered the puzzled, astonished look Oswald's bullet had frozen on the president's face. "His head was so beautiful," she said.

LESS THAN ten minutes after the third shot, the presidential limousine careened into the emergency-room driveway at Parkland Hospital, followed by the Secret Service trail car and the vice president's vehicle. Agents leaped out with guns drawn. One agent brandished an automatic rifle.

There were no medical carts awaiting the stricken president at the entrance. Some Secret Service agents ran inside and yelled for a cart. Hospital staffers rolled two to the car. They lifted the broken body of Governor Connally from his seat, and his wife exited the vehicle.

The president's bodyguards wanted to snatch him from the backseat as quick as they could and rush him inside. Time, they believed, was of the essence. Once Clint Hill got out of the car, the other agents got a clear view of the backseat for the first time. Other cars from the motorcade arrived at Parkland. Dave Powers ran up to the limousine, looked inside, and gasped: "Oh, my God, Mr. President, what did they do?"

Jackie told him, "Dave, he's dead."

Agents bent over Kennedy to lift him out of the car, but Jackie would not let him go. She had wrapped her arms tight around him and cradled his head in her lap. She curled over him in a protective embrace. She ignored the agents' request. Please, they told her, they needed to get the president inside so the doctors could treat him. Clint Hill, the agent she trusted most, beseeched her to release her husband.

"Please, Mrs. Kennedy," he said.

She would not budge.

"Please," Hill begged, "we must get the president to a doctor."

Jackie said no. "I'm not going to let him go . . . you know he's dead. Let me alone."

Lyndon Johnson arrived at Parkland. During the wild ride from Dealey Plaza, the Secret Service had already assigned some of the president's agents to Johnson. Emory Roberts said to agent Bill McIntyre, "They got him. You and Bennett take over Johnson as soon as we stop." In the car, Johnson's agent Rufus Youngblood told him, "An emergency exists. When we get to where we're going, you and me are going to move right off and not tie in with the other people."

Johnson replied, "OK, partner." As soon as the vice president's car arrived at the hospital, Youngblood and other agents hustled him inside. JFK was still in his limousine, and the agents did not allow Johnson to approach the president's car. They feared other assassins might be after Johnson. Lady Bird glimpsed a splash of color—a pink suit—as she was rushed into Parkland. Youngblood asked a hospital staffer to lead the Johnson party to a quiet, out-of-the-way room. When they got there, LBJ leaned against a wall and gazed into Lady Bird's eyes. "Lyndon and I did not speak. We just *looked* at each other, exchanging messages with our eyes."

Ken O'Donnell came to the room. "It looks pretty bleak. I think the president is dead."

When the first reporters arrived at Parkland, the president had still not been removed from the limousine. Merriman Smith was the first journalist to see the wounded John Kennedy. "We skidded around a sharp turn and spilled out of the pool car as it entered the hospital driveway." Smith ran to the side of limousine. "The President was face down on the back seat. Mrs. Kennedy made a cradle of her arms around the president's head and bent over him as if she were whispering to him. . . . I could not see the president's wound. But I could see blood splattered around the interior of the back seat and a dark stain spreading down the right side of the president's dark gray suit."

Bob Clark joined Smith beside the car. "They simply let us go up and stand as close as we could. We were standing literally a couple feet from the car, starring down at Kennedy. He was stretched out in the back seat. He was lying with the side of his head exposed and his head in Jackie's lap. I was not conscious of any wound to the head, so that part of his head was hidden, probably deliberately by Jackie. It was just a frozen scene. Jackie was sitting there, saying nothing."

Smith spoke to agent Hill, who was still leaning over the back seat. "How badly was he hit, Clint?"

"He's dead."

CLINT HILL'S intuition and close relationship with Jackie told him why she was covering her husband's head with her arms and her body. She did not want anyone to see him that way. Hill showed her he understood. O'Donnell knew too: "She did not want strangers looking at her husband's broken and bleeding head." Hill removed his suit coat.

"Hill threw his coat over Jack's head," Jackie remembered, "and I held his head to throw the coat over it." Now no one standing there would see the president's horrible wound or his eyes fixed wide open in a stare.

Jackie released her hold, and Secret Service men lifted John Kennedy's unconscious body from the backseat. They brought him inside the hospital at 12:38 P.M. (CST). It was eight minutes since he had been shot.

KBOX reporter Ron Jenkins had chased the president's limousine from the Trade Mart to Parkland. By the time his mobile unit got there, police officers had blocked the entry. Still on the air, Jenkins reported that a policeman was shouting "No. You *cannot* come in here! You *cannot* come in here!" Jenkins tried to find a back way into the building.

Back in Dealey Plaza, just a few minutes after the assassination, another local reporter broadcast his report: "This is Pierce Allman from the Texas School Book Depository building for WFAA News." Allman summarized what he had just seen and heard: "Just a few minutes ago, the president of the United States turned from Houston Street onto Elm Street on his way to a scheduled luncheon appearance at the Stemmons Trade Mart."

The journalist sounded out of breath, as though he had been running. "And as he went by the Texas School Book Depository, headed for the triple underpass, there were three loud, reverberating explosions. Nobody moved. Everyone seemed stunned. A few seemed to look around, wondering 'who has the firecrackers?' Then suddenly the Secret Service men sprang into action. The convertible bearing the president and Mrs. Kennedy sped away and officers, both plain clothes and uniformed, seemed to spring from everywhere at once, guns drawn, ordering people to lie flat."

Allman chased down and interviewed motorcade spectators. "There are two witnesses who were near the president's car at the time of the explosions who say that shots *were* fired—from which upper window we do not know. We *do* not and *cannot* confirm the reports at this time that the president has been shot. One witness says that he definitely was shot—that he was shot twice—that he saw the president slump in his seat. As I say, this is *not* confirmed at this time. From where I am the police have two witnesses. They are bringing them in now."

The journalist had managed to slip inside the building before police sealed it off. "I am in the Texas School Book Depository. . . . We will try to learn more and relay word to this station."

LEE HARVEY Oswald was on the run—and he did not have much of an escape plan. If he had owned a car, he could have driven himself out of the city and as far away from Dallas as he could get, possibly fleeing the country to Mexico. But he did not even know how to drive.

Lee Harvey Oswald was no John Wilkes Booth. After Abraham Lincoln's assassination, Booth had a fast horse standing behind Ford's Theatre, a planned escape route, and the names and locations of sympathizers who might help him during his escape south from Washington.

Oswald had no one. He walked seven blocks on Elm Street, then at 12:40 P.M. (CST), he flagged down a passing city bus that was headed in the direction he wanted to go. Although Oswald was not standing at a scheduled bus stop, the driver opened the door and let him board anyway. Its route would take Oswald past the Book Depository, back to the scene of his crime. He might have enjoyed witnessing the chaos he had created. But before the bus could get near the Depository, heavy traffic slowed its pace to almost a standstill.

Did he recall the last time he had tried to escape on a bus? The night he tried to murder General Walker, it was dark, and traffic was light. He had enjoyed a smooth ride home. That would not happen today.

It was absurd. The man who had just shot the president of the United States was stuck in traffic, trying to flee on a public bus. Realizing the ridiculousness of his position, Oswald stood up, walked to the front of the bus, took a transfer, and asked the driver to let him off. Within a few minutes, at about 12:47 P.M. (CST), he caught a taxi to his rooming house. Oswald did not want the driver to know where he lived, so he did not give him his numbered address but had the taxi drop him a few blocks away from his rooming house. From there Oswald walked to 1026 North Beckley Street. When he arrived, the proprietor of the rooming house was watching news of the assassination on television.

She told Oswald that Kennedy had been shot. He said nothing. He hurried to his room, changed jackets, picked up his revolver and some ammunition. Then, after a couple of minutes, he left at 1:03 P.M. (CST) without saying a word. No one knows what Oswald planned to do next. Perhaps he hoped to get to the bus station and buy a ticket to Mexico—he still had enough money on him for that, although not for much more. If he hoped to escape capture, he needed to flee Dallas.

Soon a roll call of employees at the Book Depository would reveal that only one man could not be accounted for—Lee Harvey Oswald.

AT THE Trade Mart, a rumor spread from table to table that President Kennedy had been shot. *New York Times* reporter Tom Wicker witnessed it. "It was the only rumor that I had ever seen; it was moving across that crowd like a wind over a wheat field."

AT PARKLAND, the Secret Service agents and hospital staffers rushed the president into trauma room one. Assistant White House press secretary Malcolm Kilduff watched as Jackie Kennedy, "her hair flying and dripping with blood," helped push the stretcher through the halls. They lifted the president from the cart and laid him on his back upon the examination table. Nurses cut away his clothing. Doctors looked for his vital signs. He had no blood pressure. He was not breathing. His heartbeat was sporadic and weak. The pupils of his eyes were dilated and fixed. The doctors inserted tubes into his veins. They gave him blood transfusions. They cut a tracheotomy in his throat to improve his breathing.

From throughout Parkland Hospital, doctors rushed to the emergency room. Some thought they could help. Others were unneeded voyeurs who wanted only to lay eyes upon the president so that one day they could say that they had been there.

Jackie spoke to Powers and O'Donnell. "Do you think he still has a chance?" The longtime, faithful aides knew better but could not say it. "I did not have the heart," O'Donnell recalled, "to tell her what I was thinking."

Merriman Smith found a telephone inside the hospital, called his office at 12:39 P.M. (CST), and dictated another bulletin: "FLASH FLASH KENNEDY SERIOUSLY WOUNDED . . . PERHAPS FATALLY BY ASSASSINS BULLETS."

Without official confirmation, Smith did not want to report what Clint Hill had told him—that the president was already dead.

A radio station interrupted a program of music with this announcement: "Ladies and gentlemen, here is a bulletin from the WQMR newsroom. An unknown sniper has fired three shots at President Kennedy in Dallas. Repeating this bulletin . . . received from the United Press. A sniper has fired at President Kennedy."

The announcer continued: "Now the remainder of the bulletin just clearing says that a sniper has seriously wounded the president in downtown Dallas. Repeating, the United Press says that a sniper seriously wounded President Kennedy in downtown Dallas today, perhaps fatally."

While doctors worked on the president, some of the Secret Service agents worked on the car. To prevent curiosity seekers from peering into the backseat or news reporters from snapping photographs of the blood and gore, agents mounted a top—a hardtop, not the clear plastic bubble top—to the convertible. Then they got steel buckets filled

The president's limousine at Parkland Hospital. Secret Service agents attached a hardtop and tried to wash away the president's blood. Note the bucket on the ground.

with water and towels and began to wash the backseat and the floor. *Time* magazine reporter Hugh Sidey watched. "It was an eerie scene. A young man, I assume he was a Secret Service man, with a sponge and a bucket of red water . . . was trying to wipe up the blood and what looked like flakes of flesh and brains in the back seat. The red roses were in the front seat."

It was a stupid thing to do. The car was a crime scene and full of evidence. Everything in it, including possible bullet fragments, should have been left as is and preserved for the investigation that was sure to follow the assassination. It was as though they believed, through some faulty logic, that if they could just wipe away the evidence of the crime, they could turn back the clock and pretend it had never happened. Try as they might, the agents could not wash away all the blood. It was like a scene from *Macbeth*, Shakespeare's violent play of murder and revenge, when "all great Neptune's ocean" would not "wash this blood clean from my hand." Jackie recalled what else lay in the car. "Every time we got off the plane that day, three times they gave me the yellow roses of Texas. But in Dallas they gave me red roses. I thought, how funny, red roses—so the seat was full of blood and red roses."

JACKIE KENNEDY approached the trauma room and tried to enter. "I'm not going to leave him. I'm not going to leave him," she told Dave Powers.

A burly nurse tried to block her way. It was against hospital policy for family members to enter the room. Jackie told her she was going in and pushed the nurse. The nurse pushed back. "I want to be with him when he dies," Jackie insisted.

A navy admiral on the president's staff rushed to her aid.

"It is her right," he commanded, "it is her prerogative."

The nurse shrank away. Jackie walked into the room where desperate surgeons worked to save her husband's life.

The appearance of the president's wife, a haunting pale figure in the bright pink suit, shocked the doctors. One of them suggested that she might want to leave and wait outside.

"But . . . it's my husband," she said, "his blood, his brains are all over me."

Blood streaked her face, saturated her white gloves, and stained her suit and stockings. She nudged one of the doctors, and without speaking she held out her cupped hands. She was holding a part of the president's brain. Dazed, in shock, perhaps she thought they would need it. Maybe they could put it back inside his head. She handed it to the doctor.

If John Kennedy had been any other patient, the doctors would have already pronounced him dead, perhaps even dead on arrival at Parkland Hospital. But this was the president of the United States—they had to try everything. As a last resort, one of the surgeons began to massage the president's heart, hoping to stimulate a rhythmic beat. It was too late. He was dead.

"I am sorry, Mrs. Kennedy," said one of the doctors, "your husband has sustained a fatal wound."

"I know," she whispered.

THE DOCTORS recorded the time of death as 1:00 P.M. (CST). One by one, members of the medical team left the trauma room. A nurse handed two paper bags to one of the Secret Service agents. They contained John Kennedy's suit coat, pants, shirt, tie, and other garments. The agent was also given Clint Hill's bloodstained suit coat. As the room emptied, Jackie Kennedy approached the table on which her husband lay dead. She pressed her cheek against his still warm face. She kissed his body. Then she removed her wedding ring and slipped it onto one of his fingers.

Assistant White House press secretary Malcolm Kilduff needed to find Vice President Lyndon Johnson to tell him that John Kennedy was dead. Kilduff went to the holding room where the Secret Service had hidden Johnson from view. LBJ's bodyguards had kept him far from trauma room one, so he had no personal knowledge of the president's condition. Kilduff spoke. "Mr. President . . ."

Stunned, Johnson did a double take. That was the first time he was called that.

Now he knew. The president of the United States was dead.

Kilduff asked Johnson if he should make an announcement to the press. No, LBJ told him, it would be better to wait until after he had left the hospital for Love Field and returned to the safety of Air Force One.

"I think I had better get out of here . . . before you announce it. We don't know if this is a worldwide conspiracy, whether they are after me as well as they were after President Kennedy, or whether they are after Speaker [John W.] McCormack, or Senator [Carl] Hayden [the two men who, after LBJ, were next in the line of presidential succession]. We just don't know."

Agent Rufus Youngblood decided he should escort the new president to the plane with as little fanfare as possible. There would be no entourage, no big motorcade accompanied by police cars and motorcycles with their screaming, attention-getting sirens. And Youngblood dared not put Johnson in SS-100-X, the presidential limousine. How could he? The backseat was still wet with blood and brains.

Instead, Johnson got into an unmarked car that gave no outward clue to the identity of its precious passenger. It was *not* a convertible.

AT 1:27 P.M. (CST), Walter Cronkite broadcast an update: "We just have a report from our correspondent, Dan Rather, in Dallas, that he has confirmed that President Kennedy is dead. There is still no official confirmation of this, however."

At 1:32 P.M. (CST) Cronkite made another announcement: "This is the bulletin that just cleared from Dallas, that the two priests who were in the emergency room, where President Kennedy lay after being taken from the Dallas street corner where he was shot, say that he is dead. Our man, Dan Rather in Dallas reported that about ten minutes ago, too."

At Parkland Hospital, at 1:33 P.M. (CST), Mac Kilduff walked into a room to make a statement to the journalists who awaited him there. His hands trembled. Then he spoke: "President John F. Kennedy died at approximately 1:00 CST today here in Dallas. He died of a gunshot wound to the brain. I have no other details regarding the assassination of the president."

Merriman Smith transmitted to UPI a three-word report: "PRESIDENT KENNEDY DEAD."

In the KLIF radio newsroom in Dallas, station owner Gordon McLendon handed off the microphone to a colleague: "Bob, do you have more?" He did: "The President is clearly, gravely, critically and perhaps fatally wounded. There are strong indications that he may have *already* expired, although that is *not* official. But the extent of the injuries to Governor Connally is a closely shrouded secret at the moment."

Then he coughed and cleared his throat. The sound of papers rustling in his hand—the latest bulletins from United Press—went over the air. Then he spoke: "President Kennedy is *dead*, Gordon."

Now the announcers spoke simultaneously over each other's sentences: "This is the official word" and "Ladies and gentlemen, the president is dead." Then one voice alone speaks: "The president, ladies and gentlemen . . . is dead."

WITHIN MINUTES, news of Kennedy's death flashed across America. At 1:38 P.M. (CST), a visibly shaken Walter Cronkite appeared on CBS television and made this announcement. "From Dallas, Texas, the flash, apparently official: President Kennedy died at 1:00 P.M., Central Standard Time, two o'clock, Eastern Standard Time—some thirty-eight minutes ago." Cronkite removed his eyeglasses, shook his head and paused. He was on the verge of tears.

Just a few weeks ago he had enjoyed the privilege of conducting a one-on-one, sit-down television interview with President Kennedy. Cronkite pulled himself together and continued. "Vice President Lyndon Johnson has left the hospital in Dallas, but we do not know to where he has proceeded; presumably he will be taking the oath of office shortly and become the thirty-sixth president of the United States."

THROUGHOUT THE nation, people at home that afternoon sat in their living rooms, riveted by television and radio alerts. Others gathered in quiet groups around office televisions sets and radios. Millions of people working that day were out to lunch when Kennedy was shot, and they heard the news when restaurants tuned their TVs and radios to news broadcasts. On the streets, many people gathered in front of appliance stores and watched the silent televisions on display behind plate-glass windows.

Drivers stopped in traffic and got out of their cars to talk to other drivers. When newspapers started publishing special editions that afternoon, frantic customers snapped up the copies as soon as they were delivered to newsstands, drugstores, and other outlets. In Chicago, one man ripped an outdoor, red metal news box for the *Chicago American* right out of the ground and drove off with a stolen stack of papers announcing the assassination.

In Nashville, Tennessee, the way that word of the assassination came to David Lipscomb High School was representative of how the news arrived at schools across the nation. A telephone rang in study hall. A teacher answered the call. She looked at her students, wondering what to do. "I don't know whether I should tell you this or not," she said. "I just don't believe it . . . President Kennedy has just been shot in Texas." A student shouted, "It couldn't have happened!" Another said, "I don't believe it."

Tommy Ingram, editor in chief of the school newspaper, the *Pony Express*, reacted fast to the breaking news. "By this time," he wrote later, "the intercom was on and the horrible message was being heard by the entire student body as it was by the entire world." In the corridors and classrooms, students spoke in hushed tones. Ingram deployed cub reporters to interview classmates and teachers. Randy McLean, a senior, said that he was "stunned" and that he couldn't "believe that anyone as alive as Kennedy was, is dead." Senior class president Bill Steensland confided, "I haven't yet, but I'm going to go home and cry." Patty Pettus, a junior, had several reactions: "Russia, Mrs. Kennedy, and the children. It's like a dream. I couldn't believe it." Lola Sue Scobey, secretary of the student body said, "We don't realize how historical this day is. Even if we were not a fan of Kennedy, it was still tragic . . . what will happen from this point on?"

Principal Damon Daniel summoned his stunned students to an assembly program. "We consider ourselves free men, yet in this country of freedom of speech and freedom of worship," he told them, "cruel tragedy still strikes." He encouraged them too. "We will lie down and bleed a while, but then we will rise again and fight." Until then, he continued, "a son, a brother, a father, and a husband has been lost, and it is our duty to weep with those that weep." Ingram had one of his staffers take photographs in the halls and at the assembly, and those reaction shots captured weeping teachers and dazed students.

The staff scrambled to get out an issue. On November 27, the day before Thanksgiving, the *Pony Express* published an ambitious, large-format, two-page extra headlined LIPSCOMB MOURNS KENNEDY. STUDENTS STUNNED AS NATION'S LEADER DIES.

Later, the National Scholastic Press Association honored the *Pony Express* by giving the paper its coveted All-American rating and naming it a Medalist, the association's highest award. "The extra edition of your paper covering the events of November 22, 1963, gave excellent, timely coverage of student reaction plus coverage of the assassination and events immediately following. Congratulations on mobilizing your staff."

Editor in chief Ingram also received a letter from the editor of the *Nashville Tennessean*, John Seigenthaler: "Dear Tommy: I have just read your *Pony Express* 'Extra' of Nov. 27 . . . it was a thoughtful and professional piece of journalism. You and your entire staff have my congratulations." Tommy did not know it, but Seigenthaler was a friend of John and Robert Kennedy.

IN DALLAS, it was time to send for a casket. The Secret Service ordered one from a local funeral home. While it was on its way, hospital staffers washed the president's body and wrapped his head in towels and sheets of plastic so his blood would not stain the silk lining of the coffin. Then they wrapped the entire body. Mrs. Kennedy did not watch this. When the coffin arrived, funeral-home workers wheeled it into the emergency room. They lifted Kennedy's corpse from the table and laid it in the coffin. Then they closed the lid. The president was ready to go home.

BACK IN Washington, a telephone in the office of Attorney General Robert Kennedy rang at 1:45 P.M. Eastern Standard Time. It was the direct line that linked the offices of the attorney general and the director of the FBI. Robert Kennedy had insisted on setting up the line so that J. Edgar Hoover could not avoid taking his calls. Hoover had great antipathy for all the Kennedys, but he disliked Bobby most of all. The director judged the Kennedys to be hypocrites and moral failures, and Hoover knew all about the president's indiscretions with women, going back to the young naval officer's World War II affair with a probable Nazi spy, the beautiful Danish blond journalist Inga Arvad.

The director, a veteran of decades of Washington intrigue and turf battles, bristled at the presence of the special phone on his desk. But the attorney general was his superior in rank, and Hoover had to submit to what he believed was a humiliation. But the phone worked both ways. This time, it was Hoover who called Kennedy. Robert Kennedy's assistant, Angie Novello, picked up the receiver.

"This is J. Edgar Hoover. Have you heard the news?"

She had. The attorney general was not in his office. Novello knew what Hoover wanted her to do, but she could not bear it.

"Yes, Mr. Hoover, but I'm not going to break it to him."

"The president has been shot," Hoover said. "I'll call him."

Robert Kennedy was in McLean, Virginia, having lunch at his expansive home, Hickory Hill, on the outskirts of Washington, less than a half-hour drive from the Department of Justice at Ninth and Pennsylvania Avenues. Bobby, his wife Ethel, and two Justice Department lawyers, United States Attorney Robert Morgenthau and his assistant, were sitting by the pool and eating sandwiches. Ethel walked away to answer the phone.

A White House operator told her that J. Edgar Hoover was calling.

Ethel tried to deflect the call. "The Attorney General is at lunch." The operator told her it was urgent. Ethel told her husband, "It's J. Edgar Hoover."

Bobby came to the phone.

Hoover spoke. "I have news for you. The president has been shot."

Kennedy asked for details.

"I think it's serious," Hoover said. "I am endeavoring to get details. I'll call you back when I find out more." Then he hung up.

Horrified, Bobby cried out, "Jack's been shot!"

AT THE United States Capitol, press liaison Richard Riedel walked onto the floor of the Senate and said, "The president has been shot. The president—he's been shot." JFK's brother, Senator Edward "Ted" Kennedy, happened to be presiding over the chamber. Whenever the president of the Senate—Vice President Lyndon Johnson—was absent, senators took turns wielding the authority of the gavel. Riedel approached the rostrum.

"The most horrible thing has happened! It's terrible, terrible!"

"What is it?" Kennedy asked.

"Your brother, the president. He's been shot."

"How do you know?"

Riedel told him, "It's on the ticker. Just came in on the ticker." Edward Kennedy fled the Senate chamber.

AT HICKORY Hill, Robert Kennedy hurried to leave. He assumed he would be flying to Dallas to be with his brother. Then a call came from the White House. Bobby listened. Then he cried out.

"Oh, he's dead!"

"He had the most wonderful life," he told Ethel. Then he walked out to the pool and broke the news to his guests.

"He died."

Soon the phone rang again. It was J. Edgar Hoover with more details. He did not know that Bobby had already heard what he was about to tell him.

Bobby interrupted him. "It may interest you to know that the president is dead."

CHAPTER 7

"I HAVEN'T SHOT ANYBODY"

The Secret Service agents wanted to remove the president's body from Parkland and take it to Air Force One. A local official had other ideas. Earl Rose, Dallas County medical examiner, told agent Roy Kellerman that the corpse could not be removed until after an autopsy was performed in Texas.

Kellerman rebuffed him. "My friend, this is the body of the President of the United States, and we are going to take it back to Washington."

"No," Rose said, "that's not the way things are. When there's a homicide, we must have an autopsy."

Forget it, Kellerman told him. "He is the President. He is going with us."

Rose would not back down. "The body stays."

The Secret Service agent was furious. "My friend, my name is Roy Kellerman. I am Special Agent in Charge of the White House Detail of the Secret Service. We are taking President Kennedy back to the capital."

Earl Rose was not impressed. "You're not taking the body anywhere. There's a law here. We're going to enforce it."

Admiral George Burkley, the president's physician, joined the argument, telling Rose that they could not keep Mrs. Kennedy waiting.

Rose snapped at him. "The remains stay."

Burkley was now as angry as Kellerman. "It's the President of the United States!"

"That doesn't matter," Rose said. "You can't lose the chain of evidence."

A justice of the peace, who had the authority to overrule Rose, showed up. He was not sympathetic. "It's just another homicide, as far as I'm concerned."

That was the tipping point. Agent Kellerman cursed the official and announced that "we're leaving." Ken O'Donnell backed him up. "We're leaving *now*. Wheel it out!"

At 2:08 P.M. (CST), the agents began to roll the coffin through the emergency room to the white hearse waiting outside. They pushed aside Rose, the justice of the peace, and a local cop. The dispute almost broke out into a fistfight. As much as JFK's Irish mafia, the agents on his detail had grown to love him. They would stand for no more delays. They marched right past the state officials, out the door, and to the hearse. John Kennedy's loyalists had no way of knowing that they had made a serious mistake. Their obstructions and the resulting failure to conduct an autopsy in Dallas within a few hours of the murder would come to haunt the history of John Kennedy's assassination for the next fifty years. The intense display of emotions by Kennedy's grieving staff and Secret Service detail at Parkland Hospital, although understandable, served the nation poorly, and would, in time and for decades to come, create widespread suspicion and mistrust about the facts of the assassination and encourage many wild theories about the murder.

• • •

WHEN LEE Oswald left his boardinghouse, he started walking. By coincidence, a Dallas policeman driving his car through the neighborhood spotted Oswald walking on the sidewalk. The police officer, J. D. Tippit, knew the president had been shot and had heard over his car radio at 12:45 P.M. (CST) a physical description of the suspect obtained from witnesses, especially Howard Brennan, who had seen the man in the Book Depository window. The radio transmission was brief: "Attention all squads, the suspect is reported to be a white male, approximately thirty, slender build, 5 feet 10 inches, weighs 165 pounds, reported to be armed with what is thought to be a thirty-caliber rifle. No further description at this time or information."

Oswald matched it in a general way, so Tippit decided to pull over to the curb and have a word with him. The thirty-nine-year-old policeman stopped the car and called out to Oswald through the open passenger-side window.

Oswald stopped walking and approached the vehicle. In a casual way, he leaned on the top edge of the passenger door, looked inside, and conversed with Tippit. They spoke for less than a minute. No one knows what they discussed. Witnesses who saw the encounter were too far away to overhear the exchange. Whatever Oswald said, it must not have satisfied Tippit, because he swung open his car door and got out. When he stood up in the street, he did not reach for his holster and place his hand on the butt of his revolver. He must have concluded that Oswald was not dangerous. Tippit started to walk around the front hood of his car, toward the curb where Oswald waited for him.

A moment later, at 1:15 P.M. (CST), Oswald pulled his revolver from his jacket pocket, aimed it at the policeman's chest, and opened fire. He had taken Tippit by surprise and shot him three times before the policeman could even draw his own pistol. Tippit collapsed to the ground. Oswald paused, and then walked over to the wounded policeman. Disabled, helpless, Tippit was still alive. Then Oswald pointed his pistol at the helpless officer, took aim, shot him in the head, and killed him.

This was a dirty killing. Yes, Oswald had already committed a horrible crime—he had murdered the president of the United States. But he had done it from a distance. The rifle was an impersonal, antiseptic weapon that allowed Lee to remain detached from the ugly reality of his crime. Oswald did not have to look Kennedy in the face before he shot him. He did not see the wounds he inflicted on Kennedy and Connally. He did not have to look into the backseat of the car at the blood and brains. And he did not see the surprised, eyes-wide-open stare his third shot had frozen on the president's face.

But now, an hour later, Lee Harvey Oswald committed murder of another kind—up close and personal. He had spoken to J. D. Tippit. When he leaned on the top edge of the passenger side door, he had bent down and chatted with the officer; he was close enough to the policeman to smell him. Oswald had looked him in the face. Then, when Tippit stepped out of the car, Oswald had shot him at close range. Not just once, to disable Tippit and then flee. Not twice, to keep him down. Three times in the chest, to kill him. Oswald was close enough to hear Tippit's reaction to being shot.

But the policeman refused to die at once. As Tippit lay dying on a city street, his lungs emitted a gurgling sound. He could not speak or call out for help. He no longer had the strength to grasp his revolver, raise his arm, and shoot his assailant. Oswald stepped across the curb and into the street. He walked over to the fallen policeman. Half an hour ago, Tippit was home having lunch with his wife. Now he was bleeding to death on a Dallas street. But not quick enough for Oswald. The assassin was in a hurry. He could not wait for the cop to die.

Now he did something truly depraved. Oswald stood over Tippit and contemplated his victim. He was at point-blank range. Oswald raised his arm and pointed the barrel of his revolver at Tippit's head. For the fourth time—once more than he had needed to kill President

Kennedy—he squeezed the trigger, and put a bullet through J. D. Tippit's brain. He walked away, swung open the cylinder of his revolver, pulled out the four empty cartridges, and tossed them on the ground. It was more evidence against him for the police to recover later.

Several witnesses saw Oswald either shoot Tippit or flee the scene. One man heard him mutter "poor dumb cop" or "poor damn cop" as he fled. Some of them followed him, and then, from a safe distance, began chasing him. Less than one hour after the assassination of President Kennedy, but without knowing they were in pursuit of the president's killer, a small posse of several citizens was chasing Lee Harvey Oswald through the streets of Dallas.

As he ran, Oswald loaded four fresh bullets into his pistol and pushed the cylinder into firing position. He was ready to kill again.

WHY DID Oswald murder Tippit? Did the officer believe he matched the description of the suspect broadcast by the police dispatcher? Did Oswald appear nervous, evasive, or suspicious? Did Tippit threaten to take him in for questioning? The only logical reason for Oswald to murder a police officer was to avoid immediate arrest. Or maybe Oswald just panicked. It was a rash and foolhardy thing to do. Several witnesses could identify him as a cop-killer. As soon as they reported the shooting, the whole neighborhood would be crawling within minutes with police officers and detectives hunting for him. The Tippit shooting dramatically reduced Oswald's chances of escape.

If Oswald had not killed Tippit, he would have had more time to carry out his escape. He had not yet been identified by name as a suspect in the assassination of the president. It would have taken time for the Secret Service and Dallas police to obtain a roster of all Book Depository employees and then conduct a roll call to discover if any of them were missing from the premises. Once they discovered Oswald's absence, they would have sent men to his boardinghouse; they would discover that he had already been there and gone. Then, within a few hours, they would have discovered strange things about him—his discharge from the Marine Corps; his interest in Russia, Cuba, and Communism; and his defection to the Soviet Union and return to the United States.

Buell Frazier would tell them about the curtain rods that morning, and he could lead them to Ruth Paine's house and to Oswald's Russian-born wife, Marina. Also, once they found the rifle, they could begin tracing its serial number to the place that sold it and to the name of the person who bought it. That would also take time. Enough to give him at least a couple hours, and maybe more, as a head start. If Oswald was lucky, the authorities might not name him as a suspect or show his picture on TV until late that afternoon or the evening. The newspapers might not even print his name and photograph until the next morning.

But Oswald destroyed all of that advantage. Not until he killed a policeman did he become the object of a furious manhunt. Back at the Book Depository, the Dallas Police, FBI, and Secret Service were still trying to fit the pieces together. No one had pursued Oswald after he left the building. No one had been searching for him. But now, forty-five minutes after he had strolled away from the Book Depository, Oswald was running for his life.

LYNDON JOHNSON waited aboard Air Force One for the hearse carrying Kennedy's body to arrive at the airport. He was already aboard the plane before the Kennedy entourage left the hospital at 2:08 P.M. (CST). Some of Johnson's political advisers were urging him to fly back to Washington right away and leave Mrs. Kennedy behind to catch Air Force Two, the vice presidential jet, later in the day. Rufus Youngblood agreed. Nothing was more important than protecting the life of the new president. The best way to do that was to get Lyndon Johnson back to the capital as soon as possible. If this was a conspiracy, and if LBJ was the next target, he needed to get out of Dallas.

In one of his first acts as chief executive, he decided they would not fly back to Washington immediately and abandon Mrs. Kennedy in Dallas. Because she refused to leave without her husband's corpse and because Johnson refused to depart without her, he ordered Air Force One to wait. All of them would fly back together. Johnson also decided that he would take the oath of office on the ground in Dallas before the plane took off.

At Parkland, JFK's agents wheeled his coffin to the hearse. Jackie refused to be separated from her husband, so she sat in back, next to the casket. Clint Hill and two military officers sat with her. Three Secret Service agents, including the driver, sat in front. The modest motorcade—the white Cadillac hearse followed by four cars—from the hospital to the airport was not as grand as the one that had left Love Field for downtown Dallas just two hours ago. During the ride, Admiral Burkley got Jackie's attention. He offered her two of the red roses from her bouquet. They were in perfect condition. "These were

Secret Service agents and members of President Kennedy's staff carry his casket aboard Air Force One.

under, were in his shirt," he told her. They had broken off the bouquet during the mayhem in the presidential limousine. She put them in one of the pockets of her pink suit.

The hearse drove onto the field. At the airport, Kennedy's aides and Secret Service agents carried the heavy burden up the stairs of an airline ramp and onto the plane. They could barely lift it—it weighed eight hundred pounds, not counting the added weight of the president's body. Then it would not fit through the airplane door—it was too wide. To make the coffin narrower, they had to break the carrying handles off its sides so they could push it onto the plane. They secured it in a small cabin near the back, where some seats had been removed to accommodate it.

Although the coffin was now on board, Lyndon Johnson did not want Air Force One to depart Dallas until he took the oath of office and was sworn in as the thirty-sixth president of the United States. Johnson, a Texan, had summoned an old friend, a local federal district court judge named Sarah T. Hughes, to rush to Love Field to swear him in. This was all a great bit of historical drama, because in fact the swearing in was a formality to confirm what had already happened. Johnson did not need to take the oath in order to assume the office—by authority of the United States Constitution he had already become president upon the death of his predecessor.

But Lyndon Johnson wanted to take the oath at once as a symbol of the continuity of the American government. A president might die, but democracy would live. And he wanted the ceremony photographed to transmit that symbol around the world. A White House photographer, Captain Cecil Stoughton, was aboard the plane. He loaded two cameras with film and planned in his head where he and the new president should stand in the cramped cabin to create the most solemn and impressive photograph. The image would convey that Johnson was now in charge of the government.

WQMR got word of what was about to happen. "Just to show you how swiftly action is taken, as it was upon the assassination of President Lincoln, Johnson—Lyndon Johnson—is expected to be sworn in *aboard* an airliner, before flying back immediately to the nation's capital. So, possibly, at this very instant as we are speaking, Vice President Lyndon Johnson is becoming the president of the United States aboard an airliner."

By tradition, when a president took the oath of office, he stood alone, raised his right hand, and repeated the words of the oath spoken to him by the chief justice of the Supreme Court of the United States. On November 22, Lyndon Johnson had something else in mind. He wanted two people standing by his side. One was his wife, Lady Bird. The other was Jackie Kennedy. It was a bold and, in the opinion of some of President Kennedy's staffers, an outrageous and offensive request. How could a woman widowed just hours ago, under the most horrible of circumstances, be expected to pose for pictures? Johnson realized that his request demanded sensitivity, delicacy, and tact.

Before he became vice president, when he was a United States senator and held the powerful post of majority leader, he was renowned for his legendary skill at persuading other people to do what he asked. But it was one thing to strong-arm a fellow politician in a backroom deal over a piece of legislation in Congress, and quite another to handle the bereaved and beloved First Lady of the United States, who was still in a state of shock after seeing her husband murdered in front of her eyes. LBJ sensed he could not delegate this request to others. He would have to appeal to Jackie himself.

Lyndon and Lady Bird Johnson went to Jacqueline Kennedy's private compartment to offer their condolences. "Dear God, it's come to this," said Lady Bird. Jackie said she was grateful her husband did not die alone: "Oh, *what* if I had not been there. Oh, I am so glad I was there."

Lady Bird asked her if she would like to change into fresh clothes. No, the new widow replied, "I want them to see what they have *done* to Jack."

Then LBJ broached the awkward subject. "Well—about the swearing in," he said.

Jackie indicated that she understood: "Oh, yes, I know, I know. What's going to happen?"

Johnson explained that he had summoned a federal judge to administer the presidential oath aboard Air Force One before the plane flew back to Washington. LBJ wanted her standing by his side. It was for history. Believing she had agreed, Johnson left her alone to compose herself until Judge Hughes arrived.

More than one person suggested to Jackie that she change clothes. Her suit, white gloves, and stockings were caked with dried blood—the bright red, wet blood spilled two hours ago had, after exposure to oxygen, solidified and taken on a darker color. Each time someone asked her, the more adamant she became. "Everybody kept saying to me to put a cold towel around my head and wipe the blood off." But she defied them. No, she insisted, she would not change. "I want them to see what they've done," she repeated more than once.

To prepare for the swearing-in ceremony, she retired to her small bathroom. There, she said, "I saw myself in the mirror; my whole face was spattered with blood and hair . . . I wiped it off with Kleenex . . . then one second later I thought, why did I wash the blood off? I should have left it there, to let them see what they've done. . . . If I'd just had the blood and caked hair when they took the picture . . . I should have kept the blood on."

EVERYTHING WAS ready for the swearing in. But Johnson was still waiting for Jackie. He did not want to proceed without her. Johnson spotted two of Kennedy's aides and said, "Do you want to ask Mrs. Kennedy if she would like to stand with us?" They hesitated. "She said she wants to be here when I take the oath," the new president told them. "Why don't you see what's keeping her?"

Ken O'Donnell went to Jackie, who told him, "I think I ought to. In the light of history, it would be better if I was there."

She emerged from her room and entered the cabin where Lyndon Johnson, Judge Hughes, the photographer, and a number of the passengers awaited her. Merriman Smith was on the plane as a pool reporter and watched the scene unfold. "The large center cabin was dark, all shades drawn, members of the Kennedy staff sitting around, some staring straight ahead, some crying softly."

Cecil Stoughton had started taking pictures of the scene even before Jackie walked in. Then she appeared. Smith saw it all. "Mrs. Kennedy walked down the narrow corridor from her bed chamber and into the lounge. She was dry eyed, but her face was a mask of shock." Her appearance startled everyone—they had assumed she would change into a fresh outfit. Then Johnson held her hand. The gesture struck Smith: "Like a man might lead a small child, Johnson took her hand and led Jackie to a place at his left side." LBJ told her, "This is the saddest moment of my life." Lady Bird Johnson stood at her husband's right. Stoughton raised one of his cameras, and then, as Judge Hughes read the oath and Johnson repeated it, the photographer took several shots before switching to his second camera. The only other sounds in the cabin were the clicking shutters of the cameras.

Lyndon Johnson being sworn in aboard Air Force One before it departs Dallas.

Jackie's appearance horrified Stoughton, who aimed his lens high to crop out the lower part of her body to hide the bloody skirt and stockings.

At 2:38 P.M. (CST), Johnson raised his right hand and was sworn in as the thirty-sixth president of the United States. It took less than a minute for him to speak the words: "I do solemnly swear that I will faithfully execute the office of President of the United States, and will to the best of my ability, preserve, protect, and defend the Constitution of the United States. So help me God."

Lyndon Johnson had already been president for almost two hours, but when the oath was done, his presidency officially began. And Stoughton had taken what remains, to this day, perhaps the most iconic, riveting, and harrowing photographs in all of American history.

After the swearing in, Jackie returned to the back of the plane, to the casket, where she found Ken O'Donnell, Larry O'Brien, and Dave Powers. She sat down and began to weep. "Oh, it's happened."

Lyndon Johnson gave an order: "Now, let's get airborne."

Before Air Force One left Dallas, Stoughton hurried off the plane with his cameras and film. It was his job to rush to develop the pictures so they could be published in the evening newspapers throughout the country.

Within a few minutes, Air Force One hurtled down the runway and took flight for Washington.

BY THE time Oswald made it to a commercial district on West Jefferson Boulevard, he had lost the people who had followed him from the Tippit murder scene. For now, he was safe. But then he made a mistake. When he heard a police siren, he turned his back to the street and pretended to study a display of footwear behind the plate-glass windows of a shoe store. That attracted the attention of the manager, Calvin Brewer. Oswald seemed furtive and suspicious. He watched Oswald and followed him down the sidewalk. Then, at 1:40 P.M., Oswald ducked inside a movie theater at 231 West Jefferson, sneaking past the ticket seller's window without paying for admission. He took a risk when he hid inside the Texas Theatre. If anyone had followed him there, one telephone call would summon a dozen police cars to the scene. Police departments were, and still are, relentless in pursuing a criminal who has shot one of their own. Even if Oswald had evaded his pursuers, his failure to buy a movie ticket might provoke theater employees to call the police to report a nonpaying customer. Either way, Oswald could find himself trapped inside a building with no escape.

Someone *had* followed Oswald. It was the shoe-store manager. He had tracked him to the theater. When he saw Oswald duck inside, he persuaded employees there to call the police.

Oswald had murdered John Kennedy less than an hour and a half ago. Now he sat in the darkness of a theater watching a matinee showing of a movie about World War II titled *War Is Hell*. The theater was nearly empty. A few men occupied scattered seats. Oswald sat near the back.

Oswald did not enjoy the movie for long. Several police cars pulled up in front of the theater, and detectives ran inside toward the screen and climbed onto the stage. The houselights went on, and a witness identified Oswald. As officers rushed him, Oswald shouted, "This is it!" Others heard him say "Well it's all over now." He punched one policeman in the face, then he reached for his pistol. He was eager to kill more cops. One of them grabbed his arm and tried to twist the revolver out of his hand. Another punched him. Another whacked him in the head with the butt of a shotgun.

Overpowered, Oswald gave up. "Don't hit me anymore. I am not resisting arrest!" Then he shouted, "I protest police brutality!" As the police tried to drag Oswald out of the theater he yelled, "They're violating my civil rights!"

In the lobby, he spotted a TV camera and shouted, "I want my lawyer. I know my rights. Typical police brutality. Why are you doing this to me?"

At 1:50 P.M. (CST), one hour and twenty minutes after the president had been shot, Oswald was in custody. But his captors did not know the importance of their prisoner. They were not sure they had caught Kennedy's killer—although they had their suspicions—but they were convinced they had just apprehended the cold-blooded murderer of Officer Tippit.

At 1:52 P.M., they shoved their prisoner into a car and drove him to police headquarters at City Hall, a building not far from the Texas School Book Depository.

In the car, Oswald demanded, "What's this all about? I know my rights. I don't know why you are treating me like this. Why am I being arrested? The only thing I've done is carry a pistol in a movie." Oswald added, "I don't see why you have handcuffed me."

One of the detectives said, "You've done a lot more. You have killed a policeman."

"Police officer been killed?" Oswald asked. "I hear they burn for murder. You fry for that."

"You might find out," a policeman replied.

"Well," Oswald countered about the electric chair, "they say it just takes a second to die."

Oswald kept refusing to identify himself, so a policeman pulled the wallet out of his pocket and searched it for identification. He found a library card bearing the name of Lee Oswald. Then he found another document with the name A. J. Hidell, one of Oswald's aliases.

When the car arrived at the City Hall garage, the police advised their suspect that he might want to turn his face away from the journalists and TV cameras waiting for him there. He was defiant: "Why should I hide my face? I haven't done anything to be ashamed of."

LEE HARVEY Oswald was about to face the first of several police interrogations. He would be pitted against a savvy homicide detective,

Will Fritz, a department legend and veteran master of extracting confessions. He would become the suspect's principal interrogator in what would become a multiday game of cat-and-mouse with the chatty assassin.

Good cops had good instincts. From his first encounter with Oswald, Fritz's intuition told him this was the man who had murdered the president. In the hours ahead, Dallas police, FBI, and Secret Service would not be the only ones allowed to question him. Others who had no business doing so, who could help spoil the legal case against Oswald, would soon have access to him too.

By 2:00 P.M., the arresting officers at the Texas Theatre had brought Oswald to police headquarters. When detectives took him into an interrogation room, a man there recognized Oswald at once. It was Bill Shelley, foreman at the Texas School Book Depository. He was there giving an affidavit to a policeman. Shelley told the police, "Well, that is Oswald. He works for us. He is one of my boys."

Oswald said his name was "Hidell." In his wallet, one piece of identification said his name was Hidell, and another said it was Oswald.

A detective asked which name was real.

"You find out," Oswald replied.

At 2:20 P.M. on Friday, November 22, Oswald was moved to the office of Captain Fritz. The Dallas district attorney, Henry Wade, had jurisdiction over the Tippit case, as he did over all murders in Dallas. Even when it became obvious that Oswald was also a suspect in the murder of President Kennedy, Wade and the Dallas police retained control of Oswald. No federal statute made killing a president a federal crime, so the U.S. attorney and the FBI had no power to seize Oswald from the custody of the Dallas Police Department.

Detective Elmer Boyd asked Oswald why he had a bruise above his eye.

"Well," Oswald explained, "I struck an officer and the officer struck me back, which he should have done."

Oswald was playing contrite for now. Captain Fritz entered the office and, by 2:30 P.M., the first interrogation of Oswald had begun. It would be the first of four.

Fritz asked for his full name.

"Lee Harvey Oswald."

Fritz asked where he worked.

"Texas School Book Depository," Oswald said.

Fritz wanted to know how he got the job.

Oswald told him that a lady he knew recommended him for the job, and that he got it through her.

Fritz was a master of the conversational style of interrogation. He wanted to warm Oswald up with a series of routine questions to get him talking.

Why, the detective asked, did he possess a card that identifies him as "Hidell"?

It's just a name that "[I] picked up in New Orleans."

The employment records of the Book Depository said Oswald lived in Irving.

No, Oswald said, he lived on Beckley in the Oak Cliff section of Dallas.

Oswald added that his wife was staying with friends in Irving. This is the first the police had heard about the Dallas rooming house. The place might be filled with evidence. Fritz sent officers to North Beckley Street to search Oswald's room.

Soon two FBI agents arrived to join the questioning. One was James Hosty, the special agent who had called on Marina Oswald a few times after she and Lee had moved to Texas. Those visits had infuriated Lee. Now, he and Hosty were meeting for the first time.

"Oh, so *you're* Hosty, the agent who's been harassing my wife!"

"You have the right to remain silent," Hosty advised him.

Oswald had remained calm while Captain Fritz questioned him. Now Oswald was enraged, screaming and cursing at Hosty. What was the story, Fritz wondered, between Oswald and this FBI agent?

"My wife is a Russian citizen who is in this country legally and is protected under diplomatic laws from harassment by you or any other FBI agent." Oswald was out of control. "The FBI is no better than the Gestapo of Nazi Germany! If you wanted to talk to me, you should have come directly to me, not my wife."

The angry exchange between Oswald and Hosty alarmed Fritz. Just when the homicide detective was having a calm chat with his prisoner and getting to know him, the FBI barged in and infuriated Oswald. It would be tough to question him now. Fritz wanted to know what was going on. What did Oswald mean by "accosted"?

"Well, he threatened her. He practically told her she'd have to go back to Russia. He accosted her on two different occasions."

Oswald was so angry about the visits that after they happened, he dropped off an anonymous menacing letter for Hosty at the local FBI office, warning that unless the special agent stopped bothering Marina, he would take action against the FBI.

Oswald's hands were cuffed behind his back. He told Fritz he wanted the handcuffs removed. That might be dangerous. Oswald was a suspect in the murder of a policeman. And he had tried to shoot more cops at the Texas Theatre. Free of the cuffs, he might spring from out of his chair, tackle one of the detectives or special agents, and grab for a gun. Fritz settled on a compromise, telling his men to remove the cuffs and recuff Oswald's hands in front of his body.

"Thank you, thank you," Oswald said.

To Hosty, he said, "I'm sorry for blowing up at you. And I'm sorry for writing that letter to you." Now that Oswald appeared to have calmed down, Fritz hit him with his first serious question.

"Do you own a rifle?"

"No, I don't."

Asked if he had ever seen a rifle in the Texas School Book Depository, Oswald said that he had seen one there two or three days before the assassination and that Roy Truly and some men were looking at it.

Oswald was trying to implicate the manager of the Book Depository in the assassination.

"Did you ever own a rifle?" Fritz repeated.

"I had one a good many years ago. It was a small rifle . . . but I have not owned one for a long time."

Hosty interrupted Fritz's line of questioning and asked Oswald if he had ever been to Mexico.

"Sure. Sure I've been to Mexico. When I was stationed in San Diego with the Marines, a couple of my buddies and I would occasionally drive down to Tijuana over the weekend."

Fritz jumped back in. He did not want the FBI agent to take the interrogation off track or anger Oswald. He asked Lee if he had a Russian wife, if he had ever been to Russia, and for how long. Oswald answered all the questions. Then Fritz asked whether he had owned a rifle in Russia.

"You know you can't own a rifle in Russia. I had a shotgun over there. You can't own a rifle in Russia."

Hosty interrupted with another question. "Mr. Oswald, have you been in contact with the Soviet embassy?" Oswald's blood started boiling again.

"Yes, I contacted the Soviet embassy regarding my wife. And the reason was because you've accosted her twice already!"

Hosty continued. "Have you ever been to Mexico City? Not Tijuana. Mexico City . . . have you ever been to Mexico City?"

"No! I've never been there. What makes you think I've been to Mexico City? I deny that!"

Captain Fritz told Oswald to calm down. It was obvious to Fritz that allowing the FBI agent to question the suspect was counterproductive. Whenever the police detective soothed Oswald and got him talking, Hosty provoked him.

"Okay," Fritz said, "let's take break."

AT THE Paine house, Ruth and Marina watched television coverage of the assassination. News reports that the shots were fired from the Book Depository troubled Marina, but the presence in the garage of what she assumed was the rifle still rolled in its blanket had calmed her. If she still had the rifle, that meant it could not have been Lee. Soon the police, following up Depository employment records, drove out to Ruth's house. When they arrived, the screen door was closed but the front one was open. They could see Marina sitting in the living room watching television. Ruth Paine invited them in.

They wanted to speak to Marina. They asked her if her husband owned a rifle. "Yes," she answered. They began to search the premises.

"When they came to the garage and took the blanket," she thought, "well, now, they will find it."

Then it happened. One of the policemen reached down to pick up the blanket and its contents. It sagged in his hand. "They opened the blanket but there was no rifle there." Until this moment, Marina did not know the rifle was gone. She could guess where it was now.

Marina was devastated. "Then, of course . . . I knew that it was Lee." So that's why he had come to Mrs. Paine's one day early.

IN THE rear of Air Force One, Jackie Kennedy sat next to her husband's coffin during the entire flight back to Washington, D.C. Top aides to John Kennedy took turns visiting her at the back of the plane, where they reminisced and told stories of happier days. They did not want her to mourn alone. Dave Powers told stories about the president's trip to Ireland that June. Ken O'Donnell said, "You know what I'm going to have, Jackie? I'm going to have a hell of a stiff drink. I think you should too." He offered to make her a scotch. "I've never had a scotch in my life," she replied. Godfrey McHugh said that didn't matter. "Now is as good a time as any to start." Jackie acquiesced and downed two. It was the beginnings of an Irish wake. Ken O'Donnell remembered how this day had begun. "You know what, Jackie? Can you tell me why we were saying that this morning? What was it he said at the hotel? 'Last night would have been the best night to assassinate a President.' Can you tell me why we were talking about that?"

Surrounded by the men who would miss Jack the most, Jackie

was overcome with emotion. "You were with him from the start and you're with him at the end."

During the flight, passengers divided into two camps, those who had served President Kennedy, and those who worked for Lyndon Johnson. For the duration of the flight, an uncomfortable tension filled the air. Some Kennedy aides had never liked LBJ, whose chief nemesis had been the president's own brother, Attorney General Robert Kennedy. Some of President Kennedy's staffers were snobs who believed that Johnson, who had attended a teacher's college in Texas, not an elite, Ivy League university as many of them had, was their social and intellectual inferior. Behind his back, they called him "cornpone" and mocked him as a crude country hick from the Texas Hill Country.

In truth, Johnson was a master politician with a record of important achievements. Johnson's earthiness was a mask that often concealed his sophisticated grasp of events and his keen understanding of the minds and behavior of others. Without Johnson as his vice presidential running mate in 1960, it is unlikely that John Kennedy would ever have been elected president in the first place. In their anguish, some Kennedy staffers now resented Johnson for becoming president. In fact, on November 22, Lyndon Johnson acted with grace and dignity. He did the best he could under the most trying of circumstances.

Forty thousand feet below the jet carrying home the body of John Kennedy and the new president, Lyndon Johnson, was a stunned nation grieving its fallen leader. By midafternoon, almost everyone in America knew about the assassination. On the flight back, Jackie Kennedy began to plan her husband's funeral. Before she was done, she would oversee the biggest, most majestic public funeral in American history since the death of Abraham Lincoln ninety-eight years earlier.

IN DALLAS, at 3:40 P.M. (CST), Captain Fritz returned to his office and continued to question Oswald.

"I asked him why his wife was living in Irving and why he was lving on Beckley."

His wife, Oswald explained, was staying with Mrs. Paine, who is trying to learn Russian. Marina teaches her, and Mrs. Paine helps her out with the baby.

"How often do you go out there?" the detective asked.

"Weekends."

Why didn't Lee stay at Mrs. Paine's full-time? Fritz wanted to know.

Oswald said that he didn't want to stay there all the time because Ruth Paine and her husband did not get along too well.

Fritz asked Oswald some questions about his schooling, his background, and whether he owned a car. No, he had no car, Oswald said. Then Fritz asked about his Marine Corps service.

"Did you win any medals for rifle shooting in the Marines?"

"The usual medals," said Oswald, admitting that he had received an award for marksmanship.

It was part of Fritz's interrogation style to shift from one subject to another, unrelated one, to keep a suspect off balance and never allow him to get too comfortable. He did that now.

"Lee, why were you registered at the boarding house as O.H. Lee?"

It was the landlady's fault, said Oswald. "The lady didn't understand me." He just left it that way.

Another of Fritz's tactics was to ask a suspect a question he already knew the answer to, and which the suspect could answer truthfully without incriminating himself. Observing how a suspect behaved while he was telling the truth would help Fritz intuit when he was telling a lie. He asked a series of the easy questions now.

Did he work at the Depository today?

Oswald said yes, and that he had worked there since October 15.

Fritz asked what part of the building he was in at the time the president was shot.

Oswald claimed he was having lunch on the first floor. They broke for lunch about noon, and he came down and ate.

"Where were you when the officer stopped you?"

On the second floor drinking a Coca-Cola. There's a soda machine in the lunchroom there. Oswald said he went up to get a drink.

Fritz asked what Oswald did after the president was shot. He said he left the building.

And where did he go?

Oswald said he went home to his room on Beckley. He was telling the truth. "I took the bus and went home, changed my clothes, and went to a movie."

That was an odd thing to do, Fritz thought. Who goes home to get a pistol before going to the movies?

Why did he do that?

" 'Cause I felt like it."

"Because you felt like it?"

"You know how boys do when they have a gun, they just carry it."

Now Captain Fritz slipped in a real question.

"Did you shoot Officer Tippit?"

It was not a silly thing for the detective to ask. He was a master at questioning murder suspects, and Oswald would not have been the first one he had lulled into confessing his crime. Fritz hoped Oswald would make it easy for him.

No, Oswald said. "The only law I violated was in the show: I hit the officer in the show, and he hit me in the eye and I guess I deserved it. That is the only law I violated. That is the only thing I have done wrong."

Fritz asked Oswald what he did in the Soviet Union, and Lee told him he worked in an electronics factory. The detective jumped back to the assassination.

Why did Oswald leave the Depository after the shooting?

"Shortly after the president was shot . . . I figured with all the confusion there wouldn't be any more work to do that day."

Fritz asked him what kind of work he did there. Oswald explained that he filled orders. What floors, Fritz continued, did Oswald have access to? The detective might have been curious to see if Lee would lie about the sixth floor. But Oswald was nonchalant.

"I was just a common laborer," he said, and "as a laborer, I have access to the entire building." Oswald told Fritz that the books were kept on the third through sixth floors. So he went to all the floors. Including the sixth.

It was time, Fritz decided, to ask another zinger of a question.

"Did you shoot the president?"

It did not work. Oswald was not ready to admit anything.

"No, I emphatically deny that."

A detective walked into Fritz's office to tell him that some Secret Service agents, including Forrest Sorrels, were waiting to speak with him. Fritz terminated the interrogation and went to see them.

The first round of Fritz's interrogation of Oswald had not gone badly. No, Oswald had not blurted out a confession. But the police detective had already learned several things about the suspect. Foremost, Oswald would talk to Fritz. It became obvious that he was a man who liked to talk. He was not one of those people who, once in custody, would clam up and refuse to say a word. He had not even demanded a lawyer before conversing with Fritz. Oswald was calm and arrogant. His disdain for the authorities oozed from his voice, facial expressions, and body language. But he could lose control when needled by an FBI agent.

Agent Hosty had gotten under Lee's skin and caused him to lose his temper. Fritz concluded that he, and not the FBI or the Secret Service, should remain Oswald's principal interrogator. Too much direct interaction between Lee and the men from those federal agencies might be counterproductive and ruin the trust that the wily old Texan cop was trying to build with Oswald. After decades on the job, Will Fritz had a sixth sense about innocence and guilt, and truth and deception.

And Oswald had already told several lies. He claimed he did not own a rifle, had never been to Mexico City, had not registered at the boardinghouse under a false name, had eaten lunch that day on the first floor of the Book Depository, and had not shot President Kennedy.

Forrest Sorrels asked Captain Fritz if Oswald had confessed. No, Fritz said, but the interrogation was far from over. Sorrels said he would like to talk to the suspect at some point. How about right now?, Fritz offered. The detective led the Secret Service agents to a room behind his office, and then had Oswald brought in. Oswald stiffened at the sight of them.

"I don't know who you fellows are, a bunch of cops."

Sorrels gave Oswald his name and showed him his identification.

"I don't want to look at it." Oswald was defiant. "What am I going to be charged with? Why am I being held here? Isn't someone supposed to tell me what my rights are?"

"Yes, I will tell you what your rights are," Sorrels said. "Your rights are the same as that of any American citizen. You do not have to make a statement unless you want to. You have the right to get an attorney."

Oswald asked, "Aren't you supposed to get me an attorney?"

Sorrels said no, but informed Oswald that he was free to be represented by the attorney of his choice. "I just want to ask you some questions," Sorrels continued. He got Oswald talking about his time in the Soviet Union, but Lee became impatient. "I don't care to answer any more questions." This round of interrogation was over.

ABOARD AIR Force One, Johnson worked the phone. There was one call he dreaded to make. It was to President Kennedy's mother, Rose. "What can I say to her?" he asked Lady Bird. Then he spoke to Rose Kennedy: "I wish to God there was something I could do." She replied, "We know how much you loved Jack and how Jack loved you." LBJ was uncomfortable and wanted to get off the phone as soon as possible. "Here's Lady Bird," he said as he handed the phone to his wife. "Oh, Mrs. Kennedy," she said, "we must all realize how fortunate the country was to have your son as long as it did." Rose Kennedy did not want to linger on the phone any more than the Johnsons did. She ended the call, "Goodbye, goodbye." Then he called political allies and congressional leaders. He wanted to meet with some of them tonight, and more tomorrow.

• • •

IN DALLAS, Captain Fritz had wanted to send Oswald downstairs for a lineup. Now was as good a time as any to do it. At 4:10 P.M., three detectives escorted Lee out of the Homicide and Robbery office and into the hall filled with reporters, photographers, and a television camera. No one had been searched. Anyone in that hallway, outraged by the president's murder, could pull a pistol from a pocket or a camera bag and shoot Oswald. But they were satisfied to fire questions at him. "Did you kill the president?" one reporter shouted.

"No, sir," Oswald said. "Nobody charged me with that."

It was a classic Oswald response, one that harked back to his New Orleans radio interviews. That August, when asked whether he had tried to renounce his American citizenship, Oswald evaded the question and did not give a direct answer. Instead, he said people who renounced their citizenship were not allowed to reenter the United States. Since he *was* back in the country, wasn't it obvious that he had not renounced his citizenship? Of course it was a lie. He had attempted to do that very thing as soon as he had arrived in the Soviet Union.

Now, in police custody, Oswald had just exhibited the same verbal and logical tic. Did he kill President Kennedy? He evaded the question. Isn't the fact that he had not been charged with that crime proof that he did not commit it? An innocent man asked if he has just murdered the president of the United States will likely respond, "Hell no!" Oswald's evasive response of "Nobody charged me with that" is his hallmark liar's "tell."

When the detectives got Oswald to the basement, they searched him. They found in his left pants pocket five bullets for his .38-caliber revolver. "What are these doing in there?" an incredulous detective asked. It was obvious that since his arrest Lee had not undergone a proper search. Of the bullets he was nonchalant. "I just had them in my pocket." On his person the detectives also discovered his bus transfer, a few insignificant documents, and thirteen dollars and eighty-seven cents.

IT WAS dark when Air Force One landed at Andrews Air Force Base outside the nation's capital at 6:05 P.M. (EST). Crowds of mourners had flocked there to watch the jet land and to see Kennedy's flag-draped coffin removed from the plane. It was not unlike the scene earlier that day when the president landed at Love Field in Dallas. This time the crowd did not cheer. Silence ruled. Television cameras broadcast live coverage of the event. Robert Kennedy, members of the cabinet, and senior government and military officials stood and watched.

The president's brother rushed to the plane and boarded it through a front door. He ran down the aisle, brushing past everyone in his path, including the new president. There was only one person in the world Bobby wanted to see, and she was at the back of the plane, sitting by a flag-draped coffin. They found each other and embraced.

"Hi Jackie, I'm here," was all he could say.

The American people were about to get their first look at Jackie Kennedy since the assassination almost five hours ago. A rear door on the plane opened. An elevated platform was put in place to receive the coffin. Then Jackie Kennedy appeared in the doorway. Standing next to her was her brother-in-law, Robert Kennedy. The image was confusing. He had not been in Dallas, so how was it that he was exiting Air Force One with Jackie?

All across the country, millions of people staring at their television screens gasped when they saw the bloodstains on her clothing. Jackie had still not changed out of the clothes she had worn in Dealey Plaza. She wanted Americans to see her pink suit. She wanted them to bear witness to the bloodstains on her jacket, skirt, and stockings.

On television screens, as she walked to the navy ambulance, viewers saw that her legs were smeared with copious amounts of blood. She wanted to sear these images into the collective memory of the American people so that they would never forget. It worked. To this day, decades after the assassination, the mere sight of an image of her in that suit triggers flashbacks in the minds of every person who remembers November 22, 1963.

President Lyndon Johnson strode toward the lights, microphones, and cameras. This time Jackie Kennedy did not stand beside him as he made his first public statement as the new president: "This is a sad time for all people. We have suffered a loss that cannot be weighed. For me, it is a deep, personal tragedy. I know the world shares the sorrow that Mrs. Kennedy and her family bear. I will do my best. That is all I can do. I ask for your help—and God's."

On the night of November 22, President Johnson composed two handwritten letters. They were not orders to important U.S. government officials or communications to world leaders. Instead, LBJ addressed them to the two children who had lost their father. "Dear John," he began his note to Kennedy's son, "It will be many years before you understand fully what a great man your father was. His loss is a deep personal tragedy for all of us, but I wanted you particularly to know that I share your grief—You can always be proud of him."

To the dead president's daughter he wrote: "Dearest Caroline—Your father's death has been a great tragedy for the Nation, as well as for you at this time. He was a wise and devoted man. You can always be proud of what he did for his country."

IN DALLAS, another plane departed Love Field after Air Force One had taken off. It was the C-130 cargo plane that had ferried the president's Lincoln Continental limousine to Texas. Secret Service agents had driven the car from Parkland Hospital back to Love Field. At the hospital, an agent removed the two flags—one bearing the presidential seal and the other the American flag—that had hung from the flagpoles on the hood during the motorcade. He handed them to Kennedy's secretary, Evelyn Lincoln.

Upon the car's arrival in Washington, agents drove it to the White House garage, where it was hidden from view. Color photographs taken of it there reveal bloodstains on the two-tone blue leather upholstery. The agents in Dallas had not been able to wash away all the blood.

FROM ANDREWS Air Force Base, President Kennedy's body was not yet ready to go home to the White House. First, accompanied by Jackie, a navy ambulance took him to Bethesda Naval Hospital, across the Maryland border from Washington. There would be an autopsy to document the official cause of death. Kennedy had been a naval officer, so Jackie, even before Air Force One had touched down, chose Bethesda Naval Hospital.

When Jackie entered the hospital, she was taken to a waiting room on the seventeenth floor. As she settled in for a long night, the president's brother Robert told her that a suspect had been arrested for her husband's murder.

"They think they found the man who did it," the attorney general said. "He says he's a Communist."

Jackie was aghast, and she said to her mother, who had joined her in Bethesda, "He didn't even have the satisfaction of being killed for civil rights. . . . It's—it had to be some silly little Communist."

IN DALLAS, at 7:40 P.M. EST, (6:40 P.M. CST) the "little communist" was back in Captain Fritz's office for more questioning. He asked for the lawyer John Abt.

At 7:10 P.M. CST, Oswald was brought before a judge to be charged with a crime. Judge David Johnston told him, "Mr. Oswald, we're here to arraign you on the charge of murder in the death of Officer J.D. Tippit."

"Arraignment! This isn't a court. You can't arraign me in a police station. I can only be arraigned in a courtroom. How do I know this is a judge?"

Alexander told him to shut up.

"The way you're treating me, I'd might as well be in Russia."

At 7:28 P.M., Oswald was questioned for ten minutes by federal agents, then was led out into a hall crammed with reporters.

"These people here have given me a hearing without legal representation."

"Did you shoot the president?"

"I didn't shoot anybody, no sir."

In front of the cameras, Oswald remained evasive and did not give a direct answer. The reporter did not ask him if he had shot *anybody*. He asked if he had shot John F. Kennedy. But Oswald would not speak the words, "No, I did not shoot the president."

At 8:30 P.M. (CST), Oswald was back in Fritz's office for more questioning.

Fritz asked if he kept a rifle in Mrs. Paine's garage in Irving.

Oswald said he did not.

The detective asked if he brought one with him when he came back to Dallas from New Orleans.

Oswald denied it.

"The people out at the Paine residence say you did have a rifle, and that you kept it out there wrapped in a blanket."

"That isn't true."

Fritz was closing in on Oswald. He tried again.

"You know you've killed the president, and this is a very serious charge."

"I haven't killed the president."

Nonetheless, Fritz reminded him, the president "had been killed."

Oswald retorted that the people will forget that within a few days, and there will be another president.

• • •

AT BETHESDA the president's casket was placed on a cart and wheeled to the room where pathologists and technicians waited to examine him. The autopsy commenced at 8:00 P.M. (EST). Attendants removed his body from the casket, laid him on a table, and unwound the plastic wrapping. They photographed his face, which had been undamaged by the bullets, and his head, which had been ravaged by one. They photographed his neck wound, then rolled him over to photograph his back wound. Then they made X-rays of his skull. Doctors did not need to saw off the top of his head to recover what remained of his brain. They just reached into the wound and extracted it. After examining it, they sealed it in a stainless-steel container with a screw-top lid.

Hours passed, but Jackie Kennedy, still waiting on the seventeenth floor, refused to leave or to sleep.

IN BETHESDA, after the pathologists finished their work, the morticians arrived to prepare the president for burial. Not knowing whether Mrs. Kennedy would choose an open- or closed-coffin viewing before the funeral, the undertakers prepared the president's corpse for an open-coffin viewing. They closed his eyes and sealed them shut. They concealed the tracheotomy incision that the Dallas doctors had made in his neck. Then they labored on their most difficult task, reconstructing the side of John Kennedy's skull that the fatal bullet had blown open. A White House courier brought a selection of eight of the president's suits, four pairs of shoes, plus shirts and ties for the morticians to dress him.

IT WAS close to midnight, November 22. Almost twelve hours since the assassination. As surgeons cut open President Kennedy's corpse, Lee Harvey Oswald was giving a press conference in a hall in the Dallas police station. It would later become known as the famous midnight press conference.

"I was questioned by a judge without legal representation."

"Well, I was questioned by a judge. However, I protested at that time that I was not allowed legal representation."

"During that, ah, that, ah very short and sweet hearing. Ah, I really don't know what this situation is about. Nobody has told me anything except that I am accused of, ah, of murdering a policeman. I know nothing more than that and I do request that, ah, someone to come forward, ah, to give me legal assistance."

"Did you kill the president?" one of the journalists in the crowd asked.

Again Oswald denied his guilt. "No, I have not been charged with that. In fact, nobody has said that to me yet. The first thing I heard about it was when the newspaper reporters in the hall asked me that question."

It was a classic Oswald verbal evasion, followed by a lie.

"You *have* been charged," one of the reporters told him.

"Sir?" Oswald replied. He seemed surprised.

"You have been charged," the reporter repeated.

The expression on Oswald's face changed in an instant, to an appearance of stunned despair. He lowered his head and bit his lip. The reporters kept firing questions.

"What did you do in Russia?"

"How did you hurt your eye?"

Oswald did not answer. Another reporter called him by name.

"Oswald, how did you hurt your eye?"

"A policeman hit me."

LEE HARVEY Oswald's first official press conference had lasted six minutes. It was odd that the prime suspect in the murder of the president of the United States was giving a press conference at all. Oswald was taken back to his cell, only to be called out again for a second fingerprinting session. Then, late in the night, Oswald was given a paraffin test to determine if he had fired a gun.

"What are you trying to do, prove that I fired a gun?" he said to the man performing the test.

"I'm not trying to prove anything. We have the test to make, and the chemical people at the lab will determine the rest of it."

Oswald told him that he was wasting his time and that he did not know anything about the murders.

At 1:35 A.M., Lee Harvey Oswald was roused from his cell and taken to see a man on the fourth floor of city hall. It was Judge Johnston again, who had arraigned him earlier for the murder of Officer J. D. Tippit. This early-morning summons could mean only one thing, and Oswald knew it.

He greeted the judge with sarcasm. "Well, I guess this is the trial."

"No sir," Johnston said. "I have to arraign you on another offense."

The judge spoke in the formal language of the law. "Lee Harvey Oswald, hereinafter styled Defendant, heretofore on or about the twenty-second day of November, 1963, in the County of Dallas and State of Texas, did then and there unlawfully, voluntarily, and with malice aforethought kill John F. Kennedy by shooting him with a gun against the peace and dignity of the State."

"I don't know what you're talking about. That's the deal is it?"

Johnston reminded Oswald of his right to an attorney.

"I want Mr. John Abt of New York," the defendant said. He was told again that he would have the opportunity to contact the lawyer of his choice.

CHAPTER 8

"WE HAD A HERO FOR A FRIEND"

At Bethesda, the work continued. The morticians did not finish until after three A.M. on Saturday, November 23, but once they had, Jackie could take the president's body home to the White House.

Jackie got in the gray navy ambulance and sat next to her husband's coffin. Robert Kennedy joined her and sat on the floor. Clint Hill followed Jackie. The driver, Bill Greer, the Secret Service agent who had driven the presidential limousine in the motorcade in Dallas, was behind the wheel. The ambulance and a small motorcade left Bethesda at 3:56 A.M., making its way down Wisconsin Avenue to Massachusetts Avenue, then to Twentieth Street on its way to the White House. In the middle of the night, those in the motorcade were surprised to see people standing vigil on the sidewalks and cars pulled over along the route. The presidential casket team was in the last car of the motorcade. Lieutenant Sam Bird turned around and saw "hundreds of automobiles following us, bumper to bumper as far as the eye could see, their headlights flashing." Impromptu as it was, this was the slain president's first funeral procession, the first of several in the days to come. At the White House, the lights were on, and the staff was waiting. Across the street, in Lafayette Park, crowds assembled.

Taz Shepard, Kennedy's naval aide, telephoned from the White House to the U.S. Marine Corps barracks at Eighth and I streets. He spoke to the duty officer: "Break out the Marines. The Commander in Chief has been assassinated, and I want a squad at the White House double-quick. You better move!"

At the barracks, a contingent of Marines awoke, donned their dress blue uniforms in record time, and arrived at the White House in seventeen minutes. Flaming torches illuminated their path as they marched through the gate and onto the White House grounds. They were in place before the ambulance carrying the president's body arrived. At 4:34 A.M. (EST), a military honor guard carried the coffin up the steps of the portico, through the hall, and into the East Room, where the body of Abraham Lincoln had once lain in repose.

It was the first public clue about what kind of funeral Jackie Kennedy had in mind. She watched the honor guard carry her husband into the East Room. This eerie scene, unfolding in silence in the middle of the night, was sad but touching in its simplicity.

JACKIE KENNEDY was exhausted. She had been awake for more than twenty hours. She needed rest. The president had been assassinated almost fifteen hours ago. In the morning, she would have much to do and many decisions to make. For now, she went upstairs to the family suite on the second floor of the White House. There in the privacy of her bedroom, she undressed, removing her suit, stockings, and other garments, stained by the tragedy she had suffered that day.

By this time, it was five A.M. Under normal conditions, the Executive Mansion would be dark and almost empty at this hour. But tonight a military honor guard in the East Room stood watch over a dead president through the night. Every guest bedroom was occupied by Kennedy family members or close friends. Jackie's mother, at her

President Kennedy lies in repose in the East Room of the White House.

daughter's insistence, slept in the president's bedroom. The president's children had been put to bed hours ago knowing that their father was dead but without the comfort of seeing their mother. Cabinet members, military officers, and friends of Jackie roamed through the halls, held quiet conversations, took catnaps on furniture, ate sandwiches, or tried to *do* something. Some planned for the day to come. Others tried to record their memories of the day that had been.

One of them, Charlie Bartlett, had introduced young Jack and Jackie all those years ago at a dinner party at his home. He was present at the beginning. Now he was here at the end. He found some White House stationery and began to write:

> *We had a hero for a friend—and we mourn his loss. Anyone, and fortunately there were many, who knew him briefly and over long periods, felt that a bright and quickening impulse had come into his life. He had uncommon courage, unfailing*

humor, a penetrating, ever curious intelligence, and over all
a matchless grace. He was our best. He will not be replaced,
nor will he be forgotten, for in truth he was a kind of cheerful
lightning who touched us all. We will remember always with
love and sometimes, as the years pass and the story is retold,
with a little wonder.

The journalist Mary McGrory hosted a dinner party at her apartment: "We'll never laugh again," she told her guests.

One of them, Daniel Patrick Moynihan said: "Oh, we'll laugh
again, Mary. But we'll never be young again."

Soon McGrory would compose her tribute to Kennedy. Like most
members of the press, she loved him. "He brought gaiety, glamor, and
grace to the American political scene in a measure never known before. That lightsome tread, that debonair touch, that shock of chestnut
hair, that beguiling grin, that shattering understatement—these are
what we shall remember. He walked like a prince and talked like a
scholar. His humor brightened the life of the Republic . . . shown his
latest nephew in August, he commented, 'he looks like a fine baby—
we'll know more later.' When the ugliness of yesterday has been forgotten, we shall remember him, smiling."

JACKIE KENNEDY awoke in her bedroom at the White House. She had
slept only a few hours. Had it really happened? Or had it all been just
a nightmare? It was real. At eight fifteen on the morning of Saturday,
November 23, she met with her children to talk to them about their
father's death. The night before, she had not told Caroline and John
the awful news. She knew she would not arrive home before their
bedtime, so Jackie's mother deputized the children's beloved nanny,
Maud Shaw, to tell them. Shaw did not want to do it and begged to be
relieved of the responsibility of telling two young children their father
was dead. In the end, she agreed to do it.

At 10:00 A.M. on the twenty-third, Jackie attended a private mass in the East Room for the president's friends and family. Elsewhere in the White House, staff members were already busy cleaning out the Oval Office in the West Wing. A tearful Evelyn Lincoln, John Kennedy's personal secretary since his years in the Senate, gathered papers and personal mementoes from his desk. It was all happening so fast. A pair of JFK's famous rocking chairs was removed at 1:31 P.M., loaded onto a handcart, and rolled across a driveway to the Old Executive Office Building, an annex of the White House. A photograph of them, one turned upside down and stacked upon the other, became a symbol of the rapid transition.

There was a lot of grumbling among some of President Kennedy's staffers about the new president. A number of officials had advised Johnson to occupy the presidential office at once as a symbol of the continuity of government, and he had done so. Johnson's actions caused Kennedy staffers to murmur that it was unseemly and in poor taste for the new president to move into the White House so quickly. But the Oval Office is in the West Wing, which is in a separate building that is connected to the main house by a colonnade.

Lyndon Johnson had not moved into the White House, which contained the historic rooms as well as the president's private living quarters. Johnson was emphatic that Jackie and her children should continue to live there as long as she wished. Mary Todd Lincoln had stayed on in the White House for more than a month after her husband's assassination, and Lyndon Johnson believed that Jackie Kennedy, and not he, should decide when she should move out. He vowed to give her all the time she needed. But she already knew that she did not want to remain for long.

As Jackie's plans for the funeral evolved, she decided her husband should be buried at Arlington National Cemetery, the historic graveyard established during the Civil War on General Robert E. Lee's estate, across the Potomac River from the White House. Thousands of soldiers from the Civil War, World War I, World War II, and the Korean War had been buried there. Arlington was also the site of the famous Tomb of the Unknowns.

Just before two o'clock on Saturday afternoon, Jackie departed the White House with a small entourage that included three of the president's siblings: his brother Robert and two of his sisters, Jean Smith and Patricia Lawford. Others followed in separate cars. On the way to Arlington, the motorcade stopped at the Pentagon to pick up Secretary of Defense Robert McNamara. It was cold and raining when Jackie Kennedy arrived at Arlington Cemetery to inspect the proposed site for her husband's grave. It was a lovely spot of ground in front of the old Lee mansion, and it enjoyed a panoramic view of Washington. Jackie remembered that when John had visited this spot a few months before, he said it was so beautiful that he could stay there forever. After fifteen minutes, she nodded her approval. "We went out and walked to that hill, and of course you knew that was where it should be," she said.

IN DALLAS, Lee Harvey Oswald woke up on Saturday, November 23, after his first night in police custody. The previous day, detectives had discovered his loaded rifle, the empty paper bag, and the three spent cartridge cases at the Book Depository. They never found any curtain rods there. Soon, through the unique serial number stamped into the weapon, the FBI would discover records proving that Oswald had ordered it by mail from a sporting goods store in Chicago. They found the order form and also the postal money order Oswald had used to pay for the rifle. The FBI also traced the purchase of the pistol he had used to shoot J. D. Tippit.

After Oswald's arrest, detectives subjected him to a total of twelve hours of questioning. He was surly, defiant, arrogant, defensive, and self-pitying. He talked a lot, but unfortunately, the Dallas Police Department failed to make tape recordings of those grueling, extensive conversations detectives had with him on November 22 and in the days that followed. He admitted nothing. He actually seemed to enjoy the attention as he toyed with the Dallas police, FBI, and Secret Service interrogators. Oswald insisted he was innocent. He denied shooting President Kennedy or Officer Tippit. He claimed he did not even own a rifle.

At 10:35 A.M. on Saturday, November 23, Oswald was brought to Captain Fritz's office, and the questioning continued.

"Lee, tell me what you did when you left work yesterday."

He said he rode a bus to his rooming house. When he got off he got a transfer and used it to take another bus to the theatre where he was arrested. A policeman took the transfer out of his pocket at that time. But that wasn't true. After Oswald got off the bus, he took a taxi to Oak Cliff. There was no reason for him to lie about taking a cab—it would have been a harmless admission. And the police did not discover the transfer until later. Perhaps he was tired. Or not thinking clearly.

Fritz asked if he brought curtain rods to work on the morning of the assassination.

"No." Oswald spoke the truth.

Fritz repeated the question.

Oswald denied it again.

Was he sure, Fritz asked, that he did not place a long package on the back seat of Wesley Frazier's car, and then carry the package into the Book Depository?

Oswald said he didn't know what he was taking about. Two lies. He *had* carried a long package, and he had *not* brought any lunch.

Oswald denied having a conversation with Wesley Frazier about curtain rods or taking them into the Depository.

"I didn't carry anything but my lunch," he insisted.

And he did not tell Frazier that he was in the process of fixing up his apartment.

Or that the reason for his visit to Irving on the night of November 21 was to obtain some curtain rods from Mrs. Paine.

"No, I never said that."

Fritz asked Oswald if he ate lunch with anyone on November 22.

Oswald said he ate with two "colored boys." One was named "Junior" and he could not remember the name of the other one. He said he had a cheese sandwich and an apple, which he got at Mrs. Paine's house before he left.

Another lie, and one easy for Fritz to disprove.

Fritz asked Oswald to tell him more about the Paines. What was Marina's living situation? What did Lee know about Mr. Paine?

Did he keep any of his belongings at the Paine residence?

Oswald said that some of the things he brought back from New Orleans in September were in Ruth Paine's garage—two sea bags, and a few boxes of kitchen articles.

And, Fritz suggested, a rifle?

Oswald denied ever storing a rifle in Mrs. Paine's garage.

Fritz asked Oswald what friends or relatives of his lived nearby and whether he'd ever had any visitors at his rooming house on Beckley Street. Then the detective returned to the rifle.

He asked if Oswald had ever ordered guns through the mail.

Oswald said he had never ordered guns, and did not have any receipts for any. There was that verbal and logical tic again. If Oswald had no receipts, he implied, then that must be proof that he had ordered no firearms.

"What about a rifle?" Fritz asked.

Oswald said he did not own a rifle, nor had he ever possessed one.

Fritz pressed him. If he never ordered a gun or purchased a gun then where did he get the pistol he had in his possession at the time of his arrest?

Oswald admitted he bought it about seven months ago but refused to answer any more questions about the revolver or any other guns until he talked to a lawyer.

Oswald told Fritz he was wasting his time.

According to one witness to the interrogation, Oswald could not resist showing off his knowledge and superiority. "Oswald stated that at various times he had been thoroughly interrogated by the FBI . . . that they had used all their standard operating procedures . . . their hard and soft approach . . . their buddy system." Oswald boasted that he was familiar with all types of questioning and would not make any statements.

Fritz figured he could keep Oswald talking anyway if he asked him to elaborate on his beef with the FBI.

"What do you mean by that?"

Oswald said that the FBI was abusive and impolite when they spoke to Marina three weeks ago. The agents were obnoxious and had frightened her.

Fritz's next question probed for a motive.

"What do you think of President Kennedy?"

"I have no views on the president. My wife and I like the president's family. They are interesting people. I have my own views on the president's national policy. I have a right to express my views but because of the charges, I don't think I should comment further."

Oswald knew where the detective was going and wanted to foreclose any hint of motive. Oswald insisted that he had nothing against John Kennedy personally.

"I am not a malcontent; nothing irritated me about the president."

Oswald refused to submit to a polygraph examination.

He said he would not take one for the FBI in 1962, and he would not do it now for the Dallas police.

Fritz confronted Oswald with one of two Selective Service cards that he carried in his wallet at the time of his arrest. The card, in the name of Alek James Hidell, was signed in that name. In Russia, "Alek" was Marina's nickname for her husband. The card also carried the photograph of Lee Harvey Oswald. But authentic U.S. military Selective Service cards did not use photographs, so this card was an obvious forgery.

Oswald refused to admit whether he signed the card with the name Hidell.

He conceded that he carried the card but refused to say for what purpose.

Fritz allowed Secret Service inspector Thomas Kelley to ask a question.

Kelley wanted to know if Oswald watched the parade yesterday. No, he had not.

"Did you shoot the president?"

Oswald said no.

"Did you shoot the governor?"

No. He claimed that he did not even know that Connally had been shot.

That might have been true. Oswald had been isolated from other prisoners and had not been allowed to watch television or see newspapers. Unless a policeman, FBI agent, or Secret Service agent had told Oswald about Connally's wounds, Oswald would not have known he had shot him too. If Oswald replayed the assassination in his head, he might have wondered how he had managed to shoot a man he had never had in his sights.

Captain Fritz ended this round of questioning.

WHILE JACKIE Kennedy was selecting a grave site for her husband, Marina Oswald was allowed to meet with her husband for the first time since the assassination. She had not seen him for about thirty hours—enough time for him to ruin their lives. Marina was afraid— afraid that her husband had finally gone and done something crazy, that he had actually murdered the president of the United States and a policeman. And having grown up in a totalitarian state, she was afraid the United States government might imprison her or deport her to the Soviet Union. Worse things than that have happened to innocent people in Communist countries who had any connection to enemies of the state.

At 1:15 P.M. (CST), Marina and her mother-in-law were taken to the fourth-floor visiting room. Her husband appeared in front of her. They could not embrace because a glass partition separated them. She had to stand—there was no chair where she could sit. They picked up the telephone handsets and began speaking in Russian. Lee glanced at his mother.

"Why did you bring that fool with you? I don't want to talk to her."

"She's your mother. Of course she came," Marina said. "Have they been beating you?"

"Oh no. They treat me fine. Did you bring Junie and Rachel?"

"They're downstairs. Alek, can we talk about anything we like?"

"Of course, we can speak about *absolutely* anything." Oswald believed the police were recording their conversation. He hoped his sarcasm served to warn Marina of the danger.

"They asked me about the gun."

"Oh, that's nothing."

"I don't believe that you did that. Everything will turn out well."

"Oh sure, there's a lawyer in New York who will help me. You shouldn't worry. Everything will be fine." Lee tried to reassure her. "Don't cry. There's nothing to cry about. Try not to think about it. And if they ask you anything, you have a right not to answer. You have the right to refuse. Do you understand?"

"Yes."

Lee and Marina began to cry. He must have known his life with Marina and the children was over forever. He had destroyed it. At best, he would spend the rest of his life in prison. At worst, he would be put to death in the electric chair, probably within a year. A few years earlier, Marina was an unsophisticated Russian teenage girl who found herself smitten with a minor immigrant celebrity who sweet-talked her into marriage and then into leaving her family and immigrating to America. But life here was not what he promised, and he was not the man he had pretended to be. Life with him had been hard, bitter, and poor.

Now, after she had entrusted him with her life and had borne him two children, his unstable nature had taken him on a quixotic psychological journey that had climaxed in the murder of the president of the United States. As the wife of the accused assassin, she was fast becoming the most notorious woman in America.

If Lee Oswald had any sense of the enormity of what he had done to *her*—setting aside for a moment the crime he had committed against the president—he gave no sign of it now. He remained calm and spoke as though nothing had happened, as though he was in jail for a traffic offense or a shoplifting charge.

"You have friends. They'll help you," Lee told her. "If it comes to that, you can ask the Red Cross for help." That was the same advice he had given her in his note before he set out to assassinate General Walker. "You mustn't worry for me. Kiss Junie and Rachel for me."

"I will," Marina said. "Alka, remember that I love you."

"I love you very much. Make sure you buy shoes for Junie."

Neither one of them realized this would be their last meeting. Marina and Lee would never see each other again. But she did not need to see him again to reach a conclusion about his guilt. Lee was too calm. He spoke vaguely. He did not complain of mistreatment by the authorities, and that was not like him. He was *always* a complainer, and now he sputtered no outraged proclamations of his innocence. When Marina looked into her husband's eyes, she knew, as she later told investigators, that he had killed the president.

AT 1:07 P.M., a few minutes before Lee and Marina began their reunion, District Attorney Henry Wade gave an impromptu interview to the press on the fourth floor of the city jail. He boasted that he was going to put Lee Harvey Oswald to death.

"Mr. Wade, do you expect to call Mrs. Kennedy or Governor Connally . . . in this trial as witnesses?"

"We will not, unless it's absolutely necessary, and at this point, I don't think it'll be necessary."

"How soon can we expect a trial?"

"I'd say around the middle of January." That was in just seven or eight weeks.

"Has Mr. Oswald expressed any hatred, ill will, toward President Kennedy or, for that matter, any regret over his death?"

"He has expressed no regret that I know of."

"It's rumored that perhaps this case would be tried by a military court because, of course, President Kennedy is our commander in chief." In the spring and summer of 1865, a military court of nine judges tried eight of John Wilkes Booth's alleged co-conspirators in the assassination of Abraham Lincoln and the attempted murder of Secretary of State William Seward. That court sentenced four of the defendants to death, and they were hanged.

Wade was skeptical.

"I don't know anything about that. We have [Oswald] charged in the state court, and he's a state prisoner at present."

"And will you conduct the trial?"

"Yes sir. I plan to."

"In how many cases of this type have you been involved, that is, when the death penalty is involved?"

"Since I've been district attorney we've asked—I've asked the death penalty in twenty-four cases."

"How many times have you obtained it?"

"Twenty-three." In Wade's mind, Oswald was as good as dead. Putting aside the assassination of the president, Wade knew he would get an easy conviction in the murder of Officer J. D. Tippit. That crime alone would send Oswald to the electric chair. He would be lucky to live out the year of 1964.

At 2:05 P.M., Captain Will Fritz followed Wade's comments with some of his own. Later it became notorious as the "cinched" interview.

"Captain, can you give us a resume of what you now know concerning the assassination of the president and Mr. Oswald's role in it?"

"There is only one thing that I can tell you without going into the evidence before talking to the district attorney. I can tell you that this case is cinched—this man killed the president. There's no question in my mind about it."

"Well, what is the basis for that statement?"

"I don't want to go into the basis. In fact, I don't want to get into the evidence. I just want to tell you that we are convinced beyond any doubt that he did the killing."

"Was it spur of the moment or a well-planned, long thought-out plot?"

"I'd rather not discuss that. If you don't mind, please, thank you."

"Will you be moving him today, Captain? Is he going to remain here?"

"He'll be here today. Yes, sir."

FINALLY, AT 3:37 P.M., Robert Oswald was allowed to see his brother, in the same room where Lee had met with Marina.

"This is taped," Oswald warned Robert.

"Well it may be or it may not be."

Robert noticed the cut above Lee's eye.

"What have they been doing to you? Were they roughing you up?"

"I got this at the theater. They haven't bothered me since. They're treating me all right."

"Lee, what the Sam Hill is going on?"

"I don't know."

"You don't know? Look, they've got your pistol, they've got your rifle, and they've got you charged with shooting the president and a police officer. And you tell me you don't know? Now, I want to know just what's going on."

"Don't believe all this so-called evidence."

Robert stared into Lee's eyes, searching for answers.

Lee shut him down. "Brother, you won't find anything there."

"Well, what about Marina? What do you think she's going to do now, with those two kids?"

"My friends will take care of them."

Oswald told Robert his daughter June needed shoes. "Junie needs a new pair of shoes." It was incredible, surreal. Lee was under arrest for a double homicide, and his mind was distracted by the trivial subject of footwear for a toddler.

"Don't worry about that. I'll take care of that."

Robert asked about an attorney and offered to get one for his brother.

"No, you stay out of it."

"Stay out of it? It looks like I've been dragged into it."

Lee said he doesn't want a local attorney. He wants Abt from New York.

Robert prepared to leave. "I'll see you in a day or two."

"Now, you've got your job and everything," Lee warned him. "Don't be running back and forth all the time and getting yourself in trouble with your boss."

"Don't worry about that," Robert assured him. "I'll be back."

Lee said good-bye. "All right. I'll see you."

The Oswald brothers would never meet again.

DALLAS POLICE paraded Lee Harvey Oswald many times before newspaper reporters and television cameras. In a crowded hallway, they allowed him to make several public statements that were filmed and broadcast across the country. Oswald played dumb.

"I really don't know what this situation is about," he told reporters, "except that I am accused of murdering a policeman. I know nothing more than that." Oswald said he wanted a lawyer. "I do request that someone . . . come forward and give me legal assistance."

When a reporter asked him point-blank, "did you shoot the president?" Oswald gave an odd, wordy, and indirect reply: "No, I have not been charged with that. In fact, nobody has said that to me yet. The first thing I heard about it was when the newspaper reporters in the hall asked me that question."

The prosecutor discussed the case against Oswald in front of reporters and pronounced him guilty. At one point, Oswald raised his handcuffed hands and, for several seconds, clenched his right fist into what appeared to be a Communist salute. Photographs and videos captured the moment. Another time, Oswald complained to journalists that his "fundamental hygienic rights" were being violated because the police would not allow him to take a shower. He told reporters he had a cut above his eye because a policeman had hit him. He asked for a lawyer several times, but the police and prosecutors ignored him. A policeman walked through a crowded hallway holding Oswald's rifle above his head like a cheap bowling trophy.

The Dallas Police Department allowed its headquarters to deteriorate into a carnival-like spectacle. Shouting, pushing reporters packed the halls and, like jackals, became frenzied whenever the police teased them with a glimpse of their prisoner. "Oswald, did you shoot the president?" yelled one journalist during one of these brief, impromptu hallway interviews.

"I didn't shoot anybody sir," he replied. "I haven't been told what I am here for."

When another reporter shouted the same question, Oswald said, "No, they've taken me in because of the fact that I lived in the Soviet Union." Then Oswald claimed, "I'm just a patsy," by which he meant that he was the fall guy for whoever committed the crime.

Oswald admitted just one thing. When asked whether he was in the Book Depository at the time of the assassination, he said yes. "I work in that building . . . naturally, if I work in that building, yes sir," he was there. But he denied everything else.

When a third reporter asked if he was the gunman—"Did you fire that rifle?"—Oswald uttered an emotional denial. "I don't know the facts you people have been given, but I emphatically deny these charges!" Oswald's denials did not surprise the detectives. Experienced policemen knew that most murderers denied their guilt.

There was no proper security at police headquarters. No one checked IDs or searched the journalists who crowded the hallway. What explains the incompetence of the police when they had Oswald in their custody? The wild atmosphere was shameful. The answer is simple: police officials wanted to curry favor with the journalists from all over the country who had descended upon Dallas. The assassination had stained the city's and its police department's reputations. There was disturbing talk that the people of Dallas shared some kind of collective guilt for the murder. The police wanted the reporters to say good things about Dallas, so they gave the press free rein. It was a fateful decision that impeded their investigation and put Oswald's life in danger.

ELSEWHERE IN Dallas on November 23, word had gotten out about Abraham Zapruder's home movie. He had already been interviewed on a local television station. Journalists desperate to purchase the rights to his film went to his office to meet with him. He had locked up the film overnight. He hoped to sell it for a lot of money, and soon he would.

BACK AT the White House, after Jackie returned from Arlington Cemetery, she remained in seclusion. Aside from the morning mass, she participated in no other events that day. She received only a few visitors—close friends and family. She needed to gather her strength for the ordeal that lay ahead. In two days, on Monday, November 25, she had to be ready to preside over two events that would test her body and soul.

The first was her husband's public funeral. With meticulous attention to detail, Jackie Kennedy threw herself into planning the event. With Abraham Lincoln's funeral as her inspiration, researchers had set to work. They uncovered historical details that had been forgotten since the Civil War, including the exact way that the White House entrances and East Room chandeliers had been draped in mourning with ribbons of black crepe paper.

Then, after the funeral, Jackie had to prepare for a second event. On Monday night she would host a birthday party for her son, John Jr. In two days, on the day of his father's funeral, he would be three years old. Jackie would not hear of canceling the party.

• • •

ON THE afternoon of Saturday, November 23, at 4:45 P.M., President Lyndon Johnson read to the nation over live radio and television his proclamation of a national day of mourning for President Kennedy.

> *To the people of the United States: John Fitzgerald Kennedy, 35th President of the United States, has been taken from us by an act which outrages decent men everywhere. . . . Now, therefore, I, Lyndon B. Johnson, President of the United States of America, do appoint Monday next, November 25, the day of the funeral service of President Kennedy, to be a national day of mourning throughout the United States. I earnestly recommend the people to assemble on that day in their respective places of divine worship . . . and to pay their homage of love and reverence to the memory of a great and good man.*

On the night of Saturday, November 23, reporters begged Police Chief Jesse Curry for a tip on when Oswald would be transferred from the jail in City Hall to the County Jail about a mile away. They were tired. Most of them had been covering the story from about one P.M. Friday, through the middle of the night, and then all day Saturday into the evening. They wanted to leave so they could take a break and rest, but they worried that the prisoner would be moved in the middle of the night when they were gone and that they would miss out on the story.

The chief told them it was safe for them to leave—and promised that the suspect would not be moved tonight. "I think if you fellows are back here by ten o'clock in the morning," Curry said, "you won't miss anything."

Curry then conferred with Captain Fritz about the timing.

"What time do you think you will be ready tomorrow," Curry asked.

Fritz said he did not know exactly when.

"Do you think about ten o'clock?"

"I believe so."

Curry stepped out of Fritz's third floor office to talk to the press to confirm that Oswald would be moved tomorrow morning. "I believe if you are back here by ten o'clock you will be back in time to observe anything you care to observe."

Asked if Oswald had admitted killing the president, Curry said no.

"I don't think we've made any progress toward a confession."

"You don't think so? "

"No."

"Why are you so pessimistic about a confession?"

"Well, you know we've been in this business a good while, and sometimes you can sort of draw your own conclusions after talking to a man over a period of time. Of course he might have a change of heart, but I'd be rather surprised if he did."

A reporter asked about security.

"Will you transfer him under heavy guard?"

"I'll leave that up to Sheriff Decker. That's his responsibility."

"The sheriff takes custody of him here?"

"Yes, that's all I have, gentlemen, thank you."

ALL THROUGH the day of November 23, and into the night, the body of John F. Kennedy lay in repose at the White House.

THE DEATH threats against Oswald began after midnight on Sunday, November 24.

Then at 2:30 A.M., an anonymous man called the Dallas FBI office and warned, "I represent a committee that is neither right nor left wing, and tonight, tomorrow morning, or tomorrow night, we are going to kill the man who killed the president. There will be no excitement and we will kill him. We wanted to be sure and tell the FBI, police department, and Sheriff's Office and we will be there and we will kill him."

The night man at the office who had taken the message informed an agent, Milton L. Newsom, of the threat. When Newsom called the Dallas County sheriff's office to report it, Deputy Sheriff C. C. McCoy said he had received a similar call. The two messages were almost identical, except the caller to the sheriff's office claimed he spoke on behalf of a secret group of about one hundred people "who have voted to kill the man who killed the president."

The purpose of the call was to warn the sheriff's office in advance so no deputies would get hurt when Oswald was shot to death tomorrow. McCoy called Sheriff Bill Decker at home to alert him to the threats.

Agent Newsom also called the Dallas Police Department and reported the threats to Captain William B. Frazier. At 5:15 A.M., Frazier called Captain Will Fritz at home, who told him to notify Chief Curry.

In the meantime, Deputy Sheriff McCoy had reported the threats to Sheriff Bill Decker, who became alarmed. He told McCoy to call the Dallas police and tell them he wanted Chief Curry to call him right away. McCoy made the call and told Frazier to have Curry call Decker.

Then McCoy conveyed a more urgent message. Decker thought the Dallas Police should drive Oswald over to the county jail right now. That way, the transfer will be over before the reporters—or potential assassins—can gather at City Hall.

But Frazier could not reach Chief Curry on the phone. When Captain C. E. Talbert showed up for duty at 6:15 A.M., Frazier brought him up to date. Talbert reached the assistant chief of police, who told him to send men to Curry's house. Around 7:30 A.M., Talbert also called FBI agent Newsom, the man who, more than four hours ago, first spread word of the threats. Talbert said that Chief Curry would not be in the office until 8:00 or 9:00 A.M.

Newsom was worried. He asked about the transfer plans.

Talbert was nonchalant. He said he did not think there would be any effort to sneak Oswald out of the city jail because the police wanted to maintain good relations with the press. The media had set up extensive coverage to cover the transfer, and he did not think that Chief Curry would want to "cross" the media.

BEFORE OSWALD went anywhere that morning, Captain Will Fritz wanted to see him for one last time. After two days of trying, Fritz knew he could not break Oswald's willpower, frighten him, or trick him into confessing. Oswald was too cool and collected for that. Yes, he had exploded in occasional flashes of anger and frustration, but he always reeled himself back in to a state of calm. He was unflappable. Not even the sight of his wife or brother weakened him or compelled him to confess.

This Sunday morning was Captain Will Fritz's last chance to speak freely with Oswald before he was transferred from the city jail, before he got a lawyer (who would no doubt advise his client not to say another word to the police), and before the next stage of the case, when the grand jury, the prosecutor, and the trial would claim center stage. There was already plenty of evidence. Indeed, the first-day evidence alone was enough to indict Oswald and put him on trial. District Attorney Henry Wade did not need a confession. He could get a conviction without one.

Fritz was sure Oswald was guilty. But the old detective wanted the psychological satisfaction of hearing it from Oswald's own lips. And he wanted to know, *why?* This was the most notorious case of Fritz's career, but the legendary lawman had failed to break the president's assassin. There was one form of leverage Fritz had not used. He could have threatened to deport Marina, and to take her children away from her (at least the one born in the United States). But that kind of threat was for the federal authorities, not Fritz, to make.

Fritz showed Oswald one of the "backyard" photographs of him holding the rifle he claimed he never owned. The detective hoped that once Oswald saw this undeniable proof, he might confess.

Oswald said he had never even seen the photograph. "I know all about photography . . . that is a picture that someone else has made. I never saw that picture in my life." Again, he denied any involvement in or knowledge of the Kennedy or Tippit shootings.

"The only thing I am here for is because I popped a policeman in the nose in a theater on Jefferson Avenue, which I readily admit I did, because I was protecting myself."

Fritz turned the questioning over to Postal Inspector Harry D. Holmes. The U.S. Post Office had been researching the identity of the person who rented the post office box to which Klein's Sporting Goods shipped the Mannlicher-Carcano rifle.

Holmes asked if Oswald had a post office box in Dallas.

"Yes."

Box 2915. Oswald said he had rented it at the main post office a few months before moving to New Orleans. He explained that he rented it in his own name and only he and his wife had access to the box.

Oswald had just admitted the rifle that killed the president was shipped to his post office box.

Holmes asked him to confirm that no one else received mail in that box. He was struck by Oswald's response. "He denied emphatically that he had ever ordered a rifle under his name, or any other name, nor permitted anyone else to order a rifle to be received in this box . . . he denied that he ever . . . bought any money order . . . to [pay] for such a rifle."

"Well," Fritz asked, "have you shot a rifle since you have been out of the Marines?"

Oswald said no.

"Do you own a rifle?" Fritz asked again.

"Absolutely not! How can I afford a rifle? I make $1.25 an hour. I can't hardly feed myself."

Inspector Holmes wanted to ask about the Dallas post office box again.

"Did you receive mail through box 2915 under the name of any other than Lee Oswald?"

"Absolutely not."

"What about a package to an A. J. Hidell?"

"No!"

"Did you order a gun in that name to come here?"

"No, absolutely not!"

"Had it come under that name, could this fellow have gotten it?"

"Nobody got mail out of that box but me, no sir."

AT ABOUT 10:20 A.M. on Sunday, while Captain Fritz and others conducted their latest interrogation of Oswald, Chief of Police Curry gave a hallway press conference. On Saturday night he had already told reporters the approximate time his men would transport Oswald to the County Jail on Sunday. Now he even revealed what special methods he would employ to keep his prisoner safe from harm.

"Chief, you say that you're going to take him . . . in an armored car. Have you ever had to do this with another prisoner?

"Not to my knowledge."

"Is it a commercial-type truck, the kind that banks use?"

"Yes, sir."

"[The] threats on the prisoner's life . . . did they come in right through the police switchboard?"

"Yes."

"Do you have details on all of them?"

"No."

The reporters wanted to know if the police had discovered any accomplices. Did anyone help Oswald?

"This is the man, we are sure, that murdered . . . assassinated the president." But Curry did not go so far as to deny the possibility of a conspiracy.

"To say that that there was no other person that had knowledge of what this man might do, I wouldn't make that statement because there is a possibility that there are people who might have known this man's thoughts and what he could do, or what he might do."

But has he confessed? the reporters yearned to know.

"Does he show any signs of breaking—to make a clean breast of this . . . to tell the truth about what happened?"

"No, sir, there is no indication that he is close to telling us anything."

The journalists kept questioning Curry, but he had nothing more to say.

INSIDE FRITZ'S office, Holmes asked Oswald about a third post-office box, one he rented in Dallas after he moved back to the city from New Orleans. What business did he list on the application? None, Oswald claimed. Then why, Holmes asked, did he state on the rental application that his business was the Fair Play for Cuba Committee and the American Civil Liberties Union?

"Maybe that's right," Oswald conceded.

Holmes asked why he did that.

"I don't know why."

Fritz asked about Oswald's involvement with the FPCC, and if he was in contact with its New York office.

"Yes, I wrote to them, and they sent me some Communist literature and a letter signed by Alex Hidell."

Hidell! That was the name on the card the police found in Lee's pocket, and the name authorized to receive mail at his New Orleans post-office box. Oswald said he never knew Hidell and never saw him in New Orleans.

Inspector Kelley asked Oswald if he believed in what the FPCC stands for.

"Yes, Cuba should have full diplomatic relations with the United States. There should be free trade with Cuba and freedom for tourists from both countries to travel within each other's borders."

Oswald denied that he had moved to Dallas to start an FPCC chapter there. He was telling the truth.

Kelley wondered if Oswald's motive was connected to Cuba.

Did Oswald think that Cuba would be better off now that President Kennedy was dead?

Oswald said he doubted that the attitude of the U.S. government would change. Lyndon Johnson belonged to the same political party, and, Oswald concluded, "His views would probably be largely the same as President Kennedy."

Fritz asked the same question Oswald answered on the New Orleans radio shows. "Are you a Communist?"

"No, I am not a Communist. I am a Marxist, but not a Marxist-Leninist." Oswald said he was a "pure Marxist."

Fritz had no idea what Oswald was talking about. "What's the difference?"

"It would take too long to explain," Oswald said, implying that Fritz lacked the sophistication to understand what he was talking about.

Fritz coaxed him to elaborate.

"Well, a Communist is a Marxist-Leninist, while I am a true Karl Marxist. I've read just about everything by or about Karl Marx." Oswald's boasts that he was a Marx scholar were absurd.

Oswald said he was an avid reader of Russian literature, "whether it's Communist or not."

Secret Service agent Kelley asked if Oswald subscribed to any Russian magazines or newspaper.

"Yes, I subscribe to the *Militant*. That's the weekly of the Socialist Party of the United States."

Oswald had a question for Kelley. "Are you an FBI agent?"

"No, I'm not. I am a member of the Secret Service."

Oswald revealed when he was standing in front of the Depository, about to leave, a young crew-cut man rushed up and said he was from the Secret Service, showed him a book of identification, and asked where the phone was.

Oswald pointed him toward the pay phone in the building, and he started toward it, and then Oswald left.

Fritz allowed Kelley to continue the questioning. The clock was ticking. Soon Oswald would be gone, taken to the County Jail.

Fritz switched to a subject guaranteed to infuriate Oswald, the circumstances of his parting with the Marine Corps. Fritz said he understood that Oswald had been dishonorably discharged. Oswald cut him off.

"I was discharged honorably." It was changed later because he attempted to renounce his American citizenship while living in Russia. Because he never changed his citizenship, he wrote a letter to secretary of the Navy John Connally to have this discharge reversed, and after considerable delay, he received a "very respectful" reply in which Connally stated that he had resigned to run for governor of Texas, and that his letter was being referred to the new secretary.

Fritz asked about a map that was found in Oswald's room at the boardinghouse.

"Lee, we found a map in your room with some marks on it. What can you tell me about those marks? Did you put them on there?"

"My God!" He knew what Fritz was implying, that the map was evidence that he had planned the assassination in advance. "Don't tell me there's a mark near where this thing happened?" Oswald remembered that he marked the location of the Texas School Book Depository on the map. But he offered an innocent explanation.

"What about the other marks? I put a number of marks on it. I was looking for work and marked the places where I went for jobs or where I heard there were jobs."

Fritz was skeptical. He asked why was there an X at the location of the Book Depository.

"Well, I interviewed there for a job, in fact, got the job. Therefore the X."

This time Oswald spoke the truth. He had used that map to plot out his route when he went from place to place in search of employment. The X that he penned in front of the Depository did not foreshadow the assassination. On November 22, he had already worked there for several weeks. Why would he need to highlight on a map the place where he already had a job?

It was 11:00 A.M. in Dallas. Soon Oswald would be taken to the basement garage of City Hall and driven to the County Jail.

Captain Fritz asked Oswald one more time about the hours preceding the assassination.

Why did he go to Irving to visit his wife on Thursday night instead of Friday, like he normally did?

"I learned that my wife and Mrs. Paine were giving a party for the children, and that they were having in a house full of neighborhood children there [on Friday], and that I just didn't want to be around at such a time." So, he went out Thursday night.

What about the next morning, Fritz asked. "Did you bring a sack with you."

"I did."

"What was in the sack?"

"My lunch."

"How big of a sack was it? What was its shape?"

"Oh, I don't recall, it may have [been] a small sack or a large sack. You don't always find one that just fits your sandwiches."

Oswald knew that Wesley and his sister had already told the police all about his large paper bag. He tried to discredit the accuracy of their recollections by creating ambiguity about its size.

Fritz pursued the subject, asking where Oswald put the sack when he got into Frazier's car.

"In my lap, or possibly in the front seat beside to me, where I always did put it because I don't want it to get crushed."

He denied putting any package in the backseat.

Fritz told Oswald that Wesley Frazier says that he brought a long parcel over to his house and put it in the backseat of his car.

"Oh, he must be mistaken, or thinking about some other time when he picked me up."

Fritz asked Oswald to tell him again where he was in the Depository at the time of the assassination. The detective was curious to see if Oswald would tell the story in the same way he had during a prior interrogation.

Oswald said that when lunchtime came he was working on one of the upper floors with one of the black employees who said "come on let's eat lunch together." Oswald told him, "you go ahead, send the elevator back up to me and I will come down just as soon as I am finished."

But, Oswald explained, before he went down the assassination happened. Then, "when all the commotion started, I just went on downstairs." He stopped in the second-floor lunchroom to get a Coke, and it was there, Oswald said, that he encountered a policeman. "I went down, and as I started to go out and see what it was all about, a police officer stopped me just before I got to the front door, and started to ask me some questions."

At that point, "My supervisor told the officer that I am one of the employees . . . so he told me to step aside. Then I just went on in the crowd to see what it was all about."

Oswald had just committed a huge mistake. On November 22, he told Captain Fritz that during the assassination, he was eating his lunch on the first floor of the Depository and *then* he went up to the second floor to buy a Coke. Now, two days later, he told a different story. In this new version, Oswald did not go down to the first floor to eat his lunch. Instead, he never ate lunch at all. Now he says the assassination and all the "commotion" happened *before* he had a chance to go down to the first floor and eat his lunch.

At first, it might sound like a trivial distinction. What difference did it make if Oswald was going up or down to the second floor to buy that Coke? Only this.

Oswald had just confessed that he was *not* on the first floor of the building when the president was shot. He had shattered his own alibi. Now he said "no" without realizing the significance of what he had just admitted: that he was somewhere else at 12:30 P.M. on November 22, 1963.

During the assassination of President Kennedy, Lee Harvey Oswald was not on the first floor. He was somewhere else inside the Texas School Book Depository. He was on one of upper floors. And he was alone.

And with that admission, he revealed something else. After the shooting, he did not go up the stairs. He ran down them.

It had taken Fritz almost two days and several rounds of interrogation to get to this moment, but Oswald had slipped up at last. The suspect had placed himself alone on one of the upper floors of the building from which the president was shot. Given several more hours of questioning, Fritz might be able to coax Oswald into making more mistakes and admissions.

Fritz heard a knock on his door. It was Chief of Police Jesse Curry. It was time to transfer the prisoner to the County Jail. "We'll be through in a few minutes," Fritz said.

Secret Service agent Forrest Sorrels asked Oswald about the mysterious Alex Hidell.

"I never used the name of Hidell," Oswald responded.

Fritz asked him if he knew anyone by the name of A. J. Hidell.

"No."

Oswald also denied ever using that name as an alias.

"No! I never used the name, and I don't know anyone by that name, and never had heard of that name before."

Fritz asked again about the fake selective service card bearing Oswald's photo and A. J. Hidell's name.

"I've told you all I'm going to about that card. You took notes, just read them for yourself, if you want to refresh your memory. You have the card. You know as much about it as I do." Oswald was vehement on the subject of Hidell. Despite the incriminating evidence of the card he kept in his wallet, he would not admit that he used the name under which he ordered the Mannlicher-Carcano rifle.

It was time to go.

CHAPTER 9

"LEE OSWALD HAS BEEN SHOT!"

The two armored trucks arrived on schedule on the Commerce Street side of City Hall. Assistant Chief of Police Charles Batchelor walked up the inclined loading ramp to the mouth of the garage entrance and selected the best truck, the one with room for two guards to ride with Oswald. Batchelor and the driver decided to leave the vehicle at the top of the ramp with the front end protruding onto the sidewalk. The position was not ideal. Oswald would have to walk through the basement garage and then up the ramp to reach the back of the armored car. But if the truck were backed all the way down the ramp, the steep incline might cause it to stall on the way up the exit ramp.

Captain Fritz told Chief Curry he was ready to bring Oswald down as soon as the seventy policemen in the basement had secured the area. But Fritz opposed the idea of using the armored car. "Chief," he said, "I don't think it's a good thing to try to move him in a money wagon."

Fritz explained. "We don't know the driver or anything about the wagon, and it would be clumsy and awkward, and I don't think it is a good idea at all."

But Fritz proposed a solution—transfer Oswald in an unmarked car.

Curry agreed. They would use the armored car as decoy. Anyone who tried to get Oswald would find themselves attacking an empty armored car.

Fritz supplied the details. "I'll transport him in one car, with myself and two detectives, and we'll have another carload of detectives as backup." Fritz planned to cut out of the caravan at Main Street, drive west on Main to Houston, then make a right turn to the county jail. The rest of the motorcade would follow a different route.

Fritz still worried about the media circus in the basement.

"You know, Chief," he said, "we ought to get rid of the television lights and cameramen so they don't interfere with our getting to the car."

It would have been a lot easier to protect Oswald if the basement was cleared of everybody but law enforcement officials.

Curry did not like that idea.

"The lights have already been moved back," Curry told his chief of detectives, "and the media have been moved back in the basement, back of the rail, and the people [outside City Hall] have been moved across the street."

Detective James Leavelle was worried too. He had suggested to Fritz that they take Oswald out on the first floor and put him in a car on Main Street. In other words, do what no one is expecting. Assign a small contingent of reliable detectives to march Oswald right out the front door.

"That way, we could be in the county jail before they even know we've left the [city] jail."

Fritz wondered if the number of men needed to guard Oswald could be squeezed into the elevator. And he was sure Chief Curry would insist on letting the press see the transfer.

At 11:10 A.M., Fritz prepared Oswald. "Lee, I want you to follow Detective Leavelle when we get downstairs and stay close to him."

Oswald was dressed in a T-shirt, and Fritz worried that Lee might get cold. The shirt he wore when he shot the president was in the crime lab.

Fritz sent a man to bring some of the clothing the police found at 1026 North Beckley Street in Oswald's little room. At first Lee asked for what he described as a black "Ivy League" shirt. Then he chose a black sweater. That, he said, might be a little warmer. The tattered garment had several holes. But Oswald cared more about its color than its condition.

Fritz did not object. It was the last small kindness he would offer Oswald. Lee pulled the sweater over his head and pushed his arms through the sleeves. Now he was dressed in all black—top, pants, and shoes—just as he was on the day Marina took that "Hunter of Fascists" photograph in their backyard eight months ago, when he wanted to show that he was "ready for anything."

At 11:18 A.M., Secret Service agent Sorrels approached Fritz. "Captain, I would not move that man at an announced time. I would take him out at three or four o'clock in the morning, when there's nobody around."

It was the last of many warnings not to expose the prisoner to unnecessary danger.

"Chief Curry," Fritz replied, "has gone along with [the press] and said he wanted to continue going along with them and cooperating with them all he can."

OUTSIDE ON the street, a pasty-complexioned, unattractive man wearing a dark suit, white shirt, dark tie, dark hat, and dark shoes approached the mouth of the driveway and the armored car parked at the top of the ramp. He slipped through the narrow space between one side of the truck and the wall. No one stopped him. He proceeded down the ramp until he found himself in the basement garage. He mingled with the scrum of journalists and others and melted into the crowd. He did not stand out.

To some people, he looked familiar in a vague way. They had seen him before. Other men in the basement, the ones who knew him, either did not spot him or did not care. This uninvited guest was just one of dozens of well-dressed men waiting for Lee Harvey Oswald.

At 11:19 A.M., Oswald was double-handcuffed in Captain Fritz's office. One pair secured his hands together. The second pair connected one of his wrists to one of Detective Leavelle's so that the two men could not be separated, and Oswald could not run and make a break for it. But the handcuff arrangement would also make it hard for Oswald to duck if someone took a shot at him. It was also dangerous for Leavelle.

"Lee, if anybody shoots at you," the detective joked, "I hope they're as good a shot as you are."

"Aw, they ain't going to be anybody shooting at me, you're just being melodramatic."

"Well, if there's any trouble you know what to do. Hit the floor."

"Captain Fritz told me to follow you. I'll do whatever you do."

"In that case," Leavelle said, "you'll be on the floor."

In the garage, NBC News reporter Tom Pettit was waiting for them: "We are standing in the basement corridor where Lee Oswald will pass through momentarily. Extraordinary security precautions have been taken for the prisoner." But Pettit was dead wrong. The Dallas police should have transferred Oswald in secrecy. Pettit's gullible comments echoed the statements that the naive reporters who had covered JFK's arrival in Dallas had made about security.

JIM LEAVELLE and his prisoner stepped into the elevator and rode it down to the garage level. The door opened. They walked toward the crowd and the transport car. Oswald spotted the press microphones and spoke: "I'd like to contact a member . . . representative of the American Civil Liberties Union." These were the last words he would ever speak.

Television cameras and microphones captured what happened next.

"Here they come!" a voice from the crowd shouted.

"Here he comes," said another voice.

"Is it OK?" a policeman asked.

Jim Leavelle got the word. "Okay, come on out, Jim."

When Oswald emerged with the trademark smirk on his face, and his body pinned between Leavelle and another detective, excited reporters and photographers pressed forward. Two television cameras recorded the scene. Only one, operated by NBC, was broadcasting live. Ike Pappas, a reporter for WNEW radio, spoke into his microphone:

"Now the prisoner, wearing a black sweater, is being moved out toward an armored car, being led out by Captain Fritz. There's the prisoner." As Oswald walked past him, Pappas asked him a question: "Do you have anything to say in your defense?"

It was 11:21 A.M. A moment after Pappas spoke, the man in the dark suit and hat who had walked down the driveway a few minutes earlier and had melted into the crowd emerged from it and took a lunging step toward Oswald. Leavelle and Oswald did not see him as he rushed them from their left. The detective and his prisoner continued to look straight ahead, oblivious to the danger. No one stood between the president's assassin and the man in the dark suit. In another moment the man would be close enough to Oswald to touch him. One witness thought that was exactly what was about to happen—an outraged cop or reporter was going to punch Oswald in the face. Another detective, Billy H. Combest, knew better. He recognized the man. It was a local Dallas police buff and hanger-on, Jack Ruby. Combest also saw the gun. "Jack, you son of a bitch," he yelled, "don't!"

Ruby's right hand gripped a pistol that he pointed at Oswald's torso. A newspaper photographer, Jack Beers, captured the moment. Oswald had no idea that he was a split second away from being assassinated. Ruby squeezed the trigger and fired one round from his .38 caliber snub-nosed revolver. The bullet struck Oswald. He screamed in pain—a sudden, deep, gutteral, animal-like moan.

A second photographer, Bob Jackson, snapped his shutter a split second after the bullet's impact. He captured Oswald's agonized face and his body writhing in pain. Two days ago, Jackson had missed his chance to photograph the barrel of Oswald's rifle protruding from the sixth-floor window of the Book Depository. Now he had just taken the most important picture of his life. Moments later, Oswald collapsed.

Ruby wanted to shoot Oswald again, but before he could get off another shot, policemen tackled him to the ground before he could pull the trigger again.

Ike Pappas dropped to his knees so he would not be hit by more gunfire, but he kept reporting. He screamed into his microphone: "There's a shot. Oswald has been shot! Oswald has been shot! A shot rang out. Mass confusion here. All the doors have been locked. A shot rang out as he was being led to the car."

Tom Pettit yelled into his microphone too: "He's been shot! He's been shot! He's been shot! Lee Oswald has been shot! There's a man with a gun. There's absolute panic, absolute panic here in the base-ment of the Dallas police headquarters. Detectives have their guns drawn. Oswald has been shot. There's no question about it. Oswald has been shot. Pandemonium has broken loose here in the basement of the Dallas Police headquarters."

"I hope I killed the son of a bitch," Jack Ruby said as detectives pried the revolver from his hand.

A cop who did not know him looked at Ruby and shouted, "Who is this son of a bitch?"

"Oh hell! You guys know me, I'm Jack Ruby."

The shooter's tone was as amiable as if he were handing out com-plimentary passes to his nightclub, or delivering a bag of fresh corned beef sandwiches to a police station or broadcast studio, two favorite tricks he employed to befriend cops and radio hosts.

DETECTIVE LEAVELLE was unwounded. His joke to Oswald had come true. Ruby's aim was good enough to hit his target without shoot-ing the detective. Leavelle reached down and unlocked the handcuffs that bound him to Oswald. Once he had freed himself, he unlocked the second pair around the prisoner's wrists. Detective Combest bent over Oswald, pulled up the black sweater, and searched for a wound. He found one hole in Lee's left chest. Although he had been shot just seconds ago, Oswald was already incoherent and appeared to be unconscious.

Combest spoke to him. "Is there anything you want to tell me?"

Oswald's eyes were open but he did not speak. Perhaps he couldn't. Perhaps he could not even hear the question. He had screamed in pain when the bullet struck him, and he was moaning now. He did not appear to be alert or lucid.

Combest asked him again, "Is there anything you want to say right now before it's too late?" The meaning behind the question was clear. He was asking Oswald to make a dying declaration.

This might be Oswald's last chance to claim credit for his infamous crime. To declare that he, Lee Harvey Oswald, lifelong loser and nobody, had finally done something worth remembering. Oswald appeared to shake his head a little from side to side, but it was hard to tell if it was in response to the question.

At this point, Oswald probably could not speak. If he could, he would have said *something*. For two days, he had enjoyed matching wits with Captain Fritz during their cat and mouse battle. He had reveled in his spontaneous hallway comments to the press. He had indulged his love of sarcasm and propensity to complain. Indeed, it had been almost impossible to *stop* Oswald from talking while in police custody.

He might not want to confess, but it was in Oswald's character to say something now, if only to tease Leavelle and congratulate him that Ruby had been such a good shot, or to ask who did it, or to complain that his guards had failed to protect him, or to send a final message to Marina. No, if Oswald did not speak now, it must have been because his grievous wound had silenced him.

Jack Ruby was doing all the talking now: "I hope I killed the son of a bitch," he boasted. "I'll save everybody a lot of trouble." Ruby justified what he had done. "Do you think I'm going to let the man who shot our president get away with it?"

One of the detectives told him, "Jack, I think you killed him."

"Somebody had to do it. You all couldn't."

A cop asked Ruby if he thought he could kill Oswald with one shot.

"Well, I intended to shoot him three times. I didn't think that I could be stopped before I got off three shots."

Police officers carried Oswald away from the excited journalists and the spot he was shot. Ike Pappas described what happened next: "Now the ambulance is being rushed in here . . . screaming red lights, people, policemen . . . they are rushing a mobile stretcher in . . . here is young Oswald now. He is being hustled in, he is lying flat. To me he appears dead. There is a gunshot wound in the lower abdomen. He is white . . . dangling, his head is dangling over the edge of the stretcher."

Police officers laid him on a gurney. Then an ambulance rushed him to Parkland Hospital.

Outside the Dallas County Jail, a crowd of several hundred people had assembled across street. They were waiting for Oswald. This crowd was bigger than the spontaneous mob that had gathered outside the Texas Theater on Friday afternoon and that was waiting for Oswald when he emerged. Today's crowd wanted to jeer him, and hoped to get a glimpse of him as he drove past. This was the second time in two days that a crowd had gathered on Houston Street to watch a famous man drive by. The first was on November 22, when President Kennedy's motorcade had driven down this street on his way to the Texas School Book Depository.

When Sheriff Bill Decker got word that Oswald had been shot before *his* motorcade had gotten under way, Decker walked into the street and announced the news. He hoped to disperse the crowd. The president's assassin was not coming. "Ladies and gentlemen, Lee Harvey Oswald has been shot and is on his way to Parkland Hospital," Decker told them.

They could go home now. The people in the crowd cheered and clapped their hands. They were happy that Oswald had gotten a taste of Texas justice.

LIKE PRESIDENT Kennedy, Lee Harvey Oswald looked dead on arrival at Parkland Hospital. But unlike the president, who had suffered a catastrophic head wound, Oswald exhibited a tenacity for life.

The wounded assassin was wheeled inside where some of the same people who treated President Kennedy would now try to save the life of his suspected assassin. It was about 11:30 A.M. The gurney approached trauma room one. No, do not bring Oswald in there, the hospital staff decided. Out of respect for President Kennedy, who had been pronounced dead in this room less than forty-eight hours ago, the staff refused to treat Oswald there. It would be obscene to try to save the assassin's life on the same table on which the president had died; to employ the same medical instruments to save the man who had stained their hospital gowns with the president's brains and blood.

So they rolled Oswald to trauma room two, where doctors had saved the life of Governor John Connally. Some of the same doctors who had treated JFK now worked on his killer. Although Jack Ruby had gotten off only one shot, he had inflicted a devastating wound. It was as bad as if he had shot Oswald three times. Because he shot his victim from at an angle, the bullet passed sideways through Oswald's entire torso, wreaking havoc with several internal organs.

Could he live? Would he recover for his trial? And at that trial— when he realized the physical evidence alone would convict him— would he accept his fate, and an almost certain death sentence, and finally mount his soapbox, find his voice, and shout to all the world, "Yes, I, Lee Harvey Oswald did this." And then tell the world *why*?

The doctors wanted him to live. It was their job. Whatever their personal feelings, they would do their utmost to save him. Chief Curry wanted him to live. Only if President Kennedy's assassin survived his wound could Curry salvage his own reputation, and that of his department. How could more than seventy Dallas policemen in one basement fail to prevent the murder of the most important criminal suspect in the world? If Oswald died, they would be stained forever as the fools who allowed him to be killed right under their very noses.

Captain Fritz wanted Oswald to live too. A dead suspect was of no use to one of the top interrogators in the history of the Dallas Police Department. Only if Oswald lived was there hope. Perhaps the suspect would confess and explain it all. The press wanted him to live— Oswald was a great story, and worth massive coverage in the weeks ahead. And, despite the howls of delight outside the county jail, most of the American people wanted Oswald to survive this day. True, it was hard not to take some pleasure in the knowledge that John Kennedy's murderer has suffered a kind of Old Testament or western vigilante justice for his great crime. But the people also wanted answers. Who was he? How did he do it? Why did he do it? If Lee Harvey Oswald died, he would take his secrets to the grave.

The bullet had damaged several vital organs, and Oswald had suffered a massive loss of blood. Surgeries to multiple organs seemed to offer hope, then failed. He had lost too much blood. The shock was irreversible. Lee Harvey Oswald expired in the same hospital where President Kennedy had died.

Without regaining consciousness, without speaking any last words, and without confessing his crime, Lee Harvey Oswald died at 2:07 P.M. (EST), just after the memorial service for President Kennedy at the Capitol had begun. Even if he had remained conscious long enough to speak, Oswald was not the sort of man to clear his conscience with a death-bed confession. Or to solve the case for the police. It would have given him pleasure to carry on the game from the grave, where his silence would torment his interrogators, and force them to fit all the puzzle pieces together by themselves.

NBC Television had broadcast the murder of Oswald on live television to the entire nation. Soon more than 100 million people learned that the assassin's assassin was a man born Jack Rubenstein, now known as Jack Ruby. He was a middle-aged club owner and low class creature from a world of sleazy nightlife, where men drank stiff drinks and paid women to dance and strip for them onstage. Like Oswald, he was another fringe character, an impulsive, moody, and violent man and, even more than Oswald, given to wild, unstable, emotional and often dangerous mood swings. He beat women, pistol-whipped customers, and called his constant companion—a little dog—his "wife." Ruby thought he would be a hero.

It was unwholesome, almost profane, that for the rest of the day, news of Oswald's shooting competed with coverage of John F. Kennedy's memorial events.

At 2:25 P.M., Marina Oswald came to Parkland Hospital to view her husband's body. She looked at his eyes. They were wet. "Look, Mama," she said to Lee's mother, "He cry."

The unstable Marguerite began to say crazy things.

"I think someday you'll hang your heads in shame," Marguerite scolded anyone in listening distance. "I happen to know, and know some facts, that maybe my son is the unsung hero of this episode. And I, as his mother, intend to prove this if I can."

Then she blabbered that Lee should be buried in Arlington National Cemetery. It was too much for Robert Oswald. He told his delusional mother to shut up.

On the day he died, Lee Harvey Oswald left behind another victim—the city of Dallas. Within hours of the assassination, Americans began to blame an entire city or state for the murder. Dallas was tarred as the "City of Hate." People remembered that Texas crowds had once spat at Lyndon and Lady Bird Johnson at the Adolphus Hotel. They remembered how in October 1963, hecklers disrupted a speech by Kennedy's ambassador to the United Nations, Adlai Stevenson.

Kennedy had been warned not to go to Texas. The political situation there was too volatile, and he was not liked there. He was too liberal for local tastes. One Texas politician had publicly implied it was not safe. Kennedy bristled at the suggestion that any part of the United States was too dangerous for a president to visit. Then there was that menacing WELCOME TO TEXAS newspaper ad and the WANTED FOR TREASON handbills. On November 20, Dallas Chief of Police Jesse Curry had to go on television to beseech the citizenry to treat the president with respect. Rumors spread that in Texas classrooms, schoolchildren cheered when they heard that Kennedy had been shot.

Whether the rumors were true or not, people wanted to believe them. Other stories claimed that adult Texans, upon hearing the news, had rejoiced in the streets.

Outside of Texas, politicians, newspaper editors, and others beat the drumbeat of collective guilt. They blamed Dallas right-wing fanaticism for the president's murder. But Oswald was not a John Bircher or right winger; if anything he was a leftist or, in Jackie Kennedy's words, "some silly little Communist." The assassination, they said, was the shame of Dallas. Of course such emotional criticism ignored the fact that most of Dallas had given the Kennedys a friendly reception. No one in that motorcade could have concluded that cheering Dallas crowds "hated" JFK and wanted him to die. The president was thrilled by the warm welcome he had received.

Two days after the assassination, at the congressional memorial service in Washington, D.C., Chief Justice of the United States Earl Warren upset the solemnity of the occasion with a wild and shocking indictment of the climate of hate he said had caused the assassination. "We are saddened; we are stunned; we are perplexed. John Fitzgerald Kennedy . . . has been snatched from our midst by the bullet of an assassin. What moved some misguided wretch to do this horrible deed may never be known to us, but we do know that such acts are commonly stimulated by forces of hatred and malevolence, such as today are eating their way into the bloodstream of American life. What a price we pay for this fanaticism." The chief justice condemned "the hatred that consumes people, the false accusations that divide us, and the bitterness that begets violence." Warren blamed the assassination on those "who do not shrink from spreading . . . venom." Of course Earl Warren meant conservatives and the right wing. But Lee Harvey

Oswald was not a right-wing fanatic. He was a leftist who hated the right wing, and who had tried to murder an ultraconservative army general. Everybody knew that Warren was also talking about Dallas, the city that boasted a big billboard that read IMPEACH EARL WARREN.

But it was the murder of Lee Harvey Oswald while in police custody that really turned the nation against Dallas. The assassination of the president came like a bolt of lightning that struck without warning from bright blue sky. The Oswald killing was different. On a dozen occasions, the police department had displayed their prisoner to the press in crowded, unsecured hallways. A Dallas law enforcement official announced that they had already "solved" the assassination and that the case against Oswald was "cinched." Death threats against the prisoner came in, but security at the jail remained lax. Nonetheless, detectives transferred Oswald to another location in the middle of a media circus in a poorly secured basement. The whole world saw what happened next.

Oswald's murder was the tipping point. How, people wondered, could the city's police and prosecutors be so inept? How could they subject Oswald to all of those hallway interviews? How could they fail to protect the life of the most hated man in the country?

America blamed Dallas for ending the play in mid-act. Negligence and incompetence had pushed the assassin offstage before he could perform his starring role—defendant in the trial of the century. The public wanted to hear Oswald's account of November 22. They hoped he might confess. They wanted to hear from the witnesses and see the evidence presented in a court of law. There would be no third act.

ON SUNDAY, November 24, at 12:34 P.M. (EST), Jackie Kennedy and her brother-in-law Robert entered the East Room and approached the closed, flag-draped coffin. At her request, the honor guard removed the American flag, and the casket was opened. For the last time, she looked upon her husband's face. At 12:37 P.M., she placed two hand-

The president's coffin at the East Front of the United States Capitol.

written farewell letters in the coffin. Then it was sealed and draped again with the national colors.

Military pallbearers in immaculate dress uniforms lifted the casket and carried the late president through the White House and outside to a waiting artillery caisson drawn by six gray horses. The soldiers secured the coffin with leather straps. Then, at 1:08 P.M., the cortege left the White House for memorial ceremonies at the U.S. Capitol. Military musicians, the usual crisp *rat-tat-tat* sound of their drums muffled by shrouds of black cloth wrapped around them, beat the mournful sound of the funeral cadence.

Formations of troops marched behind the caisson. And tens of thousands of people lined the route along Pennsylvania Avenue. It was just like Abraham Lincoln's funeral procession to the Capitol on April 19, 1865.

When the procession arrived at the East Front, the honor guard carried President Kennedy up the steps and past the very spot where, on January 20, 1961, he had taken the oath of office. They carried him into the rotunda and laid his coffin under the Great Dome, upon the very spot where Abraham Lincoln's coffin had once rested. At 2:02 P.M., the congressional memorial service began.

John Kennedy had begun his political career at this place, first as a congressman, then a senator. One of the speakers was Senator Mike Mansfield, one of the lions of the Senate. His eulogy, a tribute to both John and Jackie Kennedy, stunned listeners. Mansfield recalled the image of Jackie, at Parkland Hospital, removing her wedding ring and placing it in her husband's hands:

> *There was a sound of laughter; in a moment, it was no more. And so she took a ring from her finger and placed it in his hands.*
>
> *There was a wit in a man neither young nor old, but a wit full of an old man's wisdom and of a child's wisdom, and then, in a moment it was no more. And so she took a ring from her finger and placed it in his hands.*
>
> *There was a man marked with the scars of his love of country, a body active with the surge of a life far, far from spent and, in a moment, it was no more. And so she took a ring from her finger and placed it in his hands.*
>
> *There was a father with a little boy, a little girl and a joy of each in the other. In a moment it was no more, and so she took a ring from her finger and placed it in his hands.*
>
> *There was a husband who asked much and gave much, and out of the giving and the asking wove with a woman what could not be broken in life, and in a moment it was no more. And so she took a ring from her finger and placed it in his hands, and kissed him and closed the lid of a coffin.*

Jackie was awed. Of all the words of tribute spoken and written about President Kennedy, she loved none more. She wanted a copy

Jacqueline Kennedy and her children descend the steps of the East Front of the Capitol.

of the text. Before she could speak, Mansfield handed her his manuscript.

"How," Jackie asked him, "did you know?"

At the conclusion of the ceremony, Jackie and her daughter, Caroline, kneeled by the coffin to say good-bye.

THE SERVICE at the Capitol came to an end. Now the American people would have their turn to pay homage to the fallen president. The bronze doors of the East Front were thrown open to all comers who wanted to view the coffin. Tens of thousands of people were already in line. By 8:00 P.M., two hundred thousand stood in a line that stretched all the way down East Capitol Street and, twelve blocks away, past Lincoln Park, where a bronze sculpture of the Great Emancipator—Abraham Lincoln—beckoned a freed slave to rise.

Beyond the park, the line seemed limitless. That night at 9:04 P.M.,

Jackie and Robert Kennedy made a surprise return visit to the rotunda. She approached the catafalque and then, in full view of everyone in the rotunda, dropped to her knees. After two minutes, she rose and left the Capitol. She did not want to get into the car waiting at the base of the East Front.

"No, no, I just want to walk," she said.

Then she and her brother-in-law walked outside, among the crowd, to the north side of the Capitol, and then west along Constitution Avenue for a while, almost to the foot of Capitol Hill, before they got in their car and returned to the White House.

In the darkness, almost no one recognized her. People did not notice her because they found it so improbable that she would walk in public among them. After a few people realized it was her, Jackie walked to the street, got into her car, and returned to the White House.

IN A television interview on the evening of November 24, Daniel Patrick Moynihan, member of the Kennedy administration and future United States senator, reflected on the meaning of the last three days: "The French author Camus, where he came out at the end of his life, was that the world is absurd. A Christian shouldn't think that, but the—the utter senselessness—the meaninglessness—it—you know, we all of us down here know that politics is a tough game, and I don't think there is any point in being Irish if you don't know that the world is going to break your heart eventually. I guess we thought he had a little more time. So did he."

BY TWO A.M. on Monday, November 25, the line to get into the Capitol was three miles long, and by nine A.M., 250,000 people had filed past the coffin. There was no more time to admit everyone else still standing in the long line. People who had waited for hours were turned away.

The viewing could not continue because now it was time for the president's funeral. Jacqueline Kennedy could have waited at the White House while her husband's coffin was removed from the Capitol and brought to the mansion. But instead she went to get him. Since 12:30 P.M. (CST) on Friday, November 22, she had accompanied her husband wherever he went: on the high-speed race from Dealey Plaza to Parkland Hospital, on the slow ride with the coffin from Parkland to Love Field, aboard Air Force One for the somber flight home to Washington, in the hearse motorcade to Bethesda Naval Hospital, from there the predawn parade home to the East Room of the White House, in the Pennsylvania Avenue procession to the U.S. Capitol, and now for President Kennedy's parting from the Congress he loved.

Wherever he went, she would follow, until she would take him to the place where she could follow no more.

Her limousine arrived at the Capitol at around 10:40 A.M. Jackie, escorted by the president's two brothers, ascended the East Front steps and walked into the rotunda. She knelt before the coffin and prayed.

Mrs. Kennedy and her husband's brothers Robert and Edward lead the procession from the White House to St. Matthew's Cathedral.

Then the military casket team lifted the coffin from the catafalque, carried it out of the rotunda, and descended the stairs to the East Front. Once at ground level, they strapped the casket to the caisson. At 11:00 A.M. the cortege departed the Capitol for the White House.

From there, starting at 11:35 A.M., Jacqueline Kennedy led a procession on foot to St. Matthew's Cathedral, eight blocks away. Many heads of state from around the world—kings, presidents, prime ministers, and more—plus distinguished diplomats had traveled to Washington for the event. Mourners included French president Charles de Gaulle, British prime minister Sir Alec Douglas-Home, Irish president Éamon de Valera, and Ethiopian emperor Haile Selassie. More than 220 officials from ninety-two countries, including nineteen heads of state, traveled to Washington. The last time so many had gathered in one place was in 1910, for the funeral of British king Edward VII. Most, unless they were too old or too frail, walked to the cathedral. The Secret Service begged Lyndon Johnson to ride in a car

The caisson leaves the White House for St. Matthew's.

and not march in the procession—it was too dangerous, they warned, and they did not want to lose another president. But Johnson refused: "I would rather give my life than be afraid to give it."

At 12:15 P.M., the pallbearers carried the flag-draped coffin into St. Matthew's. More than one thousand people attended the service, a low mass conducted by the famous Catholic priest Richard Cardinal Cushing of Boston. The hour-long service confused John Jr. He fidgeted and asked his mother, "Where's my Daddy?"

The sacred music, the religious incantations, and the somber setting proved too much for Jackie. After enduring the pain for four days, she could no longer hold it in. She began to cry uncontrollably. Her sobbing body heaved and shook.

At the end of the mass, Cushing said words that broke her heart: "May the angels, dear Jack, lead you into Paradise. May the martyrs receive you at your coming. May the spirit of God embrace you, and mayest thou, with all those who made the supreme sacrifice of dying for others, receive eternal rest and peace."

Jackie thought his voice sounded like "a plea, almost a wail." That phrase, "May the angels, dear Jack . . ." unleashed in her fresh spasms of emotion, and she began to sob and shake again. Her daughter, Caroline, took her hand and said, "You'll be all right, Mummy. Don't cry. I'll take care of you."

After the service, Jacqueline Kennedy and her children, standing outside the cathedral, watched the honor guard carry the coffin down the steps. A military band played "Hail to the Chief." Jackie bent down and whispered in her little boy's ear, "John, you can salute Daddy now and say good-bye to him."

John Kennedy Jr. saluted his father's coffin just as he had seen soldiers in uniform do. It was a heartbreaking gesture that became one of the most unforgettable images of the funeral.

THAT NIGHT Jacqueline had expected to preside over a White House dinner for West German chancellor Ludwig Erhard. She had scheduled the event weeks ago and had mailed the invitations before the trip to Dallas. By all rights, she should be supervising last-minute preparations for a delightful social event, not her husband's funeral. Secretary of Defense Robert McNamara was one of the invited dinner guests. He had expected to pass through the Southwest Gate of the White House that day, not the gates of Arlington Cemetery.

At 1:30 P.M., the horse-drawn caisson bearing President Kennedy departed St. Matthew's for Arlington across the Potomac River. It took one hour and fifteen minutes to get there.

At a brief graveside service, a bugler played taps. He was so emotional, he played a wrong note. It sounded as though the instrument itself were weeping, which heightened the drama of the scene. A formation of fifty fighter jets screamed overhead. In a stunning tribute, the presidential jet, Air Force One, descended to an alarming altitude of just five hundred feet above the cemetery and dipped its wings

The coffin is carried to the grave site at Arlington National Cemetery.

A panoramic view of the scene at Arlington Cemetery.

in tribute while it whooshed past at 600 miles per hour. From the ground, the airplane looked enormous.

Artillery pieces fired a twenty-one-gun salute. Riflemen fired three volleys overhead, an eerie echo of the three shots fired in Dallas. Soldiers removed the flag from the coffin, folded it into a triangle, and presented it to Jackie.

At 3:13 P.M., Jacqueline Kennedy used a taper to light an "eternal flame" beside the grave. Two days ago, she had made a last-minute request. She said she wanted to light a flame at the climax of the service in Arlington that would burn forever in memory of her husband. She recalled the day when she and the president had toured the Civil War battlefield of Gettysburg. There, at the Eternal Light Peace Memorial dedicated by President Roosevelt in 1938, she had seen an eternal flame—a gas-powered fire that burned day and night, around the clock—that illuminated the top of the tall monument. At Arlington,

Jacqueline Kennedy watches as the flag from her husband's coffin is folded.

army engineers had one built at ground level next to President Kennedy's grave site in less than twenty-four hours, and it was ready in time for Jackie to light it on Monday afternoon.

When the service was over, she got into her limousine to return to the White House to greet the funeral guests. "It would be most ungracious of me," she said, "not to have all those people in our house."

After leaving Arlington, her car pulled away from the rest of the motorcade. It drove on alone to the Lincoln Memorial, where Jackie looked through the window and gazed up at the sculpture of Father Abraham, who had perished from an assassin's bullet ninety-eight years earlier. Then she returned to the White House, where she stood in a receiving line and greeted guests from around the world.

FAR FROM Washington, there were other funerals this day.

In Texas, Dallas police officer J. D. Tippit was buried with full honors; more than one thousand people attended the service. And that afternoon, Lee Harvey Oswald was buried in Texas in a cheap, cloth-covered pine casket. The timing was shocking—the same day as the funerals for the two men he was accused of murdering, President Kennedy and Officer Tippit. At Oswald's hurried, ten-minute graveside ceremony, the more than seventy-five reporters, photographers, and law enforcement officials far outnumbered the handful of mourners.

Aside from his family, no one who knew him wanted to be seen at his funeral. No friends came forward to serve as pallbearers. News reporters volunteered to carry the assassin's lonely casket to its resting place. At first, Associated Press correspondent Mike Cochran said no. "I was among the first that they asked to be a pallbearer. I refused, and then Preston McGraw of UPI stepped up and said, 'Certainly, I'll be a pallbearer.' So then, if there's one thing I knew . . . if UPI was going to be a pallbearer, I was damn sure going to be a pallbearer."

The minister who promised to officiate failed to show up. Only Oswald's family—his wife, Marina, and their two daughters, his mother, Marguerite, and his brother Robert—came to mourn him. *Life* magazine published photographs of the funeral, including a large, double-page image taken at dusk that made it appear that heaven itself—with dark, gloomy, and threatening skies—had cursed Oswald in the grave. That night, watchmen stood guard over the assassin's final resting place to prevent distraught citizens from, under the cover of darkness, desecrating the grave.

But Oswald did not rest in peace. After his burial, bootleg copies of gruesome photographs of his body—taken at the autopsy—leaked out and were published widely. In the days ahead, morbid curiosity seekers would flock to his grave, and someone stole his tombstone. Years later, in a bizarre attempt to verify that Oswald was really in the coffin (and not an imposter switched with the true Oswald, an outlandish claim pressed by extremist conspiracy buffs), his remains were exhumed and the corpse photographed again. It was, of course, Lee Harvey Oswald.

The Lee mansion at Arlington Cemetery after the funeral.

• • •

THAT NIGHT at the White House at 7:00 P.M., Jacqueline Kennedy hosted a little birthday party for her son. He was now three years old. She decided to hold a bigger party in a few days.

Later in the evening, Robert Kennedy asked Jackie, "Should we go visit our friend?"

Two of the many types of mourning buttons manufactured after the assassination.

A memorial banner printed in Chicago within a few days of the assassination.

Oh yes, she said.

At midnight, after everyone had gone, she returned to Arlington Cemetery to bid her husband a private and final farewell. By now the

Another of the many memorial banners.

casket had been lowered into the grave and covered with a mound of earth. It was quiet now. The eternal flame flickered and danced in the dark. The glow cast streaks of light and shadow across Jackie's face.

THE NEXT morning Lyndon Johnson found a handwritten letter from Jacqueline Kennedy on his desk. It was an affectionate farewell. Four days earlier she would have called him "Lyndon." Today she addressed the letter, "Dear Mr. President."

She thanked him for walking behind President Kennedy's coffin in the previous day's procession from the White House to St. Matthews. "You did not have to do that—I am sure that many people forbid you to take such a risk—but you did it anyway." LBJ's gesture of writing to her children had moved her: "What those letters to them will mean later—you can imagine . . . they have always loved you so much." Of all the Kennedys, Jackie had liked Lyndon Johnson the

The notorious Dealey Plaza desk set.

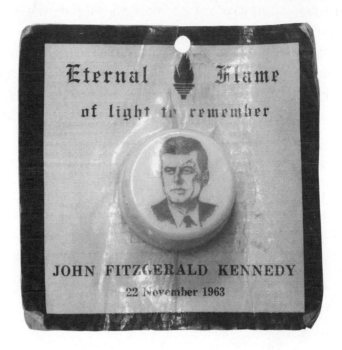

The electric Eternal Flame night-light.

most. Of course Bobby hated him and, while JFK respected him, the former rivals from Harvard and the Texas Hill Country were always men from different worlds. Now Jackie wanted to thank LBJ for his many courtesies to her: "And most of all, Mr. President, thank you for the way you have always treated me . . . before, when Jack was alive, and now as President."

As a politician's wife, Jacqueline Kennedy knew how her husband's defeat of Johnson in the race for the 1960 Democratic presidential nomination must have been a humbling experience for the powerful master of the Senate. It had not been easy for LBJ to live with the door prize of the vice presidency. And, Jackie now wrote to Johnson, she knew it: "But you were Jack's right arm—and I always thought the greatest act of a gentleman that I had seen on this earth— was how you—the Majority Leader when he came to the Senate as just another little freshman who looked up to you and took orders from you, could then serve as Vice President to a man who had served under you and been taught by you."

Jackie planned to move out of the White House in ten days, on December 6. She confided to Johnson that after she had gotten home yesterday from the funeral, and after she had greeted the last guest,

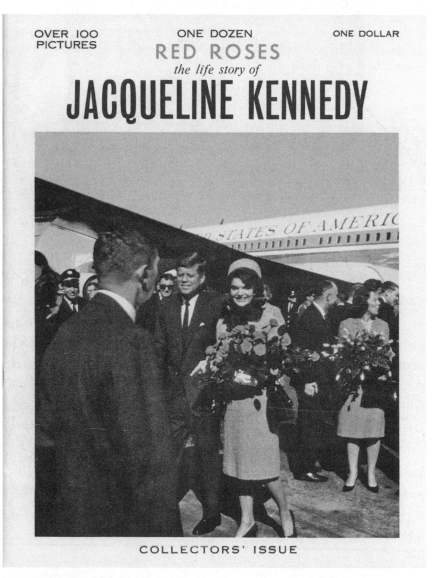

In the aftermath of the assassination, publishers issued dozens of Kennedy tribute magazines.

she went in search of memories. "It was so strange—last night I was wandering through this house." She walked through the Treaty Room, the library and more, and finally the Oval Office in which her husband had risen from his chair for the flight to Texas just four days earlier. She wanted to tell Johnson about the room: "Your office—you

Even Lee Harvey Oswald got his own magazine.

are the first President to sit in it as it looks today. Jack always wanted a re-doing and I had curtains designed for it that I thought were as dignified as they should be for a President's office."

Even in her profound sorrow, Jackie could not suppress one of the traits essential to her nature—her love of design and aesthetics. When she was a twenty-one-year-old college senior at George Washington University, her winning essay in the Prix de Paris competition for a prestigious junior editorship at *Vogue* magazine revealed her aspiration to be "a sort of Overall Art Director of the Twentieth Century, watching everything from a chair hanging in space." How, wondered this art director, hours after her husband's funeral, should LBJ decorate the presidential office? "Late last night," her letter continued, "a moving man asked me if I wanted Jack's ship pictures left on the wall for you (they were cleaning the office to make room for you)—and I said no because I remembered all the fun Jack had those first few days hanging pictures of things he liked, setting out his collection of whale's teeth, etc." She did not want to deny LBJ the same pleasure. "But of course, they are there only waiting for you to ask for them if the walls look too bare. I thought you would want to put things from Texas on it—I pictured some gleaming long horns—I hope you put them somewhere."

The letter closed on a melancholy note. Jackie had run a little White House school for a small group of children, including her own. Johnson told her that it must go on, assuring her it was not an imposition. "It mustn't be very much help to you," Jackie sympathized, "your first day in office—to hear children on the lawn at recess. It is just one more example of your kindness that you let them stay—I promise they will be gone soon." She signed her letter, "Thank you Mr. President. Respectfully, Jackie."

IT WAS done. Four days of blood and death, of mourning and drums, were over. But America would never be the same.

The people refused to forget. As after the murder of Abraham Lincoln ninety-eight years earlier, the American people coveted souvenirs and relics of the assassination. All over the country, people saved newspapers from November 22, 23, 24, 25 and 26, and in no city was any newspaper more desirable than the first edition that announced the assassination.

Entrepreneurs manufactured celluloid, pin-back memorial buttons by the millions, as well as JFK banners that were printed on black felt or on white fabric. One brazen opportunist in Dallas created a three-dimensional Dealey Plaza desk ornament, complete with ballpoint pen and paper-clip tray. In this miniature architectural model, the sixth-floor Book Depository window is always open; the Hertz clock on the roof is frozen at 12:29 P.M., one minute before the assassination; and an X marks the position of President Kennedy's limousine at the moment of the fatal shot.

Another company, inspired by the eternal flame at Arlington National Cemetery, manufactured an electric-powered, plastic JFK night-light—an "Eternal Flame of Light to Remember," guaranteed to be a "new Electronic Innovation providing 50,000 hours or more (six years) of continuous use at an approximate cost of 3 Cents per year." So the little plug-in light was not "eternal" after all.

Publishers flooded newsstands with dozens of different John Kennedy commemorative magazines. Others published Jackie tribute magazines. A record company released an album containing Lee Harvey Oswald's strange August 1963 New Orleans radio station interview (alluding to his communist ties, it was titled *Oswald: Portrait in Red*), while others issued dozens of memorial albums of JFK's speeches or of the radio and television coverage of the assassination. Even Oswald's unstable mother, Marguerite, tried to cash in. She sold a bizarre booklet featuring photographs of her son's funeral. She called it *Aftermath of an Execution: The Burial and Final Rites of Lee Harvey Oswald as Told by His Mother*. There was little demand for it, and few copies sold.

Two days after the funeral, on November 27, 1963, President Lyndon Johnson addressed a joint session of Congress.

All I have I would have given gladly not to be standing here
today. The greatest leader of our time has been struck down
by the foulest deed of our time. Today John Kennedy lives
on in the immortal words and works that he left behind. He
lives on in the mind and memories of mankind. He lives in
the hearts of his countrymen. . . . An assassin's bullet has
thrust upon me the awesome burden of the presidency. . . .
I profoundly hope that the tragedy and torment of these ter-
rible days will bind us together in new fellowship, making us
one people in our hour of sorrow. So let us here highly resolve
that John Fitzgerald Kennedy did not live—or die—in vain.

The new president vowed to carry on the dead president's work.

On November 29, Johnson appointed a special presidential commission, chaired by Chief Justice Earl Warren, to "study and report
upon all facts and circumstances relating to the assassination of the
late President, John F. Kennedy, and the subsequent and violent death
of the man charged with the assassination." LBJ instructed the special
commission to "satisfy itself that the truth is known as far as can be
discovered, and to report its findings to him, to the American people,
and to the world."

Speed was also of the utmost importance. Johnson wanted the
commission to finish its work well before November 1964, so that no
uncertainties plagued him during the presidential election.

If Lyndon Johnson wanted to *discover* the truth about John Kennedy's death, Jacqueline Kennedy wanted to *preserve* her version of
the truth about his life. On the same day that LBJ created the Warren Commission, Jackie unveiled an enduring myth that forever defined her husband's presidency. The November 29 *Washington Star*
newspaper carried an article by Jack and Jackie's old friend Charlie
Bartlett. Up until then, most journalists had focused on the late president in their editorial tributes. Yes, the press had heaped praise on
Jacqueline Kennedy for her brave and dignified behavior during the

four days from November 22 to 25. But Bartlett did not want to dwell on the widow Jackie. He wanted to write about the woman he knew.

So the man who wrote "We had a hero for a friend" published a piece titled "The Impact of Jacqueline Kennedy: Fidelity to Her Own Individuality Was Enormous Asset to President." Little did Bartlett know that Jackie's impact was just beginning. That night, during a drenching New England storm, just a week after her husband was slain in her arms, Jacqueline Kennedy had a plan to transform her husband into a legend.

CHAPTER 10

"ONE BRIEF SHINING MOMENT"

The day after Thanksgiving, on Friday, November 29, Jackie called Theodore White, Pulitzer Prize–winning author of the bestselling book *The Making of the President: 1960*. White and John Kennedy had gotten to know each other, and the president had admired him.

When Jackie called, White was not home. As he remembered, he "was taken from the dentist's chair by a telephone call from my mother saying that Jackie Kennedy was calling and needed me." He called her back. "I found myself talking to Jacqueline Kennedy, who said there was something that she wanted *Life* magazine to say to the country, and I must do it."

She told White she would send a Secret Service car to fetch him in New York and drive him up to Hyannis Port. But when White called the Secret Service he was, he wrote, "curtly informed that Mrs. Kennedy was no longer the President's wife, and she could give them no orders for cars. They were crisp."

It was impossible to fly that weekend. A northeaster or a hurricane was coming up over Cape Cod. So White hired a car and driver and headed north into the New England storm. He called his editors at *Life* to tell them about his exclusive scoop, but they told him the next issue was about to go to press. They warned him it would cost $30,000 an hour to hold the presses open for his story. It was unprecedented. But they would do it.

This meant that the most important photojournalism magazine in America would be standing still and delaying the printing of its next issue for a story that had not yet been written and would be based on an interview that had not yet even been conducted. Still, an exclusive interview with First Lady Jacqueline Kennedy was so coveted, *Life* was willing to do almost anything.

WHITE ARRIVED, he recalled, "at about 8:30 in the driving rain." Jackie welcomed him and instructed her houseguests, who included Dave Powers, Franklin D. Roosevelt Jr., and JFK's old pal Chuck Spalding, that she wanted to speak with him alone. As soon as she sat down, White began taking notes as fast as his hand could scribble: "Composure . . . beautiful . . . dressed in trim black slacks . . . beige pullover sweater . . . eyes wider than pools . . . calm voice." Then she spoke.

"She had asked me to Hyannisport," White discovered, "because she wanted me to make certain that Jack was not forgotten by history."

White was stunned. How could anyone ever forget John F. Kennedy? White was now ready to be hypnotized by a master mesmerist. Jackie complained that "bitter people" were already writing stories, attempting to measure her husband with a laundry list of his achievements and failures. Jackie hated that. They would never capture the real man.

White asked her to explain, and then, for the next three and a half hours, she delivered a jumbled, almost stream-of-consciousness narrative about Dallas, the blood, the head wound, the wedding ring, the hospital, and how she kissed him good-bye.

It was only a week after the assassination.

Then she got to the reason she had summoned White: "But there's this one thing I wanted to say . . . I kept saying to Bob, I've got to talk to somebody, I've got to see somebody, I want to say this one thing, it's been almost an obsession with me, all I keep thinking of is this line from a musical comedy, it's been an obsession with me."

She confided to White. "At night, before we'd go to sleep, . . . Jack liked to play some records . . . and the song he loved most came at the very end of this record, the last side of *Camelot*, sad *Camelot*."

She was talking about the popular Broadway musical fantasy about King Arthur's court. "The lines he loved to hear," Jackie revealed, were "Don't let it be forgot, that once there was a spot, for one brief shining moment that was known as Camelot."

In case White failed to understand, she repeated her story. "She wanted to make sure," the journalist remembered, "that the point came clear." Jackie went on: "There'll be great Presidents again—and the Johnsons are wonderful, they've been wonderful to me—but there'll never be a Camelot again."

White wanted to continue to other subjects, "But [Jackie] came back to the idea that transfixed her: 'Don't let it be forgot, that once there was a spot, for one brief moment that was known as Camelot.' "

She was determined to convince White that her husband's presidency was a unique, magical, and forever lost moment. "And," she proclaimed, "it will never be that way again."

President Kennedy was dead and buried in his grave, and she told the journalist she wanted to step out of the spotlight. "She said it is time people paid attention to the new President and the new First Lady. But she does not want them to forget John F. Kennedy or read of him only in dusty or bitter histories: For one brief shining moment there was Camelot."

Around midnight White went upstairs to write the story—*Life* needed it tonight, before he left Jackie. He came down around 2:00 A.M. and tried to dictate the story over a wall-hung telephone in her kitchen. He had already allowed her to pencil changes on the manuscript. As White spoke over the phone, Jackie overheard that his editors in New York wanted to tone down and cut some of the "Camelot" material. She glared at White and shook her head. One of his editors caught the stress in his voice and suspected Jackie. "Hey," he asked White, "is she listening to this now?"

It was Jacqueline Kennedy's tour de force, her finest hour—actually more than five hours—of press manipulation. She had summoned an influential, Pulitzer Prize–winning author to do her bidding—and like so many men she had mesmerized before, he did it. White violated all standards of journalism ethics by allowing the subject of a story to read it in advance—and edit it. But he was not acting as a journalist that night—he was serving as the awestruck courtier of a bereaved widow.

And it worked. Thanks to Theodore White's essay "For President Kennedy: An Epilogue," which ran in the December 6 issue of *Life*, Camelot and its brief shining moment became one of the most celebrated and enduring myths in American politics. To Jackie, the assassination symbolized an end of days, not just for her husband, but also for the nation.

UNBIDDEN, OTHER journalists placed themselves in the service of the newborn Kennedy legend, offering their own personal contributions to the story. In the December 2, 1963, issue of *Newsweek* magazine, editor Ben Bradlee—a close friend who had socialized with John and Jackie Kennedy—published a mythologizing essay, "He Had That Special Grace . . ."

> *History will best judge John Kennedy in calmer days when time has made the tragic and grotesque at least bearable. And surely history will judge him well—for his wisdom and his compassion and his grace. John Kennedy was a wonderfully funny man, always gay and cheerful. . . . John Kennedy was a hungry man, ravenous sometimes for the nourish-*

*ment he found in the life he led and the people he loved. . . .
John Kennedy was a graceful man, physically graceful in his
movements—walking, swimming, or swinging a golf club—
and had that special grace and the intellect that is taste. . . .
John Kennedy was a restless, exuberant man, always looking forward to the next challenge. . . . John Kennedy was a
blunt man, sometimes profane, when it came to assessing
rivals. . . . He loved his brothers and sister with a tribal love.
John Kennedy loved his children with a light that lit up his
world. . . . And John Kennedy loved his wife, who served him
so well. Their life together began as it ended—in a hospital—
and through sickness and loneliness there grew the special
love that lights up the soul of the lover and the loved alike.
John Kennedy is dead, and for that we are a lesser people in
a lesser land.*

The tributes did not stop. That month, *Look*, the *Saturday Evening Post*, and the *Saturday Review* all published worshipful JFK tributes. Today it is difficult to fathom or exaggerate the cultural influence these now vanished, oversize-format, photo-news magazines once had. But no publication became more synonymous with the "Four Days" the nation had suffered than *Life* magazine.

And in television broadcasting, no network was more celebrated for its coverage of the tragedy than CBS. On December 12, CBS presented to its employees as a memento a nineteen-page pamphlet documenting their hard work over the past three weeks. The pamphlet was illustrated with forty-eight small television screens depicting iconic images from the time of the assassination to the president's funeral. The images were taken from the CBS special *The Four Dark Days: From Dallas to Arlington*, broadcast to the nation on the night of Monday, November 25.

It was time, thought CBS president Frank Stanton, for some tasteful self-congratulation. The pamphlet contained his full-page, single-spaced letter addressed "To all CBS Employees": "The assassination of President Kennedy . . . brought to electronic journalism the most demanding challenge in its history—demanding because wholly unexpected and, while a public event of world-wide impact, profoundly moving in an exceptionally personal way to every American."

Stanton summarized what the network had done. "The first CBS News bulletin was by Walter Cronkite on the CBS Television Network at 1:40 P.M. . . . . There followed, for the next four days, what was certainly the longest uninterrupted story in the history of television and possibly all broadcasting. Before the fateful events ended on Monday night, 55 hours of news reports, bulletins, memorial concerts and special broadcasts had been presented over the CBS Television Network. . . . The cost to CBS, including loss of revenue, was $4 million; the aggregate cost to our more than 400 radio and television network affiliates about as great."

No event in history had ever called for a more supreme effort. "More than 660 CBS people—newsmen, producers, editors, writers, researchers, cameramen and technicians—worked steadily throughout the crisis. From CBS Films, jet transports carried film totaling more than 13 hours of running time to 38 countries all over the world. . . . To estimate . . . the full significance of this, one has only to recall . . . the wild rumors, the abrasive bitterness, the divisive recriminations that followed the shooting of another President on another Friday, nearly a century ago, when a nation was kept in ignorance of the full facts surrounding President Lincoln's death for weeks, and in some respects for months."

EVEN BEFORE the funeral, Jacqueline Kennedy had decided she did not want to remain living in the White House for long. Mindful perhaps of public gossip about Mary Lincoln's drawn-out, emotional, post-assassination occupancy of the Executive Mansion, Jackie was determined to avoid similar criticism. And her memories of happier times there haunted her. She wanted distance from them. Despite President Johnson's gracious invitation to remain as long as she wished, Jacqueline Kennedy wanted to be out of the mansion within two weeks of the assassination.

But Jackie had nowhere to go. She longed to return to her former neighborhood in Georgetown, but she and the president had sold the house at 3307 N Street NW after the election. On the night of November 22, Secretary of Defense Robert McNamara had offered to get it for her. "That first night [he] said he'd buy back our old house in Georgetown. That was the first thing I thought that night—where will I go? I wanted my old house back." But she chose not to take advantage of his heartfelt gesture. "I thought—how can I go back to that bedroom? I said to myself—you must never forget Jack, but you musn't be morbid." W. Averell Harriman, the famous American diplomat and elder statesman who had held several posts in the Kennedy administration, offered her the use of his elegant residence at 3038 N Street NW, a few blocks from the old Kennedy place. She accepted the offer.

But before she left the White House, there were several things she wanted to do. On December 3, just eleven days after the assassination, Jackie attended a Department of the Treasury ceremony honoring Secret Service agent Clint Hill with the Exceptional Service Award for his bravery on November 22. She had insisted that he be recognized. Her memories were fresh and painful, and it was difficult to relive that day, but she had grown close to Hill, and she wanted to show by her presence that she did not blame her trusted guardian for her husband's assassination. She also informed the Secret Service that she wanted Hill to stay on as her personal agent.

Jackie packed her children's toys, books, and clothing in cardboard boxes that she labeled herself with marking pens. She instructed artisans to carve an inscription in the marble mantel above the fireplace in the president's bedroom: "In this room lived John Fitzgerald Kennedy with his wife Jacqueline during the two years, ten months, and two days he was President of the United States."

She also handwrote thank-you notes to members of the White House staff.

Jackie Kennedy's self-imposed two-week deadline for leaving the

White House—Friday, December 6—coincided with the Presidential Medal of Freedom ceremony, which had been scheduled long before the assassination. The printed program included some of the words President Kennedy had intended to say that day. Jackie chose not to attend, but President Johnson presented the medals, and his opening remarks acknowledged the tragedy:

> *Over the past two weeks, our Nation has known moments of the utmost sorrow, of anguish and shame. This day, however, is a moment of great pride. In the shattering sequence of events that began 14 days ago, we encountered in its full horror man's capacity for hatred and destruction. There is little we do not now know of evil, but it is time to turn once more to the pursuits of honor and excellence and achievement that have always marked the true direction of the American people. . . . I want particularly to thank you for reminding us that whatever evil moments may pass by, we are and we shall continue to be a people touched with greatness called to high destiny to serve great purposes.*

Then LBJ read the citations aloud and awarded thirty-one medals. The honorees included: singer Marian Anderson, cellist Pablo Casals, architect Mies van der Rohe, photographer Edward Steichen, critic Edmund Wilson, playwright Thornton Wilder, artist Andrew Wyeth, and U.S. Supreme Court justice Felix Frankfurter.

President Johnson remembered the man in whose place he stood: "John Kennedy is gone. Each of us will know that we are the lesser for his death. But each of us is somehow larger because he lived. A sadness has settled on the world which will never leave it while we who knew him are still here." Then the president made a surprise announcement: "As a simple gesture, but one which I know he would not have counted small, it is my privilege at this moment to award the Presidential Medal of Freedom posthumously to John Fitzgerald Kennedy on behalf of the great Republic for which he lived and died."

The citation read:

> *John Fitzgerald Kennedy, 35th President of the United States, soldier, scholar, statesman, defender of freedom, pioneer for peace, author of hope—combining courage with reason, and combatting hate with compassion, he led the land he loved toward new frontiers of opportunity for all men and peace for all time. Beloved in a life of selfless service, mourned by all in a death of senseless crime, the energy, faith and devotion he brought to his extraordinarily successful though tragically brief endeavors will hereafter "light our country and all who serve it—and the glow from that fire can truly light the world."*

No one attending the event knew it, but Jackie had watched the ceremony from a spot where she was hidden from view. It was a bittersweet honor to watch another president award the medals her husband had looked forward to presenting.

On the day she left the White House, Jackie left a note and flowers for the new First Lady, Lady Bird Johnson. She and John Jr. paid a farewell call on White House usher B. C. West. They posed with him in his office for a photograph. John sat on his desk while Jackie, putting on a brave front for the camera, smiled. But she wore a black dress. In other photos taken of her that day, she did not smile.

Then Jacqueline Kennedy left the White House, returning to her beloved Georgetown, where her life with Jack had begun. She vowed never to set foot in the presidential mansion again. Now she was back on familiar ground and took comfort in it. She enjoyed private visits from her most intimate friends. Her brother-in-law Robert Kennedy was a regular caller.

On December 11, 1963, the McNamaras sent over a gift to Jackie at the Harriman house. It was an oil painting of the president by the artist Charles Fox. When Jackie unwrapped it, she was shocked. She did not want it. It was not an issue of whether or not she liked it.

She could not bear to look at it. It was too painful. She sent a hand-written note asking forgiveness for declining a gift "from the man in his cabinet who gave the most (as much as Jack's own brother Bobby gave)" to JFK.

Jackie explained: "I am in a strange locking of horns where I am sure the Secretary of Defense and his wife can outwit me. PLEASE I don't want you to give anything more for Jack—you gave him all—and my consolation is that he will be remembered as great—because of Bob McNamara."

Jackie confided that she could not even bear to display photographs of her husband. "The only photograph I have here of Jack is where his back is turned." She did not hang the oil painting. The picture was on the floor, "propped up against the wall at the little study outside my bedroom. Tonight John came out of my bedroom with a lollipop in his mouth. The picture I love was right in his way—and he took the lollipop out and kissed the picture and said Goodnight Daddy."

That broke her heart. Jackie warned the secretary of defense, "Mr. Fox may find sugary imprints he never painted in, on that picture, but you see why we could never bear to have it near us—it brings to the surface too many things."

Jackie suggested that the McNamaras take back the painting and donate it to the Kennedy Library several years down the road, after the institution was built. "So if you wish to give it to the Library and keep it till then, it would be such an honor—but what I would love most of all—is if both of you who have given so much would give nothing more—except your friendship always."

ON DECEMBER 22, three days before Christmas and a month after the assassination, President Johnson went to the Lincoln Memorial and spoke at a now forgotten candlelight tribute for John Kennedy: "Thirty days and a few hours ago, John Fitzgerald Kennedy, 35th President of the United States, died a martyr's death. The world will not forget what he did here. He will live on in our hearts, which will be his shrine. Throughout his life, he had malice toward none; he had charity for all. But a senseless act of mindless malice struck down this man of charity, and we shall never be the same."

Johnson echoed the words of Lincoln's Second Inaugural Address. "One hundred years ago . . . the 16th President of the United States made a few appropriate remarks at Gettysburg. . . . He lives on in this memorial, which is his tabernacle. As it was 100 years ago, so it is now. We have been bent in sorrow, but not in purpose. We buried Abraham Lincoln and John Kennedy, but we did not bury their dreams or their visions. They are our dreams and our visions today, for President Lincoln and John Kennedy moved toward those nobler dreams and visions where the needs of the people dwell. On this eve of Christmas, in this time of grief and unity, of sadness and continuity, let there be for all people in need the light of an era of new hope and a time of new resolve. Let the light shine and let this Christmas be our Thanksgiving and our dedication. . . . So let us here on this Christmas night determine that John Kennedy did not live or die in vain. . . ."

FOR JACQUELINE Kennedy, it was a quiet, sad, and lonely Christmas. It was the beginning of a long, dark time for her. She should have been supervising the decoration of the White House tree and the historic rooms and hosting festive receptions. Now, after only two Christmases there, she was living in a strange and empty house. But she had not forgotten the handful of people who had meant the most to her and the president.

So she selected special gifts—books, photos, personal mementos—for her intimates. To the secretary of defense, she gave a specially bound copy of *Inaugural Addresses of the Presidents of the United States from George Washington 1789 to John F. Kennedy 1961*. Jackie had the plain Government Printing Office edition custom-bound in maroon leather and the cover gilt-embossed with the presidential seal and the recipient's initials below. It was one of eighty-five copies that were bound this way.

Jackie inscribed it "To Robert McNamara—The President was going to give you this for Christmas—Please accept it now from me— With my devotion always for all you did for Jack. Jackie, December 1963."

To Dave Powers, she inscribed another copy of that book: "With my devotion always for all you did to give Jack so many happy hours— You and I will miss him most. Jackie."

She also gave him a framed set of three black-and-white photographs of Powers playing with John Jr. She inscribed the mat "For Dave Powers—Who gave the president so many of his happiest hours—and who will now do the same for his son, John Jr. With my devotion always—for your devotion to Jack. Jackie, Christmas 1963."

In January 1964, Jacqueline Kennedy went to the imposing, fortresslike headquarters of the Department of Justice, which occupied an entire city block of the south side of Pennsylvania Avenue between Ninth and Tenth streets. She rode an elevator to the fifth floor and stepped into a hallway decorated with huge, vintage WPA-style hand-painted murals depicting scenes from 1930s America. This was the executive-management floor of the department. She proceeded to a doorway marked by a wood sign painted with gold letters. It was the office of the attorney general, her brother-in-law Robert Kennedy. The film crew was waiting. She took her seat in a leather club chair placed near a fireplace hearth. The homey setting and crackling fire made it look as though she was anywhere but in a government office building.

The crew switched on the lights and aimed the cameras at Jackie, and she began to speak. It was for a short film—less than three minutes long—to thank the American people for their condolences and expressions of love for her husband. She thanked them for their letters, promised they would be archived at the Kennedy Library, and said all the usual niceties one might expect a widow in her position to say. Then she caught viewers off guard with an emotional remark that revealed how much she missed him.

In the middle of her statement, she paused and said, "All his bright light gone from the world."

In the finished film, shown at movie theaters throughout the nation, audiences read a superimposed title on the screen: "Mrs. Kennedy Speaks. Thanks 800,000 Who Sent Sympathy." Then Jackie spoke. "I want to take this opportunity to express my appreciation for the thousands of messages, nearly 800,000 in all, which my children and I have received over the past few weeks. The knowledge of the affection in which my husband was held by all of you has sustained me, and the warmth of these tributes is something I shall never forget. Whenever I can bear to, I read them. . . . All of you who have written to me know how much we all loved him, and that he returned that love in full measure."

Jackie promised to write back to her sympathizers. "It is my greatest wish that all of these letters be acknowledged. They will be, but it will take a long time to do so. But I know you will understand. Each and every message is to be treasured, not only for my children, but so that future generations will know how much our country and people in other nations thought of him. Your letters will be placed with his papers, in the library to be erected in his memory along the Charles River, in Boston Massachusetts. I hope that in years to come many of you and your children will be able to visit the Kennedy Library. It will be, we hope, not only a memorial to President Kennedy but a living center, of study of the times in which he lived, and a center for young people and for scholars from all over the world. May I thank you again, on behalf of my children, and of the president's family, for the comfort that your letters have brought to us all. Thank you."

Jackie had decided to make Georgetown her permanent home, and in February 1964 she bought a house of her own at 3017 N Street NW, near the Harriman place. She wanted to live a quiet life. But she could not find the peace she craved to heal her wounds. Both of her Georgetown homes became instant tourist attractions. Indeed, on the day she had moved into the Harriman house, news photographers and filmmakers were waiting outside to snap pictures and make movies of her. A sad image from that day shows her in a black mourning dress, her head held high as she tried to put on a brave smile, while she led her children into their new home and their new lives.

It was only the beginning. The home she purchased became another tourist attraction. In the days ahead, people stopped on the sidewalk and gaped, hoping to peek through the windows or, better yet, to glimpse Jackie entering or leaving the premises. Tour buses clogged the street and unloaded leering passengers. Photographers staked out the house day and night, hoping to take a salable photograph of the widow and her children. Some photographers even snapped images of *other* photographers photographing the house. Photojournalists not only stalked her at her home, but began following her wherever she went.

She had become an American heroine. People continued to mail condolence letters to her. Magazines would not stop publishing articles about her. She had become a public obsession. Jackie could not leave her home without taking the risk that intrusive—and possibly dangerous—strangers might accost her in the street. It happened all over again when she moved from the Harriman place to her new house across the street.

THE TRIBUTES continued through the rest of 1964. To raise support for the future Kennedy Library, Jackie helped organize a traveling exhibition of JFK's favorite mementos. She wrote the foreword to the promotional catalog brochure: "My husband had looked forward to retiring to his library at the end of his time in Washington. Now he will not see the building to be erected in his name."

Once again, *Life* magazine, just as it did so many times when the Kennedys were in the White House, put her picture on the cover and published her essay about the exhibition.

There were more tributes to come. In March 1964, *National Geographic* published a special issue devoted to the assassination. The magazine was a monthly with a long lead time, so it came out much later than the weekly December 1963 and January 1964 issues of *Life*, *Look*, and the *Saturday Evening Post*.

The *Geographic's* esteemed president and editor, Melville Bell Grosvenor, wrote, "His life was such—the radiance he shed—that if we live to be a hundred, we will remember how he graced this earth, and how he left it. Only the future can assign to John Fitzgerald Kennedy his true place in history. When men now boys are old, in distant time beyond the year 2000, they will say, 'I remember, I remember when they brought him home, the murdered President, from Dallas.' . . . Again and again the story will be told—just as I recall my Grandfather Grosvenor, at 92, telling me graphically of how, as a young student at Amherst College in Massachusetts, he traveled by horse and train to the bier of the martyred Lincoln."

IN JULY 1964, Jacqueline Kennedy announced that she was selling her Georgetown house. She had moved in just five months ago. Not only that. She was leaving Washington. Her decision shocked the political and social elites of the nation's capital. The assassination had transformed Jackie Kennedy into a national obsession. Yes, she had enjoyed great popularity as First Lady, but this was something more.

It was irrational, pathological, and even ominous. She hoped that people's fascination with her would die down. She had hoped that one day soon she could stroll the streets of Georgetown, visiting her favorite bookshops, florists, antique shops, and grocery stores as she did in the old days when Jack was a senator. But the harassment got worse every month, until she could no longer endure it. To escape, Jackie left the capital city she once loved and moved to New York City in 1964. So less than a year after the assassination, in an effort to reclaim her private life, Jacqueline Kennedy abandoned the capital. She said good-bye to her Washington days and moved to Manhattan, where she had spent many happy times before her marriage. She promised herself to honor the vow she had made when she had moved out of the White House—she would never set foot there again.

The move did not end the obsession. One relentless photographer specialized in stalking her and made a career out of taking pictures at unguarded moments. For the rest of her life, Jackie Kennedy remained an iconic figure, forever an unwilling star in the spotlight on the American stage.

FOR FOURTEEN months, between November 22, 1963, and January 20, 1965, America did not have a vice president. The Constitution made no provision to create a new one when Lyndon Johnson was elevated to the presidency. Johnson was uncontested as the Democratic nominee for president in 1964, and speculation over whom Johnson might choose as his running mate became one of the most popular guessing games in political circles. He was expected to make the announcement at the Democratic convention in late August 1964.

Robert Kennedy lusted for the spot. He viewed it as a springboard that might launch him to the presidency one day, possibly in 1968 or 1972. Bobby campaigned for it and, in a face-to-face meeting with Johnson, had almost begged for it. His loathing for LBJ had always been obvious, and he had never made much effort to hide it. And Johnson despised him for it. He had no intention of making John Kennedy's little brother his vice president.

But Johnson worried that Jackie might be promoting Bobby for the spot behind the scenes. Johnson had asked Robert McNamara to stay on as secretary of defense. The president considered him one of his most trusted advisers, and he sent him on a secret mission. Go to Jackie Kennedy, who adored McNamara, and take her temperature.

For Johnson, Jacqueline Kennedy had always been the elusive and mysterious object of his desire. Not sexually, of course, but in the sense that he craved her support, her friendship, her presence, and her essence. His frequent calls to her after the assassination, preserved when LBJ recorded them, document his pursuit. Now he wanted a sign of her blessing.

She had given it to him once, that day aboard Air Force One. But ever since the assassination, Jackie had proven resistant to his charms. She had politely declined his various invitations to return to the White House. She was elusive prey, resistant to the pressures of the "Johnson treatment." But she was not immune. Jackie well knew that the most effective way to resist Johnson was from a safe distance, before his cajoling voice or physical presence could work their magic. Jackie had always been flattered and attracted by the attentions of older, powerful men. She understood them. And they understood her. Johnson was no exception.

In his first year in office, Johnson suffered in John Kennedy's shadow. Now he was haunted by the ghosts of Camelot in exile, personified by Robert Kennedy. Thus, Johnson reasoned, he must never allow Bobby to become his running mate in the presidential election of 1964. Johnson wanted to show that he could win on his own.

But what did Jackie think? On August 3, 1964, Secretary of Defense McNamara carried out his assignment and prepared a memo to his files: "On Friday the president said he believed that Jackie had been pressuring Bobby to run for the Vice Presidency. The president had been told by one or more individuals that Jackie had repeatedly stated to Bobby he must run for the Vice Presidency and that it was at her insistence, rather than any desire of his own, that was moving him toward that objective. The President further said he felt, and had been told, that Jackie was very bitter toward him."

McNamara disagreed that Jackie had turned against LBJ. "I replied that I was certain he was wrong on both counts; that I had frequent conversations with her during the past several weeks; I knew what she thought. There was no bitterness toward the president, and she not only was not pressuring Bobby to move toward the vice presidency, but she repeatedly questioned whether it would be wise for him to accept the nomination if it were offered.

"Following my return from Newport yesterday, I called the President and repeated that I could absolutely guarantee that Jackie had no bitterness toward him, and that she had not in the past and would not in the future put pressure on Bobby to move toward the Vice Presidency."

· · ·

AT THE Democratic National Convention in Atlantic City, New Jersey, Lyndon Johnson was wise to secure his own presidential nomination and to select Minnesota senator Hubert Humphrey as his running mate before what became the unofficial "Kennedy night." Delegates were treated to a worshipful documentary about President Kennedy.

When Bobby Kennedy appeared at the podium, he received a thunderous, twenty-minute-long standing ovation. He compared his dead brother to a tragic, lost romantic hero and then quoted Shakespeare's *Romeo and Juliet*:

> *When he shall die*
> *Take him and cut him out in little stars*
> *And he will make the face of heaven so fine*
> *That all the world will be in love with night,*
> *And pay no worship to the garish sun.*

This was Robert Kennedy's coded declaration of war against Johnson. It was his announcement that from this moment on, he was the leader of a Kennedy party—or royal court—in exile.

Ten months after the assassination, on September 24, 1964, the Warren Commission completed the report of its investigation. Almost three hundred thousand words long, it was accompanied by twenty-six volumes containing seventeen thousand pages of testimony, photographs, and exhibits. In sheer bulk it was massive, staggering, overwhelming. Its central findings were that Lee Harvey Oswald had assassinated John F. Kennedy, that two of the three shots he fired had struck the president, that the assassin had acted alone, and there was no evidence that he was part of a conspiracy.

That fall, on November 3, 1964, President Lyndon Johnson got his mandate. He won a crushing victory over Republican senator Barry Goldwater, taking forty-four of fifty states and the electoral vote by a margin of 486 to 52. LBJ had won 61 percent of the popular vote,

43,127,041 to 27,175,754. His victory was total, and he had done it without Bobby Kennedy, whom he had left behind in his wake, rejected and humiliated.

A FEW weeks later, the press inundated the American people with its coverage of the first anniversary of the assassination. *Life* magazine had already given Jackie Kennedy what she wanted last year—the legend of Camelot. This year she cooperated with rival *Look* magazine for a special feature and interview.

Many journalists never got over their crush on John Kennedy. He had flattered their vanity. They could not love a new president who they believed was their inferior in culture and style. They longed for the days of the prince of Camelot.

"I feel suddenly old without Mr. and Mrs. Kennedy in the White

Chief Justice Earl Warren and members of his commission present their report to President Johnson, September 1964.

House," wrote prominent journalist James Reston. "Not only by their sheer verve and joy, the Kennedys imparted their youth to everyone, put a sheen on our life that made it more youthful than it is." Then Reston likened Lyndon Johnson to the aging sheriff in the Gary Cooper film *High Noon*: "Growing older," and reflecting a "less fresh, if no doubt a practical and effective mood." Reston waxed: "All will be well, I feel sure, but it is August, not June."

Reston's cinematic reference is telling. Like Mary McGrory, Daniel Patrick Moynihan, Charlie Bartlett and others who echoed the "things will never be the same" and the "we will never be young again" themes, the journalists who mythologized Kennedy emphasized his intangible star quality—a style, an attitude, a mood—over the details. In their swooning eulogies, none included something so practical and unromantic as a list of his failures and accomplishments.

THE YEAR 1968 was a dark time for Jacqueline Kennedy. In April, Martin Luther King Jr. was assassinated in Memphis. When King had attended President Kennedy's funeral, the civil rights leader told his wife, Coretta, that one day this would happen to him. King's prophecy had come true. Now, five years later, Jacqueline Kennedy attended *his* funeral and comforted *his* widow.

Lyndon Johnson's 1964 electoral triumph proved ephemeral. His massive escalation of the war in Vietnam turned much of America against him. Senator Eugene McCarthy challenged him for the nomination, and when Robert Kennedy saw that LBJ was vulnerable, Kennedy joined the race. In March 1968, Johnson surprised the nation by announcing that he would not seek reelection. Robert Kennedy rejoiced, believing that Johnson had reinvigorated his own campaign for the Democratic nomination and had perhaps handed him the presidency.

In June, Robert Kennedy's days ended in the pantry of the kitchen of the Ambassador Hotel in Los Angeles the night he won the California primary. His murder by an assassin's pistol shots brought back all the old feelings. Jackie was devastated and bitter. Like Jack, he had been shot in the head. Since her husband's assassination, Bobby had been her closest friend and adviser. Now, five years after she had lost Jack, Bobby had also been taken from her. She felt alone.

THINGS WERE not going well with the planning of the Kennedy Library. Jackie was not happy with the direction Harvard University was taking it. On July 16, 1968, she complained in a long, handwritten letter to Robert McNamara: "Memorials can never replace men—& one just argues to exhaustion with a lot of wooden people who will have it their way in the end anyway. . . . What is happening is that it is on its way to becoming the deadliest place in the world."

She included for McNamara's scrutiny minutes from the previous meeting of the board of trustees: "I don't know how watered down they are—if all our objections & my explosion was deleted." Jackie was worried that Harvard would have too much control over the institution. "Do you realize—we gave Harvard $20 million—plus 13 acres of land they have always coveted & could never get . . . in return they named their dismal school of Government after Jack—and in so doing, assured they'd always keep the Institute under their thumb. [Supreme Court justice] Byron White says we should have known that would happen all along—but we were in shock & in a hurry. . . . And we gave them the papers of President Kennedy—which forever will be one of their greatest treasures."

Jackie complained, "Do you know I was the only Kennedy allowed to be on the committee—because Harvard feared the Kennedys might 'use it as a springboard for the dynasty?' " Robert Kennedy had been assassinated little more than a month ago. Jackie noted the bitter irony in that. "Well there aren't many people left to use the springboard." She vowed to battle Harvard. "I am going to make such a fuss—do something so Machiavellian—once you tell me the best way to deal with these people."

Jackie fantasized about taking back JFK's papers: "I can think of nothing that would give me more pleasure than to leave the Harvard Corporation with its mouth watering—& put President Kennedy's papers in Washington—with [National] Archives—or in an adequate working Library-Warehouse somewhere in Massachusetts—Other sites were offered us—or in Ireland." Jackie relished the thought of pulling the plug on Harvard. "And Jack and Bobby would not mind the discomfort at the Corporation—it would give them a laugh in heaven."

In 1968, Jacqueline Kennedy remarried. Her choice of husband, the older, dodgy Greek shipping millionaire Aristotle Onassis, horrified many of her friends and shocked the American people. To her fans, it was a desecration of JFK's memory—the death of Camelot. She became prey to the international media.

Once upon a time she had been one of them, the "Inquiring Camera Girl" from the *Washington Star* newspaper. Once, she had enticed some of the best photographers in America—Lowe, Avedon, Shaw, and others—to employ their art in her service. She had manipulated the most prestigious photojournalism magazines in the country to present her and Jack in the best possible light. Now the tables had been turned.

Jacqueline Kennedy never again spoke publicly about November 22, 1963. She gave no interviews about the assassination, appeared in no television specials, and refused to commemorate the event. Every five years, when the national media marked the fifth, tenth, fifteenth, twentieth, twenty-fifth, and thirtieth anniversaries of her husband's death, she remained in seclusion and maintained a sphinx-like silence. Despite her substantial talent as a writer and her subsequent work as an editor, she wrote no books or memoirs. She survived the president by nearly thirty-one years, dying in May 1994 at the age of only sixty-four. In Arlington National Cemetery, she lies buried beside her husband, near the eternal flame she lit in November 1963.

EPILOGUE

"ALL HIS BRIGHT LIGHT GONE FROM THE WORLD"

In Dallas, the Texas School Book Depository still stands at the corner of Elm and Houston Streets. The building was fortunate to survive the aftermath of the assassination of President Kennedy. Ashamed that the murder had been committed in their city and embarrassed that their police department had allowed Lee Harvey Oswald to be shot to death right under their noses, many leading citizens of Dallas wanted the Book Depository to be torn down.

To them, the building was an ugly landmark of the day that Dallas could never forget, one that they feared would scar the city's reputation forever. But cooler heads prevailed, and the Book Depository was preserved for history. Its iconic silhouette looms over Elm Street, but the famous Hertz clock atop the roof—dismantled long ago—does not flash the bottom of the noon hour each day at 12:30 P.M. The Depository no longer serves as a warehouse for textbooks. Like Ford's Theatre in Washington, where John Wilkes Booth assassinated Abraham Lincoln, the Texas School Book Depository is now a museum.

Today an institution named the Sixth Floor Museum occupies the place that the Book Depository's most famous employee, Lee Harvey Oswald, made infamous. Once a controversial and unwelcome reminder of Dallas's worst day, today the museum is an important part of the city's cultural landscape and has attracted millions of visitors. It is not a shrine to an assassin and does not sensationalize the crime. Instead it is a responsible museum that frames the events in Dallas within the broader context of President Kennedy's life story, American politics, and the history of 1960s America. The museum honors Kennedy, not his assassin. What a mistake it would have been, fifty years ago, in the heat of passion, to have torn the building down.

As Oswald did, you can take an elevator to the sixth floor, and there you can retrace his footsteps to the wall of windows facing Elm Street. But you can no longer gaze out the window from which he shot the president. To protect the sniper's nest from vandals and souvenir hunters, a Plexiglas barrier now surrounds Oswald's corner window. You may, however, stand at the window beside it, look down to the street, and imagine what Oswald must have seen on that beautiful fall afternoon of November 22, 1963. How close President Kennedy must have appeared to him in the eyepiece of the rifle's scope. Indeed, in person, all of Dealey Plaza is smaller than it appears to be in photographs.

There are other things to see in Dallas: Oswald's escape route from the Book Depository to his rooming house; from there his path to the street where he shot police officer J. D. Tippit; and from there his footsteps to the Texas Theatre, where he was captured. There is another site to see: the haunted place where on the night of Thursday, November 21, Oswald decided to carry out his plan—Ruth Paine's house, still a private home, where Oswald slept on the eve of the assassination and from which he emerged the next morning with his rifle, determined to kill a president.

But it is the Texas School Book Depository and Dealey Plaza that exert the most powerful gravitational force over visitors. From Elm Street, tourists gaze up at the sixth-floor window, calculating the trajectory of the shots. The conspiracy-minded lurk behind the fence on the Grassy Knoll, speculating about a second gunman and whether he could have fired upon the presidential limousine from there.

At the decorative roadside pergola along Elm Street, they mount the low concrete pedestal where Abraham Zapruder once stood as they pan with their own cameras from left to right while they replay the famous assassination home movie in their heads. Then they step into the middle of Elm Street, dodging traffic in order to stand on the painted X that marks the exact spot where President Kennedy was shot in the head.

Inside the museum, they approach the sniper's nest, listening for the echo of the three rifle shots and the hollow ping of three empty cartridge cases bouncing on the wood floor, sounds heard by several of Oswald's coworkers that day.

Once a year, on November 22 at 12:30 P.M., on the date and time of the anniversary, Dealey Plaza resembles a flea market or a street bazaar. Vendors push trinkets and souvenirs, including bootleg autopsy photographs of John F. Kennedy's corpse. Authors peddling conspiracy theories hawk books and magazines to passersby. Assassination buffs make annual pilgrimages to attend conspiracy-oriented conferences, as if these annual rituals—through a kind of harmonic convergence—will finally reveal the truth.

There is nothing else like it in America—Dealey Plaza is the liveliest assassination site in the nation.

In contrast, in Washington, D.C., on every April 14 at around 10:15 P.M., the anniversary of the murder of Abraham Lincoln, the street in front of Ford's Theatre is deserted. Tourists make no pilgrimage there. Only a handful of people—no more than five or ten—come to maintain a nighttime vigil at the place where Lincoln was shot or to sit on the steps of the Petersen House where he died. One person, in homage to Walt Whitman's poem, usually leaves a bouquet of fresh lilacs there for Father Abraham.

• • •

IN THE fall of 2013, America's basements, attics, and closets will disgorge millions of mementoes of the Kennedy assassination. Long-hidden souvenirs overlooked for decades will be resurrected for the fiftieth anniversary.

On November 22, 1963, the American people experienced the assassination of John F. Kennedy as a shared event. On the same day at the same time, an entire nation read the same stories, saw the same photographs, listened to the same radio broadcasts, and watched the same images on television. For four days straight, the three national television networks—CBS, NBC, and ABC—immersed the American people in a shared moment of national grief. For the first time in U.S. history, the medium of television unified a nation through its coverage of a historic event. Similarly, the great weekly picture magazines, *Life* and *Look*, published photos and stories seen by tens of millions of people. The nation's newspapers, some printing several editions per day, published several hundred million copies.

Once the story was over, people did not throw away their old newspapers, magazines, and commemorative publications. Instead, they preserved them as iconic family heirlooms, as time capsules for future generations. There is no better way to experience the utter shock, disbelief, and horror caused by the assassination of President Kennedy than by returning to these original sources and imagining what it was like to be alive and reading the afternoon editions of November 22, 1963.

Other relics of the Kennedy assassination—those suppressed for the last half century—are unlikely to see the light of day for the fiftieth anniversary. Unlike the major relics of the Lincoln assassination, which the National Park Service displays in its museum at Ford's Theatre and which the American people have been able to see since the early 1900s, the principal relics of the Kennedy assassination have been hidden from the American people by the National Archives for fifty years. At Ford's, millions of Americans have seen John Wilkes Booth's revolvers, repeating carbine, pocket calendar, handwritten diary, the knife he used to stab Major Rathbone in the president's theater box, and the Deringer pistol he used to assassinate Abraham Lincoln. The museum also displays Lincoln's coat and bloodstained pillows from the Petersen House. The U.S. Army Medical Museum displays the bullet that killed Lincoln, medical instruments used on the president, and bloodstained shirt cuffs worn by one of the doctors who treated him.

The National Archives has, in contrast, thrown a veil of secrecy over the artifacts of the Kennedy assassination. The evidence of the assassination remains buried in the vaults of the Archives in an annex building in suburban Maryland. There, stored in acid-free archival boxes to ensure their long-term preservation, are the clothes President Kennedy wore on the afternoon of November 22: his suit jacket, scarred in the back with the bullet hole from Oswald's second shot, his necktie, his striped dress shirt, and other garments and accessories, still stained with dried blood.

The other Warren Commission exhibits are stored there too, including Lee Harvey Oswald's rifle, his pistol, his letters and writings, and his clothing, including the bullet-damaged sweater he wore when Jack Ruby shot him. None of these items has ever been put on public display. Also hidden away in the Archives is another relic, perhaps the most iconic symbol of the assassination, Jackie Kennedy's pink suit, and her other clothing and accessories.

The Archives seeks to suppress these from public view for another century. On November 22, 1963, Mrs. Kennedy said that she wanted the world to see her suit. Today, the National Archives wants to ban it from the sight of the American people for another century, until 150 years have passed since the assassination.

In the mid 1960s, Jackie or her mother, Janet Auchincloss, sealed in a cardboard box the suit, the other garments from that day, and a list of the contents, and mailed the container to the National Archives. For the next three decades, Jacqueline Kennedy never sought to regain possession of the clothes, supporting the conclusion that they were donated to the Archives in the 1960s. But in 2003, the Archivist of the United States yielded to Caroline Kennedy's claim of ownership, and executed an agreement banning all access to the material.

Not all of the evidence from the assassination is at the National Archives. One unique, macabre item from the collection is missing—President Kennedy's brain. During the autopsy, doctors removed it and sealed it in a leakproof stainless-steel cylindrical container with a screw-top lid. They failed to place his brain back in his head for the funeral, and so John Kennedy was buried without it. For a time, the steel container was stored in a file cabinet in the office of the Secret Service. Then it was put in a footlocker with other medical evidence and transferred to the National Archives, where it was placed in a secure room designated for the use of JFK's former secretary, Evelyn Lincoln, while she organized his presidential papers.

Then, on some day prior to 1966—no one knows when—the locker, with all its contents, disappeared. Lyndon Johnson's attorney general, Ramsey Clark, ordered an investigation, which failed to recover the brain but did uncover compelling evidence suggesting that former attorney general Robert Kennedy, aided by his assistant Angie Novello, had stolen the locker and its contents, including not only his brother's brain but also a number of medical slides and tissue samples. They have never been seen since.

Robert Kennedy did not abscond with these materials to hide evidence of a conspiracy to kill the president. It is much more likely that he took them to conceal from the American people any evidence of the hitherto unknown extent of JFK's serious health problems, illnesses, and medications.

LIKE ARTIFACTS from the Lincoln assassination, relics from the Kennedy assassination have become prized collector's items. When the Secret Service sent the Lincoln Continental limousine back to the manufacturer to be refurbished, one or more souvenir hunters sliced bloodstained swatches of blue upholstery from the backseat and sold them to eager buyers, complete with letters of authenticity. The limousine is now in a museum, but first it had been altered substantially after November 22 for use by the new president, Lyndon Johnson. Soon he stopped riding in it.

A car blanket from JFK's limousine, embroidered with the presidential seal, has been offered for private sale. And recently the American and presidential flags that flew from the front fenders during President Kennedy's last motorcade were sold to a private collector for several hundred thousand dollars. The revolver Jack Ruby used to slay Lee Harvey Oswald, valued at several hundred thousand dollars, reposes in a private collection, as do the suit, hat, and even the shoes Ruby wore the day he committed the murder.

On December 1, 1963, Jacqueline Kennedy sent a handwritten letter to Nellie Connally in care of Parkland Hospital, where the wounded governor remained bedridden. "We loved them every way that a woman can love a man, haven't we," Jackie wrote, and we were "so fortunate to have them in our arms at that terrible time." The letter was so intimate that Mrs. Connally refused to publish the rest of it in her 2003 memoirs of Dallas. Later her family sold the letter for several hundred thousand dollars.

THROUGH IT all, the myth of Camelot endures. In time, Theodore White questioned his role in creating the legend. "Quite inadvertently, I was her instrument in labeling the myth." He knew he had crossed the ethical line that divided journalist from hagiographer. "So the epitaph of the Kennedy administration became Camelot—a magic moment in American history, when gallant men danced with beautiful women, when great deeds were done, when artists, writers and poets met at the White House, and the barbarians beyond the walls held back. Which, of course, is a misreading of history. The magic Camelot of John F. Kennedy never existed . . . no Merlins advised John F. Kennedy, no Galahads won high place in his service."

And yet White could not let go. Fifteen years later, in the epilogue to his memoir, *In Search of History*, he was still in Camelot's thrall.

Writing about himself in the third person, he reflected on his historic meeting with Jackie: "The storyteller was unaware of passing a divide as he left the Kennedy compound that night. It was still raining as he reached the main highway to New York, and there he was on familiar ground. Except for the sadness and the personal ache, all seemed as it had been before. He did not know then that he and everyone else in America had, that week, passed through an invisible membrane of time which divided one era from another, and that Jacqueline Kennedy's farewell to Camelot was farewell to an America never to be recaptured."

But was it only a dream? Or does its embrace by millions of Americans—even today—make it true? Passing time and a litany of unwholesome revelations about President Kennedy's private life have not extinguished the myth. Like the eternal flame at Arlington Cemetery, it burns today. The truth once hidden behind the wizard's curtain may have tarnished the legend, but the forces that Jacqueline Kennedy unleashed to preserve and romanticize the memory of her husband are too powerful to ever put the "one brief shining moment" back into the bottle.

IN THE aftermath of the Warren Commission, many people found it hard to believe such an inconsequential man as Oswald could change history in such a monumental way. Many thought, and still think, this crime was too great to be explained by random chance. They wanted a more profound and complicated explanation. This was not unusual. For more than two and a half centuries, Americans have turned to numerous conspiracy theories to explain catastrophic events or troubled times. In the 1960s, many found Oswald's journey to the Soviet Union and his interest in Cuba suspicious. Was his murder just two days after the president's a coincidence? Or was it the result of a plot to silence him?

These and other questions provoked some critics to doubt the conclusions in the Warren Report, and to question even the most simple, obvious, and persuasive evidence of Oswald's guilt. Many of the conspiracy theorists have devoted their lives to proving that John F. Kennedy was the victim of one plot or another. Many of their theories—and there are a dizzying number of them—contradict one another. According to the most popular ones, the president was killed by a Russian or Cuban Communist conspiracy, anti-Castro Cuban exiles, the anticommunist American right wing, organized crime— the Mafia, the CIA, the FBI, the U.S. military, Texas millionaires in the oil business, the "military-industrial complex," or even Vice President Lyndon Johnson.

Some conspiracy theorists claim that Oswald fired no shots in Dealey Plaza. They argue—despite the considerable evidence against him—that he was framed. Others admit that Oswald fired the shots but insist he was not the lone gunman, and that additional snipers— two, three, four, or more—fired as many as sixteen rounds, even though most witnesses said they heard only three shots. One theory asserts that there were *two* Oswalds—the real one, and the other an imposter sent back from Russia. A few critics accuse Kennedy's Secret Service agents, the U.S. Navy doctors at Bethesda hospital, and even Dallas police officer J. D. Tippit of being part of the conspiracy.

Some of the theories rely on falsified evidence. Others are based on lies. Some theorists believe the same master conspiracy behind the Kennedy assassination controls other important and nefarious events in American life, including other subsequent assassinations. But all of the theories have one thing in common. They reject the proven role that chance, luck, randomness, coincidence, or mistake have played in human history for thousands of years. To them, there are no accidents in life. Everything that happens can be explained by conspiracy.

Just as the conspiracy theorists have questioned everything about the assassination, so must a reader question *their* writings with equal skepticism. Today we know much more about the assassination of President Kennedy than the members of the Warren Commission did. More information and sophisticated advances in science and technology have illuminated the crime and its evidence in new ways.

No one, after all these years, has yet disproved the key conclusion of the Warren Commission: Lee Harvey Oswald was the assassin and he acted alone.

Indeed, in the future—fifty or one hundred years from now—it is more likely that the discovery of any new evidence, along with further scientific advances, will only strengthen the case against Lee Harvey Oswald as the lone gunman.

One great mystery remains: Why? Why did Oswald assassinate John F. Kennedy? Oswald did not tell his wife, mother, or brother when they visited him in jail. He did not reveal his motives to the Dallas police, the Secret Service, or the FBI when they questioned him. Perhaps, embittered, he killed the president to impress Soviet officials who placed so little value on him after his defection. But by the fall of 1963, Oswald had long soured on life in Russia and had renounced the corruptions of Soviet Communism. Could he have wanted to impress Fidel Castro and seek political asylum in Cuba, fantasizing that he would become a revolutionary hero? Or could it be possible that Oswald came under the influence of others—an individual or a group—not as a knowing, paid assassin or agent of a conspiracy, but as someone who listened to whisperings in his ear telling him that any man who killed a president would go down in history?

Perhaps his motive was not politics but fame. Anyone who remembers John Kennedy remembers the man who murdered him. Or maybe Oswald was one of America's first glory killers, obsessed with Kennedy's glamorous, movie-star-like celebrity. By killing the president, Oswald's deluded mind sought to merge their identities, hoping that some of the magic Oswald never possessed—effervescence, popularity, wealth, success, and even greatness—might rub off on him. Oswald longed to possess the traits that were never meant to be his.

Or in the end, perhaps the reason is much simpler and more fundamental and lies beyond rational human understanding: Lee Harvey Oswald was evil.

It is impossible to know. Whatever his motives, Oswald took them with him to the grave. If he could return today to the scene of his crime, he would be pleased to see that, half a century later, he remains the subject of endless fascination and speculation. He taunts us still, defying us to solve the mystery of the *why* that he left behind. Unlike John Wilkes Booth, the assassin of Abraham Lincoln, Oswald did not leap to the stage, boast of his crime, and wave a bloody dagger before our eyes. No, Oswald struck from the shadows. Then he robbed us of the rest of the story.

The assassination of President John F. Kennedy on November 22, 1963, is as compelling as any drama written by William Shakespeare. It is the great American tragedy.

JOHN F. KENNEDY'S unfinished life was cut short before he could fulfill his potential. He was just forty-six years old. The nation mourned the death of its young president, not only for the loss of what he was—but also for the loss of what he might have become. JFK loved America. He was an optimist about the country's future. He had shown signs of greatness. If he had lived and won reelection in 1964, he would have served until January 20, 1969. One can only speculate what he might have accomplished if he had had more time.

Many people, especially those who lived through it, saw the Kennedy assassination as a dividing line in our history and November 22, 1963, as the day when something went terribly wrong in American life. They believe the murder ushered in a dark era and set in motion a series of awful events: the escalation of the Vietnam War, civil unrest, racial violence, and five years later, the assassinations of Martin Luther King Jr. and of the late president's own brother, Senator Robert Kennedy. We can look back in wonder, but we will never know the ways in which the death of President Kennedy altered the future course of American history.

. . .

A YEAR after the assassination, reflecting on her husband's life, Jacqueline Kennedy said, "I realized that history was what made Jack what he was. You must think of him as this little boy, sick so much of the time, reading in bed, reading the knights of the Roundtable. . . . For Jack, history was full of heroes. And if it made him this way—if it made him see the heroes—maybe other little boys will see."

She recognized that the assassination had transformed him into a hero too. "Now, I think I should have known that he was magic all along—but I should have guessed that it would be too much to ask to grow old with [him] and see our children grow up together. So now, he is a legend when he would have preferred to be a man."

Jackie contemplated the meaning of his life. "John Kennedy believed so strongly that one's aim should not just be the most comfortable life possible—but that we should all do something to right the wrongs we see—and not just complain about them. We owe that to our country . . ."

"He believed," Jacqueline Kennedy said, "that one man can make a difference—and that every man should try."

IN GEORGETOWN the house from which John and Jacqueline Kennedy set out on the journey that began on January 20, 1961, and ended on November 22, 1963, still stands. Over the last half century, little about it has changed. If you go there today, perhaps on a chilly fall evening in late November, when the crisp, fallen leaves crinkle underfoot, the twin lamps beside the front door still burn, still glowing in the darkness of the night.

ACKNOWLEDGMENTS

I thank my most important reader, my wife Andrea Mays. Despite her busy life as a professor, author, and mother, Andrea helped shepherd all three books in my presidential trilogy—*Manhunt, Bloody Crimes,* and *End of Days*—to publication. She read every page of this book more than once, from first to final draft, and her editorial comments, from overall storytelling advice to meticulous line edits, were indispensable and improved the book in countless ways. Our boys, Harrison and Cameron, also helped me tell this story by assisting on my book for young adults—*The President Has Been Shot!*—that preceded *End of Days*.

It is impossible to think about the death and funeral of John F. Kennedy without recalling the death of another president ninety-eight years earlier. The upheaval that followed the assassination of Abraham Lincoln, and the unprecedented national mourning unleashed by the events of April 14 and 15, 1865, was not seen in this country again until November 22, 1963. These two profound national tragedies have much in common. Indeed, the parallels connecting these stories are so striking that I began to think of them as bookends to one great sad, American tale of loss, legend, and myth. My previous books on the assassination of Abraham Lincoln, the chase for his killer, his funeral, and the national death pageant that transformed him from man to martyr made it possible for me to write about John Kennedy.

A number of people helped bring my version of John—and Jacqueline—Kennedy's end of days to print. Regarding Jackie it is easy to forget that, although she survived the gunfire in Dealey Plaza, the life she knew ended that day. This book is about her, too.

My "first readers" Michael F. Bishop, Ronald K.L. Collins, and David Lovett read the manuscript early on and offered many helpful comments and insights. Michael, former Congressional staffer, White House veteran, and former executive director of the national Abraham Lincoln bicentennial commission, brought his extensive knowledge of the presidency and the modern executive mansion to bear to my advantage. Ron, one of the nation's preeminent experts on the First Amendment, is a prolific author of books not only about law, but also on American legends of another kind, including Lenny Bruce and the Beats.

David Lovett, Washington lawyer, lobbyist, and association executive, scrutinized the manuscript with exquisite care and went far above and beyond the call of duty to assist me. He is an expert on the Kennedy and Lincoln assassinations, and he owns what I believe is the finest and largest private library in the world on the murder of JFK. In his spare time, he is one of the top historical researchers in Washington and one of the city's best-kept secrets. He threw open the doors and gave me total access to his vast archive. Whatever I needed—books, pamphlets, documents, images, recordings, ephemera, or objects of material culture—he provided. He is more knowledgeable than anyone alive about the bibliography and historiography of the Kennedy assassination. We have been friends for almost thirty years, and he is the only person I know who matches my obsessive zeal for tracking down obscure historical rarities. His uncanny research skills made *End of Days* a better book. In the words of Abraham Lincoln, David gave his "last full measure of devotion" to this project, and for that I thank him.

My friend Richard Thomas narrated all three audio books in this series. He is one of the finest actors of his generation and, whether he was busy in theater, television, or film, Richard always took the time to lend his great American voice to my words. His artistic choices made the text come alive. No one could have done it better, and I thank him for his generosity.

The late Wesley J. Liebeler, my professor and mentor at the UCLA School of Law, served as an assistant counsel on the Warren Commission. His insights provided a rare insider's perspective on the murder. Jim had an irreverent and irrepressible sense of humor, and I wish he were around today to read how some of the more outré conspiracy theorists have theorized that I, as Jim's protégé, must be part of the government conspiracy to cover up the true history of the Kennedy assassination. I miss Jim's maniacal, cackling laugh. I owe him much. Not only did Jim pass on to me the secrets of the Warren Commission, he gave me something far more precious. He introduced me to my wife.

Fellow UCLA Law alumnus Vincent Bugliosi is one of the finest prosecutors in American history. Vince, a three-time Edgar Award winner, is the author of one of the best nonfiction crime books ever written, one that is also one of the most frightening books of the twentieth century, *Helter Skelter: The True Story of the Manson Murders*. His monumental *Reclaiming History: The Assassination of President John F. Kennedy*, is one of the most important books about the event. He is a remarkable and generous friend, and I have enjoyed our long conversations about the events surrounding November 22.

Thanks to former editor of the *Los Angles Times* Shelby Coffey III and the great investigative journalist Edward Jay Epstein for an unforgettable conversation about Lee Harvey Oswald and the evidence against him. Ed is a brilliant historian of the assassination and his books are essential reading.

I am indebted to Clinton J. Hill, United States Secret Service, for several conversations about what he saw and did on the afternoon of November 22. He is a great American, and a brave but humble man. No one misses President Kennedy more than Clint Hill. He was there, and he knows.

John Seigenthaler and Charles Overby extended many courtesies to me in Nashville and Washington, D.C. John talked about what it was like to know John and Robert Kennedy, and he shared stories about being alone with them in JFK's last house in Georgetown. His insights on Jackie Kennedy were priceless. John speaks with touching eloquence about what it felt like to live through November 22.

Thanks to Tom Ingram, former editor in chief of Nashville's David Lipscomb High School newspaper, for providing me with an original issue of the *Pony Express* extra that covered the Kennedy assassination.

Jessica Kline assisted me with gracious expertise and good humor whenever I needed help with computer problems. Amy Hart was invaluable in producing all the printed materials I needed to revise the manuscript from first to final draft.

My literary agent and good friend Richard Abate worked tirelessly to bring *End of Days* to publication. The great agents believe that once they sell a book, their work is not done—it has just begun. This is our sixth book together. On all of them, Richard has stood beside me and he has, with great taste, humor and a historian's eye, made them better books. I always look forward to our strategy sessions at our favorite classic New York City steakhouses.

My editor Henry Ferris has been with me on all three of my books about presidential history. By now I have inflicted upon him a century's worth of American tragedy, death, and mourning. I promise that someday I will write a happier book. Until then, I owe him my thanks for his patience, kindness, and invaluable contributions along the way.

Henry's lieutenant Cole Hager assisted in gathering the photos, bringing the manuscript to final draft for publication, and getting it into production. During a hectic process he was always cool under fire, and I thank him.

Sharyn Rosenblum has been with me on all my books and remains the best publicist in the business. I have fond memories of a memorable dinner at Martin's Tavern, where we outlined the campaign for this book, and then went on a midnight walk past John and Jackie Kennedy's last Georgetown house a few blocks away, before strolling down to the old C & O Canal. The spirit of the Kennedys still lingers in their old neighborhood.

Martin's still serves customers at the same tables once occupied by Lyndon Johnson, Sam Rayburn, Richard Nixon, and a young unmarried senator named John F. Kennedy. When Kennedy dined there alone, he often sat at the first table to the right—the half booth or "rumble seat." Legend has it that he proposed to Jacqueline Bouvier at another window booth in the restaurant. My friends at Martin's—owner Billy Martin and manager Joseph Filosa—have for years been hospitable hosts to me, and to the ghosts of Camelot who still dine there.

At the Monocle, the famous Capitol Hill restaurant and watering hole for several generations of American political and government leaders, owner John Valanos and manager Nick Selimos gave me a home away from home during the time I wrote *End of Days,* and all my other books.

Thanks to my first-class legal team of Eric S. Brown and Michael I. Rudell, and also Jonathan D. Lupkin. I can always rely upon their counsel.

Thanks to my friend and fellow Washingtonian Mark Vargas for his energetic promotion of my books. We have spent hours in conversation discussing our mutual fascination with 1960s American politics, John Kennedy, Lyndon B. Johnson, and Jacqueline Kennedy. Sean Langille explained to me the mysteries of social media and other new forms of communication.

It was Douglas Brinkley who first suggested, one day in the summer of 2007 as we stood together on a Washington, D.C., street corner, that I write about the Kennedy assassination someday. He also gave valuable suggestions on how to think about this story. On another occasion, we enjoyed a marathon dinner where we talked about the life and legend of Jackie Kennedy.

I can trace the origins of this book back further, to my childhood. In the fall of 1963, I was almost five years old. I remember nothing of November 22, although that afternoon I must have been at home, watching news coverage with my mother. I do remember this. Across the street from our house lived my two favorite childhood play-mates, Ourania and Evanthia Malliris. The girls were a few years older than I. Their father was a conservative Greek grocer who did not allow them to watch television. His daughters, he dreamed, would attend prestigious universities one day—Harvard was one of the ones he mentioned—and idling in front of a TV screen had no part in that plan. On Sunday, November 24, my mother told me that the girls were coming over to our house because they had received special permission to watch something on tele-vision. It must very important, I thought, and I asked why. My mother said that the president of the United States had died, and that we were going to watch a horse-drawn carriage carry his coffin to the U.S. Capitol and then watch the memorial service there.

Several years later my mother, Dianne, led me led me to what she called her "morgue": a tall, floor-to-ceiling closet with a sliding door that concealed several shelves piled with vintage newspapers, magazines, picture books, photographs, and ephemera. She was a painter, and these were some of her references and sources for ideas. When I was eight or nine years old, I discovered a treasure trove in that closet—her time cap-sule of materials that she had collected about the assassination of president Kennedy. Mesmerized, I paged through old *Life* and *Look* magazines from the fall of 1963. With care, I opened long-folded newspapers, their pages browned and brittle, and read their frightening headlines. I did not know much about President Kennedy, and did not un-derstand the significance of everything I saw, but I knew from my mother's tears that something terrible had happened.

Every year, when late November came around, my father, Lennart Swanson, shared with me his memories of the day President Kennedy was shot and entranced me with stories about where he was at 12:30 P.M.—in Larsen's restaurant on the northwest side of Chicago—and what had happened there and in the city that afternoon and evening. He bought me the Kennedy "Eternal Flame" plug-in, electric night-light pictured in this book. He also gave me the bright-red steel *Chicago American* newspaper vending rack that I wrote about in the book and that remains, to this day, filled with a stack of copies of the edition from that unforgettable Friday afternoon.

And so I thank my parents, who inspired me long ago to tell the story that you now hold in your hands.

James L. Swanson
Edgartown, Massachusetts
August 1, 2013

BIBLIOGRAPHY

———>•◦•<———

A NOTE ON SOURCES

The bibliography of the Kennedy assassination is enormous. There are thousands of books, magazines, pamphlets, and articles on the subject. No bibliography has ever listed them all. No library owns them all. One of the finest and most extensive collections in the world is not even in a public institution, and is the private library assembled by Washington, D.C., attorney, lobbyist, and association executive David Lovett. In addition to these published sources, millions of pages of documents in government files and elsewhere pertain to the assassination of President Kennedy. No person has read them all. No one ever will. Thus, any general reader who wishes to learn more about the subject must be selective.

On November 22, 1963, the American people experienced the assassination of John F. Kennedy as a shared event. People learned the news in a limited number of ways: by word of mouth; by telephone; by listening to the radio; by watching small, black-and-white television sets that received no more than four or five channels; and by reading daily newspapers and weekly newsmagazines. On the same day and at about the same time, almost the entire nation read the same stories, saw the same photographs, and watched the same televised images. For four days straight, the three national television news networks—CBS, NBC, and ABC—immersed the American people in a shared moment of national grief. For the first time in U.S. history, the medium of television unified a country through saturation coverage of a historic event. Similarly, the great weekly photojournalism magazines, *Look, Life,* and the *Saturday Evening Post,* published images and stories seen by tens of millions of people.

Once the story was over, people did not throw away their old newspapers and magazines. Instead, they preserved them as iconic family heirlooms, as time capsules for future generations. All over the country, people put them away in basements, closets, and attics, where they can still be found today. There is no better way to learn how America experienced the assassination of President Kennedy than by going back to these original sources. As you turn the pages of these old publications, now unparalleled artifacts of how the death of the president was reported and experienced, you will travel back in time and know what it was like to be alive on November 22, 1963. Later, the two major wire services released oversized, commemorative hardcover books that seemed to find their way into half the households in America. From United Press International came *Four Days: The Historical Record of the Death of President Kennedy,* and from the rival Associated Press came *The Torch Is Passed: The Associated Press Story of the Death of a President.* Across the nation, local newspapers mailed millions of copies of these low-priced books to their subscribers.

The first major book on the Kennedy assassination was the *Report of the President's Commission on the Assassination of President John F. Kennedy* (known popularly as the "Warren Report"). Published in 1964 by the official U.S. government commission appointed by President Lyndon B. Johnson to investigate the Kennedy assassination, and chaired by Chief Justice Earl Warren, the *Report* consists of one volume of findings supported by twenty-six volumes of testimony, evidence, and exhibits. The Warren Commission concluded that Lee Harvey Oswald killed President Kennedy, that he acted alone, and that there was no evidence that he was part of a conspiracy. The publication of the Warren Report inaugurated a deluge of books on the Kennedy assassination that has not subsided to this day.

The following bibliography lists most of the books that I consulted while researching *End of Days.* For readers who want to learn more, I recommend several titles. The twenty-seven volumes published by the Warren Commission remain an essential source. Often dismissed by critics as out of date or ridiculed by the conspiracy minded as part of the cover-up, these volumes cannot be ignored, and contain vital information. Few people have ever actually read them. The unwieldy set occupied an entire shelf, and the books were expensive and difficult to cross-reference. Today, the work of the Warren Commission is available in searchable, electronic form.

For more on the sites connected to the assassination, including the Texas School Book Depository, and for a brief introduction to the story, see Conover Hunt's *Dealey Plaza National Historic Landmark Including the Sixth Floor Museum.*

The majority of books published on the Kennedy assassination advance one conspiracy theory or another. The two best non-conspiratorial books are Gerald Posner's *Case Closed: Lee Harvey Oswald and the Assassination of JFK* and Vincent Bugliosi's *Reclaiming History: The Assassination of President John F. Kennedy.* For readers deterred by the titanic size of the Bugliosi book, I recommend his shorter account, *Four Days in November.* For a brief and excellent introduction to the subject, on how it was reported, and for insightful commentary on the history and psychology of JFK assassination conspiracy theories, see Peter Knight's *The Kennedy Assassination.* For an incisive analysis that places these theories within the larger context of the modern obsession with conspiracy interpretations of a number of twentieth-century events, see Knight's *Conspiracy Culture: From Kennedy to the X-Files.*

Investigative journalist Edward J. Epstein remains one of the most important scholars of the Kennedy assassination. Beginning in 1966 with his landmark book *Inquest: The Warren Commission and the Establishment of Truth,* Epstein—neither a conspiracy theorist nor an apologist for the Warren Commission—has asked skeptical questions about the events of November 22, 1963. His three books on the subject have been collected into one volume, *The Assassination Chronicles: Inquest, Counterplot, and Legend.*

The two pioneering and encyclopedic books on November 22 to 25, William Manchester's *Death of a President* and Jim Bishop's *The Day Kennedy Was Shot,* are as compelling today as they were when they were published in the 1960s.

For more on the life of JFK, begin with James N. Giglio's *The Presidency of John F. Kennedy* and Robert Dallek's *An Unfinished Life: John F. Kennedy, 1917–1963*. For a compelling narrative on the odd life of Lee Harvey Oswald, written by a journalist who knew Oswald and his wife, Marina (and, strangely, John F. Kennedy too), see Priscilla Johnson McMillan's *Marina and Lee*. For another arresting account of Oswald, see Norman Mailer's *Oswald's Tale: An American Mystery*. *Marina and Lee* and *Oswald's Tale* remain two of the best five or six books ever written on the Kennedy assassination.

Thomas Mallon's *Mrs. Paine's Garage and the Murder of John F. Kennedy* remains one of my favorite books on the subject. More than almost any work on the subject, it resurrects the emotions of that terrible day in Dallas, fifty years ago.

GENERAL REFERENCES

Adler, Bill. *The Eloquent Jacqueline Kennedy Onassis: A Portrait in Her Own Words.* New York: William Morrow & Company, 2004.

———. *The Uncommon Wisdom of Jacqueline Kennedy Onassis.* Secaucus, NJ: Citadel Press, 1996.

Allison, Graham, and Philip Zelikow. *Essence of Decision: Explaining the Cuban Missile Crisis.* New York: Longman, 1999.

Alsop, Stewart. *The Center: People and Power in Political Washington.* New York: Harper & Row, Publishers, 1968.

———. *Jack and Jackie: Portrait of an American Marriage.* New York: William Morrow, 1996.

———. *Jackie After Jack.* New York: William Morrow & Company, 1998.

Amrine, Michael. *This Awesome Challenge: The Hundred Days of Lyndon Johnson.* New York: G. P. Putnam's Sons, 1964.

Anderson, Christopher. *Jack and Jackie: Portrait of an American Marriage.* New York: William Morrow, 1996.

Beschloss, Michael, ed. *Jacqueline Kennedy: Historic Conversations on Life with John F. Kennedy.* New York: Hyperion, 2011.

Bradford, Sarah. *America's Queen: The Life of Jacqueline Kennedy Onassis.* New York: Viking Press, 2000.

Bradlee, Benjamin. *That Special Grace.* Philadelphia: J. B. Lippincott Company, 1964.

Branch, Taylor. *Pillar of Fire: America in the King Years, 1963–1965.* New York: Simon & Schuster, 1998.

Brown, Thomas. *JFK: The History of an Image.* Bloomington: Indiana University Press, 1988.

Caro, Robert A. *The Years of Lyndon Johnson; The Passage of Power.* New York: Alfred A. Knopf, 2012.

Collier, Peter, and David Horowitz. *The Kennedys: An American Drama.* New York: Summit Books, 1984.

Dallek, Robert. *An Unfinished Life: John F. Kennedy, 1917–1963.* Boston: Little, Brown & Company, 2003.

Davis, John H. *Jacqueline Bouvier: An Intimate Memoir.* New York: John Wiley & Sons, 1996.

———. *The Kennedys: Dynasty and Disaster, 1848–1983.* New York: McGraw-Hill Book Company, 1984.

Dobbs, Michael. *One Minute to Midnight: Kennedy, Khrushchev, and Castro on the Brink of Nuclear War.* New York: Alfred A. Knopf, 2008.

Donovan, Robert J. *PT 109: John F. Kennedy in World War II*. New York: McGraw Hill, 1961.

Fairlie, Henry. *The Kennedy Promise: The Politics of Expectation*. Garden City, NY: Doubleday, 1971.

Frankel, Max. *High Noon in the Cold War: Kennedy, Khrushchev, and the Cuban Missile Crisis*. New York: Ballantine Books, 2004.

Garside, Anne. *Camelot at Dawn: Jacqueline and John Kennedy in Georgetown, May 1954*. Baltimore: Johns Hopkins University Press, 2001.

Giglio, James N. *The Presidency of John F. Kennedy*. 2nd ed. Lawrence: University Press of Kansas, 2006.

Gillette, Michael L. *Lady Bird Johnson: An Oral History*. New York: Oxford University Press, 2012.

Goodman, Jon. *The Kennedy Mystique: Creating Camelot*. Washington, DC: National Geographic, 2006.

Hamilton, Nigel. *JFK: Reckless Youth*. New York: Random House, 1992.

Hellmann, John. *The Kennedy Obsession: The American Myth of JFK*. New York: Columbia University Press, 1997.

Heymann, C. David. *A Woman Named Jackie*. New York: Carol Communications, 1989.

Hill, Clint, and Lisa McCubben. *Mrs. Kennedy and Me: An Intimate Memoir*. New York: Gallery Books, 2012.

Jones, Howard. *The Bay of Pigs*. New York: Oxford University Press, 2008.

Kelley, Kitty. *Capturing Camelot: Stanley Tretick's Iconic Images of the Kennedys*. New York: Thomas Dunne Books, 2012.

Kennedy, John F. *Public Papers of the Presidents of the United States: John F. Kennedy: 1961, 1962, 1963*. 3 vols. Washington, DC: U.S. Government Printing Office, 1962–64.

Kennedy, Robert. *Thirteen Days: A Memoir of the Cuban Missile Crisis*. New York: W. W. Norton, 1969.

Klein, Edward. *All Too Human: The Love Story of Jack and Jackie Kennedy*. New York: Pocket Books, 1996.

———. *Just Jackie: Her Private Years*. New York: Ballantine Books, 1998.

Kuhn, William. *Reading Jackie: Her Autobiography in Books*. New York: Nan A. Talese, 2010.

Lawrence, Gregg. *Jackie as Editor: The Literary Life of Jacqueline Kennedy Onassis*. New York: Thomas Dunne Books, 2011.

Leamer, Laurence. *The Kennedy Women*. New York: Villard Books, 1994.

Leaming, Barbara. *Jack Kennedy: The Education of a Statesman*. New York: W. W. Norton, 2006.

———. *Mrs. Kennedy: The Missing History of the Kennedy Years*. New York: Free Press, 2001.

Lubin, David M. *Shooting Kennedy: JFK and the Culture of Images*. Berkeley: University of California Press, 2003.

Manchester, William. *One Brief Shining Moment*. New York: Little, Brown & Company, 1983.

Matthews, Chris. *Jack Kennedy: Elusive Hero*. New York: Simon & Schuster, 2011.

———. *Kennedy and Nixon: The Rivalry That Shaped America*. New York: Simon & Schuster, 1996.

May, Ernest R., and Philip D. Zelikow. *The Kennedy Tapes: Inside the White House During the Cuban Missile Crisis*. Cambridge, MA: Belknap Press of Harvard University Press, 1997.

Miller, Merle. *Lyndon: An Oral Biography*. New York: G. P. Putnam's Sons, 1980.

Minow, Newton N., and Craig L. Lamoy. *Inside the Presidential Debates*. Chicago: University of Chicago Press, 2008.

Nasaw, David. *The Patriarch: The Remarkable Life and Turbulent Times of Joseph P. Kennedy*. New York: Penguin Press, 2012.

O'Donnell, Kenneth P., David F. Powers, and Joseph McCarthy. *"Johnny, We Hardly Knew Ye": Memories of John Fitzgerald Kennedy*. Boston: Little, Brown & Company, 1972.

Parmet, Herbert. *Jack: The Struggles of John F. Kennedy*. New York: Dial Press, 1980.

———. *JFK: The Presidency of John F. Kennedy*. New York: Penguin Books, 1984.

Perry, Barbara A. *Jacqueline Kennedy: First Lady of the New Frontier*. Lawrence: University Press of Kansas, 2004.

Piereson, James. *Camelot and the Cultural Revolution: How the Assassination of John F. Kennedy Shattered American Liberalism*. New York: Encounter Books, 2007.

Pottker, Jan. *Janet & Jackie: The Story of a Mother and Her Daughter, Jacqueline Kennedy Onassis*. New York: St. Martin's Press, 2001.

Rather, Dan, and Mickey Herskowitz. *The Camera Never Blinks: Adventures of a TV Journalist*. New York: William Morrow & Company, 1977.

Reeves, Richard. *Portrait of Camelot: A Thousand Days in the Kennedy White House*. New York: Harry N. Abrams, 2010.

———. *President Kennedy: Profile of Power*. New York: Simon & Schuster, 1993.

Schieffer, Bob. *This Just In: What I Couldn't Tell You on TV*. New York: G. P. Putnam's Sons, 2003.

Schlesinger, Arthur M., Jr. *A Thousand Days: John F. Kennedy in the White House*. Boston: Houghton Mifflin Company, 1965.

Shaw, Mark. *The John F. Kennedys: A Family Album*. Rev. ed. New York: Rizzoli International Publications, 2000.

Shaw, Maud. *White House Nannie: My Years with Caroline and John Kennedy, Jr.* New York: New American Library, 1966.

Shesol, Jeff. *Mutual Contempt: Lyndon Johnson, Robert Kennedy and the Feud That Defined an Era*. New York: W. W. Norton & Company, 1997.

Smith, Sally Bedell. *Grace and Power: The Private World of the Kennedy White House*. New York: Random House, 2004.

Steel, Ronald. *In Love with Night: The American Romance with Robert Kennedy*. New York: Simon & Schuster, 2000.

terHorst, J. F., and Col. Ralph Albertazzie. *The Flying White House: The Story of Air Force One*. New York: Coward, McCann & Geoghegan, 1979.

Wicker, Tom. *JFK and LBJ: The Influence of Personality Upon Politics*. New York: William Morrow & Company, 1968.

———. *Kennedy Without Tears: The Man Beneath the Myth*. New York: William Morrow & Co., 1964.

Williams, Juan. *Eyes on the Prize: America's Civil Rights Years, 1954–1965*. New York: Viking Press, 1987.

Widmer, Ted, ed. *Listening In: The Secret White House Recordings of John F. Kennedy*. New York: Hyperion, 2012.

Wolff, Perry. *A Tour of the White House with Mrs. John F. Kennedy*. Garden City, NY: Doubleday, 1962.

Youngblood, Rufus W. *20 Years in the Secret Service: My Life with Five Presidents*. New York: Simon & Schuster, 1973.

THE ASSASSINATION

[Associated Press]. *The Torch Is Passed: The Associated Press Story of the Death of a President.* New York: Associated Press, 1963.

Belin, David W. *Final Disclosure: The Full Truth About the Assassination of President Kennedy.* New York: Charles Scribner's Sons, 1988.

———. *November 22, 1963: You Are the Jury.* New York: Quadrangle, 1973.

Belli, Melvin M., and Maurice C. Carroll. *Dallas Justice: The Real Story of Jack Ruby and His Trial.* New York: David McKay Company, 1964.

Bishop, Jim. *The Day Kennedy Was Shot.* New York: Funk & Wagnalls, 1968.

Blaine, Gerald, and Lisa McCubben. *The Kennedy Detail: JFK's Secret Service Agents Break Their Silence.* New York: Gallery, 2010.

Bloomgarden, Henry S. *The Gun: A Biography of the Gun That Killed John F. Kennedy.* New York: Grossman Publishers, 1975.

Brener, Milton E. *The Garrison Case: A Study in the Abuse of Power.* New York: Clarkson N. Potter, 1969.

Brennan, Howard L., and J. Edward Cherryholmes. *Eyewitness to History: The Kennedy Assassination, as Seen by Howard Brennan.* Waco, TX: Texian Press, 1987.

Bringuier, Dr. Carlos. *Red Friday: Nov. 22nd, 1963.* Chicago: Chas. Hallberg & Company, 1969.

Bugliosi, Vincent. *Four Days in November: The Assassination of President John F. Kennedy.* New York: W. W. Norton & Company, 2007.

———. *Reclaiming History: The Assassination of President John F. Kennedy.* New York: W. W. Norton & Company, 2007.

Clarke, James W. *American Assassins: The Darker Side of Politics.* Princeton, NJ: Princeton University Press, 1990.

Connally, Nellie, and Mickey Herskowitz. *From Love Field: Our Final Hours with President John F. Kennedy.* New York: Rugged Land Books, 2003.

Cottrell, John. *Assassination: The World Stood Still.* London: New English Library, 1964.

Curry, Jesse. *Retired Dallas Police Chief Jesse Curry Reveals His Personal JFK Assassination File.* Dallas: American Poster & Printing Company, 1969.

David, Jay, ed. [pseud. for Bill Adler]. *The Weight of the Evidence: The Warren Report and Its Critics.* New York: Meredith Press, 1968.

Davison, Jean. *Oswald's Game.* New York: W. W. Norton, 1983.

Epstein, Edward J. *Assassination Chronicles: Inquest, Counterplot, and Legend.* New York: Carroll & Graf Publishers, 1992.

Fagin, Stephen. *Assassination and Commemoration: JFK, Dallas, and the Sixth Floor Museum at Dealey Plaza.* Norman: University of Oklahoma Press, 2013.

Fine, William M., ed. *That Day with God.* New York: McGraw-Hill Book Company, 1965.

Fitzpatrick, Ellen. *Letters to Jackie: Condolences from a Grieving Nation.* New York: Ecco Press, 2010.

Ford, Gerald R., and John R. Stiles. *Portrait of the Assassin.* New York: Simon & Schuster, 1965.

Gertz, Elmer. *Moment of Madness: The People vs. Jack Ruby.* Chicago: Follett Publishing Company, 1968.

Gillon, Steven M. *The Kennedy Assassination—24 Hours After: Lyndon B. Johnson's Pivotal First Day as President.* New York: Basic Books, 2009.

Glikes, Erwin A., and Paul Schwaber. *Of Poetry and Power: Poems Occasioned by the Presidency and Death of John F. Kennedy.* New York: Basic Books, 1964.

Greenberg, Bradley S., and Edwin B. Parker, eds. *The Kennedy Assassination and the American Public: Social Communications in Crisis.* Stanford, CA: Stanford University Press, 1965.

Grosvenor, Melville Bell. *The Last Full Measure: The World Pays Tribute to President Kennedy.* Washington, DC: National Geographic, 1964.

Hampton, Wilborn. *Kennedy Assassinated!: The World Mourns: A Reporter's Story.* Cambridge, MA: Candlewick Press, 1997.

Hanson, William H. *The Shooting of John F. Kennedy: One Assassin, Three Shots, Three Hits—No Misses.* San Antonio, TX: Naylor Company, 1969.

Harris, Patricia Howard. *An Austin Scrapbook of John F. Kennedy.* Austin, TX: Pemberton Press, 1964.

Hartogs, Renatus, and Lucy Freeman. *The Two Assassins.* New York: Thomas Y. Crowell Company, 1965

Hayes, Harold, ed. *Smiling Through the Apocalypse: Esquire's History of the Sixties.* New York: McCall's Publishing Company, 1969. (Reprint of articles, include: "Kennedy Without Tears"; "Lee Oswald's Letters to his Mother"; "You All Know Me! I'm Jack Ruby!"; and "Sixty Versions of the Kennedy Assassination.")

Henderson, Bruce, and Sam Summerlin. *1:33.* New York: Cowles, 1968.

Hlavach, Richard, and Darwin Payne, eds. *Reporting the Kennedy Assassination: Journalists Who Were There Recall Their Experiences.* Dallas: Three Forks Press, 1996.

Holland, Max. *The Kennedy Assassination Tapes: The White House Conversations of Lyndon B. Johnson Regarding the Assassination, the Warren Commission, and the Aftermath.* New York: Alfred A. Knopf, 2004.

Hosty, James P., Jr. *Assignment: Oswald.* New York: Arcade Publishing, 1996.

Hunt, Conover. *Dealey Plaza National Historic Landmark Including the Sixth Floor Museum.* Dallas, TX: The Sixth Floor Museum, 1997.

Hunter, Diana, and Alice Anderson. *Jack Ruby's Girls.* Atlanta: Hallux, 1970.

Itek Corporation. *John Kennedy Assassination Film Analysis.* Lexington, MA: Itek Corporation, 1976.

———. *Life-Itek Kennedy Assassination Film Analysis.* Lexington, MA: Itek Corporation, 1967.

———. *Nix Film Analysis.* Lexington, MA: Itek Corporation, 1967.

Kaplan, John, and Jon R. Waltz. *The Trial of Jack Ruby.* New York: Macmillan Company, 1965.

Kirkwood, James. *American Grotesque: An Account of the Clay Shaw–Jim Garrison Affair in the City of New Orleans.* New York: Simon & Schuster, 1970.

Knight, Peter. *The Kennedy Assassination.* Oxford: University Press of Mississippi, 2007.

Lattimer, John K. *Kennedy and Lincoln: Medical and Ballistic Comparisons of Their Assassinations.* New York: Harcourt Brace Jovanovich, 1980.

Leslie, Warren. *Dallas: Public and Private.* New York: Grossman Publishers, 1964.

Lewis, Richard Warren, and Lawrence Schiller. *The Scavengers and Critics of the Warren Report: The Endless Paradox, Based on an Investigation by Lawrence Schiller.* New York: Delacorte Press, 1967.

Loken, John. *Oswald's Trigger Films: The Manchurian Candidate, We Were Strangers, Suddenly.* Ann Arbor, MI: Falcon Books, 2000.

Mailer, Norman. *Oswald's Tale: An American Mystery.* New York: Random House, 1995.

Mallon, Thomas. *Mrs. Paine's Garage and the Murder of John F. Kennedy.* New York: Pantheon Books, 2002.

Manchester, William. *The Death of a President: November 20–November 25, 1963*. New York: Harper & Row, Publishers, 1967.

Mayo, John B., Jr. *Bulletin from Dallas: The Story of John F. Kennedy's Assassination as Covered by Radio and TV*. New York: Exposition Press, 1967.

McMillan, Priscilla Johnson. *Marina and Lee*. New York: Harper & Row, Publishers, 1977.

Moore, Jim. *Conspiracy of One; The Definitive Book on the Kennedy Assassination*. Fort Worth, TX: Summit Group, 1990.

Morin, Relman. *Assassination: The Death of President John F. Kennedy*. New York: Signet Books, 1968.

Mossman, Billy C., and B. C. Stark. *The Last Salute: Civil and Military Funerals, 1921–1969*. Washington, DC: Department of the Army, 1971.

Mulvaney, Jay, and Paul De Angelis. *Dear Mrs. Kennedy: The World Shares Its Grief—Letters, November 1963*. New York: St. Martin's Press, 2010.

Myers, Dale K. *With Malice: Lee Harvey Oswald and the Murder of Officer J. D. Tippit*. Milford, MI: Oak Cliff Press, 1998.

[NBC News]. *Seventy Hours and Thirty Minutes, as Broadcast on the NBC Television Network by NBC News*. New York: Random House, 1966.

———. *There Was a President*. New York: Ridge Press, 1966.

Oswald, Robert, Myrick Land, and Barbara Land. *Lee: A Portrait of Lee Harvey Oswald by His Brother Robert Oswald*. New York: Coward-McCann, 1967.

Posner, Gerald. *Case Closed: Lee Harvey Oswald and the Assassination of JFK*. New York: Random House, 1993.

Rajski, Raymond B., ed. *A Nation Grieved: The Kennedy Assassination in Editorial Cartoons*. Rutland, VT: Chares Tuttle Company, 1967.

Roberts, Charles. *The Truth About the Assassination*. New York: Grosset & Dunlap, 1967.

Savage, Gary. *JFK First Day Evidence: Stored Away for 30 Years in an Old Briefcase, New Evidence Is Now Revealed by Former Dallas Police Crime Lab Detective R. W (Rusty) Livingstone*. Monroe, LA: Shoppe Press, 1993.

Seigenthaler, John. *A Search for Justice*. Nashville, TN: Aurora Publishers, 1971.

Semple, Robert B., Jr., ed. *Four Days in November: The Original Coverage of the John F. Kennedy Assassination by the Staff of the New York Times*. New York: St. Martin's Press, 2003.

Sites, Paul. *Lee Harvey Oswald and the American Dream*. New York: Pageant Press, 1967.

Sneed, Larry A. *No More Silence: An Oral History of the Assassination of President Kennedy*. Dallas, TX: Three Forks Press, 1998.

Sparrow, John. *After the Assassination: A Positive Appraisal of the Warren Report*. New York: Chilmark Press, 1967.

Stewart, Charles J., and Bruce Kendell, ed. *A Man Named John F. Kennedy: Sermons on His Assassination*. Glen Rock, NJ: Paulist Press, 1964.

Sturdivan, Larry M. *The JFK Myths: A Scientific Investigation of the Kennedy Assassination*. St. Paul, MN: Paragon House, 2005.

Thornley, Kerry W. *Oswald*. Chicago: New Classics House, 1965.

Trask, Richard B. *National Nightmare on Six Feet of Film: Mr. Zapruder's Home Movie and the Murder of President Kennedy*. Danvers, MA: Yeoman Press, 2005.

———. *Pictures of the Pain: Photography and the Assassination of President Kennedy*. Danvers, MA: Yeoman Press, 2004.

———. *That Day in Dallas: Three Photographers Capture on Film the Day President Kennedy Died*. Danvers, MA: Yeoman Press, 1998.

Trost, Cathy and Susan Bennett. *President Kennedy Has Been Shot: The Inside Story of the Murder of a President.* Naperville, IL: Sourcebooks, 2003.

[United Press International]. *Four Days: The Historical Record of the Assassination of President Kennedy.* N.p.: American Heritage Publishing Company, 1964.

U.S. House of Representatives. *Final Report of the Select Commission on Assassinations,* and the accompanying 12 hearing and appendix volumes on the JFK Assassination, 95th Congress, 2nd Session. Washington, DC: U.S. Government Printing Office, 1979.

U.S. Senate. *Memorial Addresses in the Congress of the United States and Tributes in Eulogy of John Fitzgerald Kennedy, a Late President of the United States.* Washington, DC: U.S. Government Printing Office, 1964.

Vagnes, Oyvind. *Zaprudered: The Kennedy Assassination Film in the Visual Culture.* Austin: University of Texas Press, 2011.

Walsh, William G. *Children Write About John F. Kennedy.* Brownsville, TX: Springman-King, 1964.

[Warren Commission.] *Hearings Before the President's Commission on the Assassination of President John F. Kennedy.* 26 vols. Washington, DC: U.S. Government Printing Office, 1964.

———. *Report of the President's Commission on the Assassination of President John F. Kennedy.* Washington, DC: U.S. Government Printing Office, 1964.

White, Stephen. *Should We Now Believe the Warren Report?* New York: Macmillan Company, 1968.

Wills, Garry, and Ovid Demaris. *Jack Ruby: The Man Who Killed the Man Who Killed Kennedy.* New York: The New American Library, 1968.

Wise, Dan and Marietta, Maxfield. *The Day Kennedy Died.* San Antonio, TX: The Naylor Co., 1964.

Wolfenstein, Martha, and Kliman Gilbert, eds. *Children and the Death of a President: Multi-Disciplinary Studies.* Garden City, NY: Doubleday & Company, 1965.

Wrone, David R. *The Assassination of John Fitzgerald Kennedy: An Annotated Bibliography.* Madison: State Historical Society of Wisconsin, 1973.

Zelizer, Barbie. *Covering the Body: The Kennedy Assassination, the Media, and the Shaping of Collective Memory.* Chicago: University of Chicago Press, 1992.

CONSPIRACY LITERATURE

Adams, Don. *From an Office Building with a High-Powered Rifle: One FBI Agent's View of the JFK Assassination.* Walterville, OR: TrineDay, 2012.

Anson, Robert Sam. *They've Killed the President: The Search for the Murderers of John F. Kennedy.* New York: Bantam Books, 1975.

Armstrong, John. *Harvey & Lee; How the CIA Framed Oswald.* Arlington, TX: Quasar, 2003.

Baker, Judyth Vary. *Me and Lee: How I Came to Know, Love and Lose Lee Harvey Oswald.* Walterville, OR: TrineDay, 2010.

Bane, Bernard M. *The Bane in Kennedy's Existence.* Boston: BMB Publishing Company, 1967.

Belzer, Richard. *Hit List: An In Depth Investigation into the Mysterious Deaths of the Witnesses in the JFK Assassination.* New York: Skyhorse Publishing, 2013.

Benson, Michael. *Who's Who in the JFK Assassination: An A to Z Encyclopedia.* Secaucus, NJ: Carol Publishing Group, 1993.

Blakey, G. Robert, and Richard N. Billings. *The Plot to Kill the President.* New York: Times Books, 1981.

Bonner, Judy Whitson. *Investigation of a Homicide: The Murder of John F. Kennedy.* Anderson, SC: Droke House, 1969.

Brown, Walt. *Treachery in Dallas.* New York: Carroll & Graf Publishers, 1995.

———. *The Warren Omission; A Micro-Study of the Methods and Failures of the Warren Commission.* Wilmington, DE: Delmax, 1996.

Buchanan, Thomas. *Who Killed Kennedy?* London: Secker & Warburg, 1964.

Callahan, Bob. *Who Shot JFK? A Guide to the Major Conspiracy Theories.* New York: Simon & Schuster, 1993.

Chambers, Paul. *Head Shot: The Science Behind the JFK Assassination.* Amherst, NY: Prometheus Books, 2010.

Chapman, Gil, and Ann Chapman. *Was Oswald Alone?* San Diego: Publishers Export Company, 1967.

Crenshaw, Charles A., M.D., Jens Hansen, and J. Gary Shaw. *JFK: Conspiracy of Silence.* New York: Signet Books, 1992.

Cutler, Robert B. *The Flight of CE 399: Evidence of Conspiracy.* Beverly, MA: Omni-Print, 1969.

———. *Two Flightpaths: Evidence of Conspiracy.* Manchester, MA: Cutler Designs, 1971.

———. *The Umbrella Man: Evidence of Conspiracy.* Manchester, MA: Cutler Designs, 1975.

DiEugenio, James. *Destiny Betrayed: JFK, Cuba, and the Garrison Case.* New York: Sheridan Square Press, 1992.

Douglass, James W. *JFK and the Unspeakable: Why He Died and Why It Matters.* New York: Orbis Books, 2008.

Duffy, James R. *Who Killed JFK? Kennedy Assassination Cover-Up.* New York: Shalpolsky Publishers, 1989.

Eddowes, Michael. *The Oswald File.* New York: Clarkson N. Potter, 1977.

Evica, George Michael. *And We Are All Mortal: New Evidence Analysis in the John F. Kennedy Assassination.* West Hartford, CT: University of Hartford, 1978.

Fenster, Mark. *Conspiracy Theories: Secrecy and Power in American Culture.* Minneapolis: University of Minnesota Press, 1999.

Fensterwald, Bernard, Jr., and Michael Ewing, eds. *Coincidence of Conspiracy?* New York: Zebra Books, 1977.

Fetzer, James H. *Assassination Science: Experts Speak Out on the Death of JFK.* Chicago: Catfeet Press, 1998.

Flammonde, Paris. *The Kennedy Conspiracy: An Uncommissioned Report on the Jim Garrison Investigation.* New York: Meredith Press, 1969.

Fonzi, Gaeton. *The Last Investigation: A Former Federal Investigator Reveals the Man Behind the Conspiracy to Kill JFK.* New York: Thunder's Mouth Press, 1993.

Fox, Sylvan. *The Unanswered Questions About President Kennedy's Assassination.* New York: Award Books, 1965.

Garrison, Jim. *A Heritage of Stone.* New York: G. P. Putnam's Sons, 1970.

———. *On the Trail of the Assassins: My Investigation and Prosecution of the Murder of President Kennedy.* New York: Sheridan Square Press, 1988.

Gershenson, Alvin H. *Kennedy and Big Business.* Beverly Hills, CA: Book Company of America, 1964

Groden, Robert J. *The Killing of a President: The Complete Photographic Record of the JFK Assassination, the Conspiracy, and the Cover-Up.* New York: Viking Studio Books, 1993.

———. *The Search for Lee Harvey Oswald: The Comprehensive Photographic Record.* New York: Penguin Studio Books, 1995.

Groden, Robert J., and Harrison Edward Livingston. *High Treason: The Assassination of President John F. Kennedy: What Really Happened.* New York: Conservatory Press, 1989.

Gun, Nerin E. *Red Roses from Texas.* London: Frederick Muller, 1964.

Hancock, Larry. *Someone Would Have Talked.* Southlake, TX: JFK Lancer, 2010.

Hepburn, James. *Farewell America.* Belgium: Frontiers, 1968.

Hinckle, Warren, and William W. Turner. *Deadly Secrets: The CIA-Mafia War Against Castro and the Assassination of J.F.K.* New York: Thunder's Mouth Press, 1992.

Horne, Douglas P. *Inside the Assassination Records Review Board: The U.S. Government's Final Attempt to Reconcile the Conflicting Medical Evidence in the Assassination of JFK.* 5 vols. N.p.: Author, 2009.

Hurt, Henry. *Reasonable Doubt: An Investigation in the Assassination of John F. Kennedy.* New York: Holt, Rinehart & Winston, 1986.

Joesten, Joachim. *The Dark Side of Lyndon Baines Johnson.* London: Peter Dawnay, 1968.

———. *The Garrison Enquiry: Truth & Consequences.* London: Peter Dawnay, 1967.

———. *How Kennedy Was Killed: The Fully Appalling Story.* London: Dawnay/Tandem, 1968.

———. *Marina Oswald.* London: Peter Dawnay, 1967.

———. *Oswald: Assassin or Fall Guy?* London: Merlin Press, 1964.

———. *Oswald: The Truth.* London: Peter Dawnay, 1967.

Jones, Penn, Jr. *Forgive My Grief:* Vols. 1–4. Midlothian, TX: Midlothian Mirror, 1966, 1967, 1969, 1974.

Kaiser, David E. *The Road to Dallas: The Assassination of JFK.* Cambridge, MA: Harvard University, Belknap Press, 2008.

Kantor, Seth. *Who Was Jack Ruby?* New York: Everest House, 1978.

Kelin, John. *Praise from a Future Generation: The Assassination of John F. Kennedy and the First Generation Critics of the Warren Report.* San Antonio, TX: Wings Press, 2007.

Krusch, Barry. *Impossible: The Case Against Lee Harvey Oswald.* 3 vols. Asheville, NC: ICI Press, 2012.

Kurtz, Michael L. *Crime of the Century: The Kennedy Assassination from a Historian's Perspective.* Knoxville: University of Tennessee Press, 1982.

———. *The JFK Assassination Debates: Lone Gunman versus Conspiracy.* Lawrence: University Press of Kansas, 2006.

Lane, Mark. *A Citizen's Dissent: Mark Lane Replies.* New York: Holt, Rinehart & Winston, 1968.

———. *The Last Word: My Indictment of the CIA in the Murder of JFK.* New York: Skyhorse Publishing, 2011.

———. *Plausible Denial: Was the CIA Involved in the Assassination of JFK?* New York: Thunder's Mouth Press, 1991.

———. *Rush to Judgment: A Critique of the Warren Commission's Inquiry into the Murders of President John F. Kennedy, Officer J. D. Tippit and Lee Harvey Oswald.* New York: Holt, Rinehart & Winston, 1966.

Lifton, David S. *Best Evidence: Disguise and Deception in the Assassination of John F. Kennedy.* New York: Macmillan Publishing Company, 1980.

Livingstone, Harrison Edward. *Killing Kennedy and the Hoax of the Century.* New York: Carroll & Graf Publishers, 1995.

McClellan, Barr. *Blood, Money and Power: How LBJ Killed JFK.* New York: Hanover House, 2003.

McFarlane, Ian. *Proof of Conspiracy in the Assassination of President Kennedy.* Melbourne, Australia: Book Distributors, 1975.

McKnight, Gerald D. *Breach of Trust: How the Warren Commission Failed the Nation and Why.* Lawrence: University Press of Kansas, 2005.

Marcus, Raymond. *The Bastard Bullet: A Search for Legitimacy for Commission Exhibit 399.* Los Angeles: Rendell Publications, 1966.

Marrs, Jim. *Crossfire: The Plot That Killed Kennedy.* New York: Carroll & Graf, 1989.

Meagher, Sylvia. *Accessories After the Fact: The Warren Commission, the Authorities, and the Report.* Indianapolis: Bobbs-Merrill Company, 1967.

Melanson, Philip H. *Spy Saga: Lee Harvey Oswald and U.S. Intelligence.* New York: Praeger, 1990.

Mellen, Joan. *A Farewell to Justice: Jim Garrison, JFK's Assassination, and the Case That Should Have Changed History.* Dulles, VA: Potomac Books, 2005.

Menninger, Bonar. *Mortal Error: The Shot That Killed JFK.* New York: St. Martin's Press, 1992

Miller, Tom. *The Assassination Please Almanac.* Chicago: Henry Regnery Company, 1977.

Model, F. Peter, and Robert J. Groden. *JFK: The Case for Conspiracy.* New York: Manor Books, 1977.

Morris, W. R., and Robert B. Cutler. *Alias Oswald.* Manchester, MA: GKG Partners, 1985.

Morrow, Robert D. *Betrayal.* Chicago: Henry Regnery Company, 1976.

———. *First Hand Knowledge: How I Participated in the CIA-Mafia Murder of President Kennedy.* New York: S.P.I. Books, 1992.

Moss, Armand. *Disinformation, Misinformation, and the "Conspiracy" to Kill JFK Exposed.* Hamden, CT: Archon Books, 1987.

Murray, Norbert. *Legacy of an Assassination.* New York: Pro-People Press, 1964.

Nechiporenko, Oleg M. *Passport to Assassination: The Never-Before-Told Story of Lee Harvey Oswald by the KGB Colonel Who Knew Him.* New York: Birch Lane Press Books, 1993.

Nelson Phillip F. *LBJ: The Mastermind of the JFK Assassination.* New York: Skyhorse Publishing, 2011.

Newman, Albert H. *The Assassination of John F. Kennedy: The Reasons Why.* New York: Clarkson N. Potter, 1970.

Newman, John. *Oswald and the CIA.* New York: Carroll & Graf Publishers, 1995.

North, Mark. *Act of Treason.* New York: Carroll & Graf Publishers, 1991.

Oglesby, Carl. *The JFK Assassination: The Facts and the Theories.* New York: Signet Books, 1992.

———. *The Yankee and Cowboy Wars.* Kansas City: Sheed Andrews and McMeel, 1976.

Oswald, Marguerite. *Aftermath of an Execution; The Burial and Final Rites of Lee Harvey Oswald as Told by His Mother.* Dallas: Challenge Press, 1965.

O'Toole, George. *The Assassination Tapes: An Electronic Probe into the Murder of John F. Kennedy and the Dallas Coverup.* New York: Penthouse Press, 1975.

Popkin, Richard H. *The Second Oswald.* New York: Avon Books, 1966.

Prouty, L. Fletcher. *JFK: The CIA, Vietnam, and the Plot to Assassinate John F. Kennedy.* New York: Birch Lane Press Books, 1992.

Ramparts Magazine, ed. *In the Shadow of Dallas: A Primer on the Assassination of President Kennedy.* San Francisco: Ramparts, 1967.

Roffman, Howard. *Presumed Guilty: Lee Harvey Oswald in the Assassination of President Kennedy.* Rutherford, NJ: Farleigh Dickinson University Press, 1975.

Russell, Bertrand. *16 Questions on the Assassination.* Passaic, NJ: Minority of One, 1964.

Russell, Dick. *The Man Who Knew Too Much.* New York: Carroll & Graf Publishers, 1992.

Russo, Gus. *Brothers in Arms: The Kennedys, the Castros, and the Politics of Murder.* New York: Bloomsbury, 2008.

——. *Live by the Sword: The Secret War Against Castro and the Death of JFK.* Baltimore: Bancroft Press, 1998.

Sauvage, Leo. *The Oswald Affair: An Examination of the Contradictions and Omissions of the Warren Report.* Cleveland: World Publishing Company, 1966.

Scheim, David E. *Contract on America: The Mafia Murders of John and Robert Kennedy.* New York: Shalpolsky Publishers, 1988.

Schotz, Martin. *History Will Not Absolve Us: Orwellian Control, Public Denial, and the Murder of President Kennedy.* Brookline, MA: Kurtz, Olmer & Delucia, 1996.

Scott, Peter Dale. *Crime and Cover-up: The CIA, the Mafia, and the Dallas-Watergate Connection.* Berkeley, CA: Westworks Publishers, 1977.

Scott, Peter Dale, Paul L. Hock, and Russell Stetler, eds. *The Assassinations: Dallas and Beyond.* New York: Random House, 1976.

Shaw, J. Gary, and Larry R. Harris. *Cover-up: The Governmental Conspiracy to Conceal the Facts about the Public Execution of John Kennedy.* Cleburne, TX: Authors, 1976.

Sloan, Bill, and Jean Hill. *JFK: The Last Dissenting Witness.* Greta, LA: Pelican Publishing Company, 1992.

Stafford, Jean. *A Mother in History.* New York: Farrar, Straus & Giroux, 1966.

Stone, Oliver, and Zachary Sklar. *JFK: The Book of the Film.* New York: Applause Books, 1992.

Summers, Anthony. *Conspiracy.* New York: McGraw-Hill Book Company, 1980.

Tague, James T. *Truth Withheld: A Survivor's Story—Why We Will Never Know the Truth About the JFK Assassination.* Dallas: Excel Digital Press, 2003.

Thomas, Ralph D. *Missing Links in the JFK Assassination Conspiracy.* Austin, TX: Thomas Investigative Publications, 1992.

——. *Photo Computer Image Processing and the Crime of the Century: A New Investigative & Photographic Technique.* Austin, TX: Thomas Investigative Publications, 1992.

Thompson, Josiah. *Six Seconds in Dallas: A Micro-study of the Kennedy Assassination.* New York: Bernard Geis Associates, 1967.

Waldron, Lamar. *Ultimate Sacrifice: John and Robert Kennedy, and the Murder of JFK.* New York: Carroll & Graf Publishers, 2005.

Weisberg, Harold. *Case Open: The Omissions, Distortions and Falsifications of Case Closed.* New York: Carroll & Graf, 1994.

——. *Never Again.* New York: Carroll & Graf Publishers, 1995.

——. *Oswald in New Orleans: Case of Conspiracy with C.I.A.* New York: Canyon Books, 1967.

——. *Whitewash; Whitewash II; Photographic Whitewash; Whitewash IV; and Postmortem.* Hyattstown and Frederick, MD: Author, 1965, 1966, 1966, 1974, and 1975.

Wilber, Charles G. *The Medicolegal Investigation of the President John F. Kennedy Murder.* Springfield, IL: Charles C. Thomas Publisher, 1978

Wrone, David R. *The Zapruder Film: Reframing JFK's Assassination.* Lawrence: University Press of Kansas, 2003.

Zirbel, Craig I. *The Texas Connection: The Assassination of President John F. Kennedy.* Scottsdale, AZ: Wright & Company, 1991.

JFK ASSASSINATION FICTION

Aubrey, Edmund. *Sherlock Holmes in Dallas.* New York: Dodd, Mead & Company, 1980.

Ballard, J. G. *Love and Napalm: Export U.S.A.* New York: Grove Press, 1969. (Chapters include: "The Assassination of John F. Kennedy—Considered as a Downhill Motor Race" and "Plan to Assassinate Jacqueline Kennedy.")

Balling, L. Christian. *The Fourth Shot.* Boston: Little, Brown & Company, 1982.

Bealle, Morris A. *Guns of the Regressive Right or How to Kill a President.* Washington, DC: Columbia Publishing Company, 1964.

Berry, Wendall, and Ben Shahn. *November Twenty Six Nineteen Hundred Sixty Three.* New York: George Braziller, 1964. (poetry and art)

Braver, Adam. *November 22, 1963: A Novel.* Portland, OR: Tin House Books, 2008.

Brown, Walt. *The People v. Lee Harvey Oswald.* New York: Carroll & Graf Publishers, 1992.

Condon, Richard. *Winter Kills.* New York: Dial Press, 1974.

DeLillo, Don. *Libra.* New York: Viking, 1988.

DiMona, Joseph. *Last Man at Arlington.* New York: A. Fields Books, 1973.

Ellroy, James. *American Tabloid.* New York: Alfred A. Knopf, 1995.

Freed, Donald, and Mark Lane. *Executive Action: Assassination of a Head of State.* New York: Dell Publishing Company, 1973.

Freedman, Nancy. *Joshua Son of None.* New York: Delacorte Press, 1973.

Garrison, Jim. *The Star Spangled Contract.* New York: McGraw-Hill Book Company, 1976.

Garson, Barbara. *MacBird!* New York: Grassy Knoll Press, 1966. (play)

Harrington, William. *Columbo: The Grassy Knoll.* New York: Forge Books, 1993.

Hastings, Michael. *Lee Harvey Oswald: A Far Mean Streak of Indepence (sic) Brought on by Negleck (sic).* Baltimore, Penguin Books, 1966. (play)

Heath, Peter. *Assassins from Tomorrow.* New York: Prestige Books, 1967.

Hunter, Stephen. *The Third Bullet: A Bob Lee Swagger Novel.* New York: Simon & Schuster, 2013.

Jensen, J. Arthur. *The Kennedy Assassination: A Historical Novel.* Philadelphia: Xlibris Corporation, 2001.

King, Stephen. *11/22/63: A Novel.* New York: Scribner Book Company, 2011.

La Fountaine, George. *Flashpoint: A Novel.* New York: Coward McCann & Geoghegan, 1976.

Lawn, Donald James. *The Memoirs of John F. Kennedy: A Novel.* Seattle, WA: Castlefin Press, 2010.

McCarry, Charles. *The Tears of Autumn.* New York: Saturday Review Press, 1975.

Malzberg, Barry. *The Destruction of the Temple.* New York: Pocket Books, 1974.

———. *Scop.* New York: Pyramid Books, 1976.

Mayer, Robert. *I, JFK.* New York: E. P. Dutton, 1986.

Meltzer, Brad. *The Fifth Assassin.* New York: Grand Central Publishing, 2013.

Morris, Wright. *One Day.* New York: Atheneum, 1965.

O'Donnell, M. K. *You Can Hear the Echo.* New York: Simon & Schuster, 1966.

Shapiro, Stanley. *A Time to Remember.* New York: Random House, 1986.

Sloan, Bill. *The Other Assassin.* New York: S.P.I. Books, 1992.

Sondheim, Stephen, and John Weidman. *Assassins.* New York: Theatre Communications Group, 1991. (musical)

Stevens, James, and David Bishop. *Who Killed Kennedy: The Shocking Secret Linking a Time and a President (Doctor Who Series).* London: Doctor Who Books, 1996.

Swanson, Doug. *Umbrella Man: A Jack Flippo Mystery.* New York: G. P. Putnam's Sons, 1999.

Thomas, D. M. *Flying into Love.* New York: Bloomsbury Publishing, 1992.

Thornley, Kerry W. *The Idle Warriors.* Avondale Estates, GA: IllumiNet Press, 1991.

Thurston, Wesley S. *The Trumpets of November.* New York: Bernard Geis Associates, 1966.

Vincent, E. Duke. *The Camelot Conspiracy: A Novel of the Kennedys, Castro, and the CIA.* New York: Overlook Press, 2011.

Wilden, Theodore. *To Die Elsewhere.* New York: Harcourt Brace Jovanovich, 1976.

Woolley, Brian. *November 22.* New York: Seaview Books, 1981.

SOURCE NOTES

PROLOGUE

2 Capturing the Kennedys in photographs. See Mark Shaw, *The John F. Kennedys: A Family Album* (New York: Farrar, Straus, 1964). Shaw was a freelance photographer whose Kennedy photos were made while he was on assignment by *Life* magazine. The Mark Shaw Photographic Archive is in Drummerston, Vermont, with a website at http://markshawphoto.com. Also see Jacques Lowe, *Kennedy: A Time to Remember* (New York: Quartet Books, 1983), as well as *New York Times*, ed., *The Kennedy Years* (New York: Viking, 1964), illustrated by Lowe and others. See http://jacqueslowe.com. There were several other books written, edited, or illustrated by this photographer. Unfortunately, forty thousand of his negatives documenting the Kennedy presidency were destroyed in the September 11, 2001, attack on the World Trade Center.

CHAPTER 1: "SUCH DANGEROUS TOYS"

3 The most definitive accounts of Oswald's activities and his relationship with his wife, Marina, can be found in her Warren Commission testimony, her statements to government officials, and her intimate recollections in a book: Priscilla Johnson McMillan, *Marina and Lee* (New York: Harper & Row, 1977).

3 Oswald's loose-leaf notebook on Walker. See the testimony of Marina Oswald, *Hearings before the President's Commission on the Assassination of President John F. Kennedy* (Warren Commission, hereinafter WC), 26 vols. (Washington, DC: U.S. Government Printing Office, 1964), vol. 11, p. 292; also see her statement to the FBI, WC 22, commission exhibit (hereinafter CE) 1156, p. 1945.

4 Revolver purchase. The Smith & Wesson .38-caliber revolver was purchased through a mail-order coupon from Seaport-Traders, Inc., a mail-order division of George Rose & Co., Los Angeles. See the testimony of Heinz W. Michaelis, employee of the company, WC 7, pp. 372–78 and United States, *Report of the President's Commission on the Assassination of President John F. Kennedy* (Warren

Report, hereinafter WR; Washington, DC: U.S. Government Printing Office, 1964) p. 121, and WC 16, CE 135, p. 511.

4 Rifle purchase. Although it was ordered under an alias, handwriting analysis confirmed that the purchase order was in Oswald's handwriting. The rifle purchase actually was confirmed by the FBI on the evening of November 22, 1963. The rifle was purchased from Klein's Sporting Goods in Chicago. See WR, pp. 118–21; WC 17, CE 773, p. 635; and CE 790, p. 678. Also see WC 17, CE 788–89, pp. 677–78 (postal money order); CE 790, p. 679 (order form); and the affidavit of Louis Feldsott, WC 11, p. 205 (verifying purchase). In addition, see the testimony of postal clerk Harry D. Holmes, WC 7, pp. 289–308, and the testimony of William J. Waldman, vice president of Klein's Sporting Goods, WC 7, p. 360, and WC 7, Waldman Exhibits 1–10, pp. 692–707 (complete record of the transaction). Also, see Henry S. Bloomgarden, *The Gun: A "Biography" of the Gun that Killed John F. Kennedy* (New York: Grossman, 1975).

5 "What do you need a rifle for?" See the testimony of Marina Oswald, WC 1, p. 13.

Oswald always had a fascination with guns. For instance, he read *American Rifleman, Field and Stream, Argosy,* and *Guns and Hunting,* which were lying around on the coffee table in the waiting room of the Cresent City Garage next to the Reily Coffee Company, where Oswald once worked. This was according to the owner of the garage, Adrian Alba. See the testimony of Adrian Alba, WC 10, pp. 221–24, and WC 23, CE 1933–34.

5 Backyard photos. See the testimony of Marina Oswald, WC 1, pp. 15–16, and WC 16, CE 133A, 133B, 134, p. 510. The House Select Committee on Assassinations verified the authenticity of these photographs taken from Oswald's camera by Marina Oswald. See U.S. House of Representatives, *Report of the Select Committee on Assassinations* (hereinafter HSCA), 95th Congress, 2nd Session (Washington, DC: U.S. Government Printing Office, 1979), 6, pp. 138–230. For another analysis confirming the authenticity of these photographs, see Dino A. Brugioni, *Photo Fakery: The History and Techniques of Photographic Deception and Manipulation* (Dulles, VA: Brassy's, 1999), pp. 81–84, 89–91, 96.

6 Oswald's subscriptions to the *Militant* and the *Worker.* As early as his service in the military, one fellow Marine remembered seeing in the mailroom Oswald's subscription to the *Worker.* See the affidavit of Paul Edward Murphy, WC 8, p. 320. Another remembered seeing Oswald read this publication. See the testimony of Donald Lewis, WC 8, p. 323.

At the time of the Walker shooting, Oswald had subscriptions to both of these publications. These newspapers can be seen in the backyard photos with dates before the time the photographs were taken. See Gerald Posner, *Case Closed: Lee Harvey Oswald and the Assassination of JFK* (New York: Random House, 1993), p. 107. The Warren Commission also confirmed his subscription to these publications. See WR, appendix 14, "Analysis of Lee Harvey Oswald's Finances from June 13, 1962, through November 22, 1963," pp. 742–44. Oswald also wrote letters to these publications (and his letter actually appears in the issue of the *Militant* he is holding in the photograph). See the testimony of Arnold Samuel Johnson, director of the Information and Lecture Bureau of the Communist Party, confirming Oswald's subscription to the *Worker,* WC 10, pp. 95–107, and WC 20, Arnold Exhibits 1–7, pp. 257–75. Also see the testimony of James J. Tormey, executive secretary of the Gus Hall–Benjamin Davis Defense Committee, WC 10, pp. 107–8, and WC 21, Tormey Exhibits 1–2, pp. 674–77. In addition, see the testimony of

Farrell Dobbs of the *Militant*, WC 10, pp. 109–16, and WC 18, Dobbs Exhibits 1–13, pp. 567–80.

6   "I thought he had gone crazy" and "dangerous toys." See the testimony of Marina Oswald, WC 1, pp. 15–16.

7   "Russian men beat their wives." Ibid.

7   "I'll *make* you shut up." See McMillan, *Marina and Lee,* p. 263.

7   Dinner fight. See the testimony of Marina Oswald, WC 1, p. 6.

9   Other fights and encounters. Ibid., pp. 6, 10. The Oswalds fought often, and their relationship can best be described as abusive. In *Marina and Lee*, McMillan documents numerous quarrels, including those before friends and acquaintances.

9   Oswald practices with rifle. See the testimony of Marina Oswald, WC 1, p. 15.

9   Photographs of strange house (Walker's home). Ibid., p. 14, and WC 16, CE 2–5, pp. 3–9.

9   Fired from printing company. See the testimony of John G. Graef, Oswald's supervisor at Jaggers-Chiles-Stovall, WC 10, pp. 189–91 (complete testimony, pp. 174–94), and WC, CE 1886, pp. 691–92 (FBI verification of Oswald's last day of employment).

10   Let Marina know he lost job. She thought he was looking for work. Ibid., p. 282.

11   JFK honorary citizenship for Churchill, April 9, 1963. See *Public Papers of the Presidents of the United States, John F. Kennedy: January 1, 1963 to November 22, 1963* (Washington, DC: U.S. Government Printing Office, 1964), pp. 315–17.

12   Walker's biography. See Chris Cravens, *Edwin A. Walker and the Right Wing in Dallas, Texas*, M.A. thesis, Southwest Texas State University, 1991. Walker's papers are held at the Eugene C. Barker Texas History Center at the University of Texas at Austin.

12   Oswald's interest in Walker. Even after the assassination attempt, Oswald continued to be fascinated with General Walker's activities. In July 1963, Oswald was invited by his cousin, Eugene Murret, to speak about his experiences in Russia, before fellow seminarians at Spring Hill College in Alabama. The notes by Oswald in his own handwriting, which he made to prepare for this speech, mentioned General Walker's right-wing activities. See WC 16, CE 102, p. 441.

Also, Oswald wrote several letters to the *Worker,* including one undated, but postmarked November 1, 1963, claiming that he attended a meeting where General Walker spoke on October 23, 1963. See WC 10, CE—A. Johnson, exhibit 7, pp. 271–75.

Michael Ralph Paine remembered Oswald going to this speech and specifically was questioned about it. See the testimony of Michael Paine (the husband of Ruth Paine, at whose house Marina and her children were living at the time of the assassination), WC 2, pp. 398–401. Marina Oswald also remembered that her husband went to see this Walker speech, even remembering their conversation, in which Lee said, "Paine knows that I shot him." See the testimony of Marina Oswald, WC 5, pp. 395–96.

12   *Worker* quotes. *Worker*, Oct. 2, 1962, pp. 1, 7.

12   Other *Worker* articles quoted. *Worker*, Oct. 7, 1962, p. 1, and Apr. 2, 1963, p. 4.

13   Marina worried and saw note. The testimony of Marina Oswald, WC 1, p. 17.

13   Quotes from note left for Marina. WC 16, CE 1, pp. 1–2. Marina discusses the note in a Secret Service interview, Dec. 5, 1963. See WC 23, CE 1785, pp. 392–94. The translation from Russian also appears at WC 23, CE 1786, pp. 395–97, where she confirms the note.

15 Walker's comments. Eddie Hughes, "Close Call: Rifleman Takes Shot at Walker," *Dallas Morning News*, April 11, 1963, p. 1. To view his television interview, go to www.youtube.com/watch?v=yCjahRnkQfk.

16 Hides rifle and rides bus home. McMillan, *Marina and Lee,* p. 284.

17 Oswald tells Marina he shot Walker. Testimony of Marina Oswald, WC 1, pp. 16–17.

17 *Dallas Morning News* article. The day after the shooting, April 11, 1963, the first, early city edition of the *Dallas Morning News* contained no mention of the shooting at all. That edition probably was put to bed well before news of the shooting emerged the evening of April 10. A front-page story—Hughes, "Close Call"—appeared in the "Three Star" final morning edition. The next day, two days after the shooting, the story was reprinted on page five. Oswald expressed disappointment over the lack of coverage of his actions. However, it is possible, because of his habit of reading day-old morning newspapers, that Oswald might have not have read about the shooting until two days later. Marina Oswald testifies that her husband heard nothing on the radio that night but did buy a paper the next day and found out what he had missed. Testimony of Marina Oswald, WC 1, p. 17.

18 Oswald's comments of chasing car and "what fools." McMillan, *Marina and Lee,* p. 287.

19 Walker interaction with police. See the complete Dallas police report on the shooting (at the time, the police did not have Oswald as a suspect), WC 24, CE 2001, pp. 36–48. Also see Eddie Hughes, "Close Call: Rifleman Takes Shot at Walker," *Dallas Morning News*, April 11, 1963 (final morning edition), p. 1.

19 "Where is the rifle?" Testimony of Marina Oswald, WC 1, p. 17.

19 Oswald admits he was stalking Walker and shows her his notebook. Ibid., 17–18.

20 Oswald destroys notebook. Testimony of Marina Oswald, WC 1, p. 17, and WC 11, pp. 292–94.

20 More discussion between Marina and Oswald. Testimony of Marina Oswald, WC 1, pp. 17–18.

21 Visit with George de Mohrenschildt, Oswald's friend. Testimony of George de Mohrenschildt, WC 9, pp. 166–284. The House Select Committee on Assassinations prepared a staff report on de Mohrenschildt and his activities: HSCA 12, pp. 47–310. Of particular interest was de Mohrenschildt's unpublished manuscript and his recollection of Oswald's comment concerning the Walker shooting, pp. 200–202.

21 "[H]ow come you missed?" Testimony of George de Mohrenschild, WC 9, pp. 314–17, and testimony of Marina Oswald, WC 1, p. 18. This statement was later denied by de Mohrenschildt. See HSCA 12, p. 202.

22 "Hunter of Fascists—ha-ha-ha!!!" On one of the backyard photos, Oswald wrote a handwritten inscription, "For George, Lee Harvey Oswald" and gave it to de Mohrenschidt. HSCA 8, pp. 339–40. Years later, de Mohrenschidt gave this inscribed photograph to the HSCA. Several experts confirmed that all the backyard photos were genuine, taken by Oswald's own camera, and that the handwriting was indeed Oswald's. HSCA 6, pp. 138–225, and HSCA 8, pp. 227–389. (In fact, HSCA consulted several experts, who confirmed that sixty-three handwritten documents and fingerprints were Oswald's.) Marina Oswald jotted down in Russian on the back of this same photograph given to de Mohrenschidt, "Hunter of Fascists—ha-ha-ha!!!" See Edward J. Epstein, *Legend: the Secret World of Lee Harvey Oswald* (New York: Reader's Digest Press, 1978).

In summary, in the final report, the Warren Commission was certain that Lee Harvey Oswald fired the shot that almost killed General Walker, based on this evidence: (1) The note that Oswald left for his wife on the evening of the shooting, (2) the photographs found among Oswald's possessions after the assassination of President Kennedy, (3) firearm identification of the bullet found in Walker's home, and (4) admissions and other statements made to Marina Oswald by Oswald concerning the shooting. WR, pp. 183–87.

CHAPTER 2: "THE GLOW FROM THAT FIRE"

23 Bartletts hosting dinner party where JFK and Jackie meet. Robert Dallek, *An Unfinished Life: John F. Kennedy, 1917–1963* (Boston: Little, Brown, 2003), p. 192.

24 Wedding in Newport. See Kenneth P. O'Donnell, David F. Powers, and Joe McCarthy, *"Johnny, We Hardly Knew Ye": Memories of John Fitzgerald Kennedy* (Boston: Little, Brown, 1970), pp. 94–96.

24 JFK declares his candidacy. Senate Caucus Room, Jan. 2, 1960. Complete text of this speech, American Presidency Project of the University of California at Santa Barbara, www.presidency.ucsb.edu/ws/index.php?pid=25909. For a contemporary account of this campaign, see Theodore H. White, *The Making of the President, 1960* (New York: Atheneum, 1961).

24 For a comparison of Nixon and Kennedy in the presidential campaign of 1960, see Chris Matthews, *Kennedy & Nixon: The Rivalry That Shaped Postwar America* (New York: Free Press, 2011), and David Pietrusia, *1960: LBJ vs. JFK vs. Nixon, the Epic Campaign That Forged Three Presidencies* (New York: Union Square Press, 2008). For an early politically motivated work published during the 1960 election, see Arthur M. Schlesinger Jr., *Kennedy or Nixon: Does It Make Any Difference?* (New York: Macmillan, 1960).

27 Presidential debates. See U.S. Senate, *Final Report of the Committee on Commerce: The Joint Appearances of Senator John F. Kennedy and Vice President Richard M. Nixon, Presidential Campaign of 1960*, Senate report 994, part 3, 87th Congress, 1st Session (Washington, DC: U.S. Government Printing Office, 1961). Also see Sidney Kraus, ed., *The Great Debates: Background, Perspective, Effects* (Bloomington: Indiana University Press, 1962), and Newton N. Minow and Craig L. Lamay, *Inside the Presidential Debates: Their Improbable Past and Promising Future* (Chicago: University of Chicago Press, 2008).

28 Close election. See U.S. House of Representatives, *Statistics of the Presidential and Congressional Election of November 8, 1960* (corrected to Aug. 15, 1960) (Washington, DC: U.S. Government Printing Office, 1960), p. 51.

28 Inaugural address of JFK. See *Public Papers of the Presidents of the United States, John F. Kennedy; January 20, 1961 to December 31, 1961* (Washington, DC: U.S. Government Printing Office, 1962), pp. 1–3. The Kennedy Presidential Library online maintains a complete video of the inaugural address: www.jfklibrary.org/Asset-Viewer/BqXIEM9F4024ntFl7SVAjA.aspx?gclid=CJzYuZL7i7QCFahQOgodH34A5Q.

For an overview on the writing of this inaugural address, see Richard J. Tofel, *Sounding the Trumpet: The Making of John F. Kennedy's Inaugural Address* (Chicago: Ivan R. Dee, 2005). For a firsthand account of the contributions of Kennedy's principal speechwriter, see Theodore C. Sorensen, *Counselor: A Life of the Edge of History* (New York: Harper, 2008), pp. 200–27.

30 Bay of Pigs. For an overview of the Bay of Pigs incident, see Howard Jones, *The Bay of Pigs* (New York: Oxford University Press, 2008), and Peter Wyden, *The Bay of*

Pigs: The Untold Story (New York: Simon & Schuster, 1979). Also see Haynes Johnson, et al., *The Bay of Pigs: The Leader's Story of Brigade 2506* (New York: Norton, 1964). For a ten-minute Universal Newsreel release reporting the aftermath of the failed invasion, go to the C-SPAN website at www.c-spanvideo.org/program/Ba.

31 Cuban Missile Crisis. For a firsthand account of the Cuban Missile Crisis, see Robert F. Kennedy, *Thirteen Days: A Memoir of the Cuban Missile Crisis* (New York: W.W. Norton & Co., 1969). For a thorough discussion of the incident, see Graham T. Allison and Philip Zelikow, *Essence of Decision: Explaining the Cuban Missile Crisis* (New York: Longman, 1999). Also see Michael Dobbs, *One Minute to Midnight: Kennedy, Khrushchev, and Castro on the Brink of Nuclear War* (New York: Knopf, 2008); Ernest R. May and Philip D. Zelikow, *The Kennedy Tapes: Inside the White House During the Cuban Missile Crisis* (Cambridge, MA: Belknap, 1997); and Max Frankel; *High Noon in the Cold War: Kennedy, Khrushchev, and the Cuban Missile Crisis* (New York: Ballantine, 2004). For a ten-minute Universal Newsreel release report on this crisis, go to the C-SPAN website at www.c-span video.org/program/301730-1.

32 Space race. For a short summary of the space race between the Soviet Union and the United States, go to the History Channel's website at www.history.com/topics/space-race. Also see Von Hardesty and Gene Eisman, *Epic Rivalry: The Inside Story of the Soviet and American Space Race* (Washington, DC: National Geographic, 2007).

33 JFK address to the Joint Session of Congress on the decision to go to the moon, May 25, 1961. See *Public Papers of the Presidents of the United States, John F. Kennedy: January 1, 1961 to December 31, 1961* (Washington, DC: U.S. Government Printing Office, 1962), pp. 403–5 (complete address, pp. 396–406).

33 JFK speech at Rice University, Sept. 12, 1962. Ibid., pp. 668–71. NASA's website has a complete video of this speech: http://er.jsc.nasa.gov/seh/ricetalk.htm.

33 Civil rights and *Brown v. Board*. See generally, Richard Kruger, *Simple Justice: The History of Brown v. Board of Education and Black Americans' Struggle for Equality* (New York: Vintage, 1977), and James T. Patterson, *Brown v. Board of Education: A Civil Rights Milestone and Its Troubled Legacy* (New York: Oxford University Press, 2001).

34 JFK's actions regarding civil rights. See generally Nick Bryant, *The Bystander: John F. Kennedy and the Struggle for Black Equality* (New York: Basic Books, 2006); Juan Williams, *Eyes on the Prize: America's Civil Rights Years, 1954–1965* (New York: Viking, 1987), and Taylor Branch, *Pillar of Fire: America in the King Years, 1963–1965*, second work in the trilogy by this author (New York: Simon & Schuster, 1998).

34 Kennedy's Oval Office TV address on civil rights, June 11, 1963. See *Public Papers of the Presidents of the United States, John F. Kennedy: January 1, 1963 to November 22, 1963* (Washington, DC: U.S. Government Printing Office, 1964), pp. 468–71.

36 Kennedy's speech in the Rudolf Wilde Platz in front of the Brandenburg Gate in Berlin, June 26, 1963. See *Public Papers of the Presidents of the United States, John F. Kennedy: January 1, 1962 to December 31, 1962* (Washington, DC: U.S. Government Printing Office, 1963), pp. 524–25. On the C-SPAN cable television website, there is newsreel footage of JFK trip: www.c-spanvideo.org/program/153127-1.

37 Oswald's early childhood. There are numerous works that touched upon Oswald's early childhood, adolescence, Marine service, life in the Soviet Union, and mar-

riage to Marina. The best primary sources were Oswald's own relatives, such as the book by his brother: Robert L. Oswald, Myrik Land, and Barbara Land, *Lee: A Portrait of Lee Harvey Oswald* (New York: Coward-McCann, 1967). All his immediate relatives extensively testified before the Warren Commission: Marina Oswald (wife), WC 1, pp. 1–126, WC 5, pp. 387–408, 410–420, 588–620, and WC 11, pp. 275–301; Marguerite Oswald (mother), WC 1, pp. 126–264; Robert Oswald (brother), WC 1, pp. 264–469; and John Edward Pic (half-brother), WC 11, pp. 1–82. There also were numerous interviews, affidavits, and testimonies of acquaintances, and a biography on Marguerite Oswald: Jean Stafford, *A Mother in History* (New York: Farrar, Straus & Giroux, 1966). In addition, there was a brief summary of his life in the Warren Report: WR, appendix 13, "Biography of Lee Harvey Oswald," pp. 669–740, and Marina Oswald provided journalist Priscilla Johnson McMillan with much personal insight for her book *Marina and Lee* (New York: Harper & Row, 1977).

37 Oswald's interest in Soviet Union. In an interview with McMillan in Russia on November 16, 1959, Oswald recalled when he first became interested in the Russians, Marx, Engels, and the communists, after being handed a pamphlet on the streets of New York on the atomic spies Julius and Ethel Rosenberg. See McMillan, *Marina and Lee*, p. 62. Also see the testimony of Priscilla Post Johnson (later McMillan), WC 11, pp. 446–60, and WC 20, Johnson (Priscilla) Exhibits 1 and 2, pp. 277–89 (written notes and article she wrote for the North American Newspaper Alliance on this interview). In addition, see WR, appendix 18, "Biography of Lee Harvey Oswald," p. 694, and an interview with Aline Mosby in Moscow in November 1959, WC 22, CE 1385, pp. 701–10.

His fellow Marines called him Oswaldskovich. See the affidavit of James Anthony Bothelo, WC, p. 315. Also see the testimony of fellow Marines Nelson Delgado, WC 8, pp. 228–65, and Kerry W. Thornley, WC 11, pp. 82–115. Thornley would later publish two works—one fiction—on his relationship with Oswald: *Oswald* (Chicago: New Classics House, 1965) and *The Idle Warriors* (Avondale Estates, GA: IllumNet Press, 1991).

38 Oswald's enlistment in the Marines. See Oswald's military record of service, WC 19, pp. 656–768 (Folsom Exhibit 1, pp. 1–131). Also see the testimony of Allison G. Folsom, Lt. Col., USMC, WC 8, pp. 303–10.

38 Shooting himself. See WR, p. 683, and WC 19, pp. 747–52 (Folsom Exhibit No. 1, pp. 109–15). Also see the affidavit of his fellow Marine Paul Edward Murphy, who was in an adjoining cubicle, WC 8, pp. 319–20 (Oswald says, "I believe I shot myself"), and his Marine medical record, WC 19, Donabedian Exhibit 1, pp. 581–617.

39 Rifle scores. See WR, pp. 191–92, and appendix 13, pp. 681–82. Also see WC 16, CE 239, pp. 639–79 (Oswald's Marine Corps Score Book); testimony of Lt. Col. Folsom, WC 8, pp. 310–11; testimony of Maj. Eugene D. Anderson, WC 11, pp. 301–5 (complete testimony 301–6); and the testimony of Sgt. James A. Zahn, WC 11, pp. 308 (complete testimony pp. 306–10).

39 Dependency discharge. Oswald used his mother to request a dependency discharge from the Marines. Marguerite sent her own affidavit attesting to her disabilities. Although he returned to his mother's home, he did not stay long. See WC 19, Folsom Exhibit 1. Robert Oswald stated that his mother indeed had an accident reaching for a glass of candy. See Robert Oswald and Myrick Land, and Barbara Land, *Lee: A Portrait of Lee Harvey Oswald by His Brother Robert Oswald* (New York: Coward-McCann, 1967), pp. 93–95.

39 Travels to Soviet Union. In addition to McMillan, who happened to interview Oswald in the Soviet Union, one of the best sources for information on Oswald's activities in Russia was Norman Mailer, who interviewed some of Oswald's acquaintances and identified some of his interactions in Russia. See Norman Mailer, *Oswald's Tale: An American Mystery* (New York: Random House, 1995). Also, Oswald kept a "Historic Diary," which detailed his activities. See WC 16, CE 29, pp. 94–105. In addition, he wrote letters to his relatives. For another useful assessment of Oswald's activities in the Soviet Union, see Edward J. Epstein, *Legend: The Secret World of Lee Harvey Oswald* (New York: Reader's Digest Press, 1978).

39 Attempted suicide. See Oswald's "Historic Diary," WC 16, CE 24, pp. 94–95 (written in Russian in Oswald's handwriting and then translated).

39 Tries to renounce citizenship. See WR, appendix 13, "Biography of Lee Harvey Oswald," pp. 692–94, and appendix 15, "Oswald's Attempts to Renounce His U.S. Citizenship," pp. 747–51. Also see WC 16, CE 24, "Oswald's Historic Diary," pp. 96–98, and the diplomatic correspondence, WC 18, CE 908–10, pp. 97–105. In addition, see Oswald's handwritten declaration, WC 18, CE 912 and 913, p. 109; testimony of Richard Edward Snyder, who at the time of the interview with Oswald was second secretary and consul at the American Embassy in Moscow, WC 5, pp. 261–70 (complete testimony, pp. 260–99); testimony of John A. McVickar, assistant to Richard Snyder, WC 5, p. 300 (compete testimony pp. 299–306); and memorandum to files summarizing conversation with Oswald, WC 8, CE 910, pp. 106–7.

39 Marries Marina and has child. See McMillan, *Marina and Lee*, pp. 88–99, 136–45.

39 Returns to United States. Oswald's repatriation is a tale of bureaucratic complexities. See generally, WR, appendix 15, "Transactions between Lee Harvey Oswald and Marina and the U.S. Department of State," pp. 752–731, and McMillan, *Marina and Lee*, pp. 100–135, 146–58.

39 Struggles with low-paying jobs. From the time Oswald arrived back in the United States on June 13, 1962, he was unable to keep steady employment. He lied on job applications, and his job performances were poor. He quit or was fired from the four low-paying menial jobs he held up until November 22, 1963. While he was out of work, he appeared at both the Texas Employment Commission and the Louisiana Department of Labor offices in his efforts to find a job, and he collected unemployment compensation. His acquaintances tried to help him find work, and he was interviewed for jobs he did not obtain. His job history is as follows:

- July 17, 1963, through October 8, 1962: sheet metal worker at Louv-R-Pak Division of the Leslie Welding Company in Fort Worth, Texas; hired through a lead by the Texas Employment Commission; quit employment.

  See the testimony of foreman Tommy Bargas, WC 10, pp. 160–66. There were numerous witnesses from the Texas Employment Commission, and Commission exhibits related to Oswald's employment history. See especially the testimony of Mrs. Helen P. Cunningham, WC 10, pp. 117–36, and the testimony of placement interviewer R. L. Adams, WC 10, pp. 117–36. Oswald's inability to hold a job is also evident in the testimony of Marina Oswald, WC 1, pp. 6–29.

- October 12, 1962, through April 6, 1962: cameraman trainee at Jaggars-Chiles-Stovall Co. (printing company) in Fort Worth, Texas, after a lead from the Texas Employment Commission; hired and fired by John Graef (because Oswald could not do the work "although he was trying" and could not get along with fellow employees).

See the testimony of the director of the photographic department, John G. Graef, WC 10, pp. 174–94, and the testimony of the president of the company, Robert L. Stovall, WC 10, pp. 167–74.

- May 10, 1963, through July 19, 1963: machinery greaser at the Reily Coffee Co. in New Orleans; in response to a newspaper advertisement; fired for his inefficiency and inattention to work.

  See the testimony of maintenance man Charles Joseph Le Blanc, WC 10, pp. 213–19.

- October 16, 1963, through November 22, 1963, order filler at the Texas School Book Depository in Dallas; after a lead from Ruth Paine, who was informed by Linnie Mae Randle (the sister of Buell Wesley Frazier, who drove Oswald home on weekends to see his wife); abandoned job and left work early on November 22, 1963.

  See testimony of Ruth Paine, WC 3, pp. 33–34; testimony of Marina Oswald, WC 1, p. 29; and testimony of Roy Sansom Truly, superintendent and member of the Board of Directors, Texas School Book Depository, WC 3, pp. 213–14, as well as WR, p. 738.

39 Writes to Fair Play for Cuba Committee. See WC 20, Lee (Vincent T), Exhibits 1–9, pp. 511–32, and testimony of Vincent T. Lee, WC 10, pp. 86–95.

39 Stores rifle, sits with rifle, practices aiming. See the testimony of Marina Oswald, WC 1, p. 21.

40 Arrest in New Orleans. Oswald was interviewed after his arrest, and a police officer, Lt. Francis L. Martello, testified before the Warren Commission. See WC 10, pp. 51–62. At the police station, Oswald also was interviewed by FBI agent John Lester Quigley. See WC 4, pp.431–40. In his own handwriting, Oswald discussed his activities in New Orleans. See WC 16, CE 93, pp. 341–43 (Oswald's notes on his background).

40 Marina's scorn. See testimony of Marina Oswald, WC 1, p. 23.

40 Oswald's brief TV interview by Bill Slatter of WDSU-TV as rebroadcast on NBC shortly after the assassination, www.youtube.com/watch?v=wYcylHB7Z9k.

41 Oswald's two radio interviews in New Orleans. See testimony of William Kirk Stuckey, WC 11, pp. 156–78.

41 Tracking Oswald down. Ibid., p. 160.

41 Stuckey's first interview. For a complete transcript of the interview, see WC 21, CE Stuckey Exhibit 2, pp. 621–32. Stuckey interviewed Oswald for almost thirty-seven minutes, but the interview edited for broadcast was only four and a half minutes. Also see testimony of William Stuckey, WC 11, pp. 162–66.

44 Second interview (debate). For a complete transcript of the debate, see WC 21, CE Stuckey Exhibit 3, pp. 633–41. Two of the participants in this debate also authored books: Ed Butler, *Revolution Is My Profession* (New York: Twin Circles, 1968), and Carlos Bringuier, *Red Friday: November 22nd, 1963* (Chicago: Chas. Hallberg, 1969). For audio of these two interviews, go to www.youtube.com/watch?v=Jd_JChyrXkU&list=SPDA8542A28B0E6185. There were also several commercially released long-playing phonographic recordings of these interviews.

47 Oswald's mispronunciations and misspellings. Several professionals, including those who interacted with him, attempted to develop a psychological profile and assess his capacity for violence. See the testimony of clinical psychologist Renatus

Hartogs, WC 8, pp. 214–24, and his book: Renatus Hartogs and Lucy Freeman, *The Two Assassins* (New York: Crowell, 1965). Also see Abrahamson David, *Our Violent Society* (New York: Funk & Wagnalls, 1970), "Lee Harvey Oswald: Psychological Capacity for Violence and Murder," pp. 129–60.

50 Stuckey takes Oswald for a drink. See testimony of William K. Stuckey, WC 11, p. 171.

50 Oswald's description of his activities with the Fair Play for Cuba Committee (FPCC), printing of handbills, proud of media coverage. See WC 16, CE 93, p. 343 (notes by Oswald on his background).

51 Oswald's description of himself. See WC 16, CE 93 (notes of Oswald on his background). Also see his manifesto, entitled "The Collective," wherein he writes: "Lee Harvey Oswald was born in Oct 1939 in New Orleans La. the son of a Insuraen [*sic*] Salesman whose early death left a far mean streak of indepence [*sic*] brought on by negleck [*sic*]," WR, p. 395, and WC 16, CE 92, p. 285 of "Oswald's Typed Narrative Concerning Russia" (complete document, pp. 285–336).

52 Marina's comments of Oswald sitting with rifle. See the testimony of Marina Oswald, WC 1, p. 65, and McMillan, *Marina and Lee*, p. 362.

52 Oswald cleaning his rifle. WC 1, p. 14.

53 Oswald wanted to go to Cuba, and Marina's response. Ibid., p. 22.

53 Oswald's fantasy of killing Nixon. Robert Oswald testified that Marina stated that Lee Harvey Oswald had the "intention to shoot Richard M. Nixon and that Marina Oswald had locked Lee in the bathroom for the entire day." See WC 1, p. 335. In her second appearance before the Warren Commission, Marina confirmed this threat. She mentioned a pistol and also stated that the entire incident took about twenty minutes and he was in the bathroom for about five minutes until he "quieted down." See WC 5, pp. 392–93. Her memory was less solid as to whether there were locks on the door or whether Lee could get out when he wanted. See WC, p. 389. This testimony was confirmed by FBI interviews: Robert Oswald, WC 22, CE 1357, pp. 596, and Marina, WC 22, CE 1404, p. 786.

53 Oswald wanting to hijack aircraft and studying airline schedules. Testimony of Marina Oswald, WC 1, pp. 22–23.

53 "I cannot conclude he was against the President." Ibid., p. 22.

54 Marina on Lee's "sick imagination" and delusions of grandeur. Ibid.

54 Oswald's interest in biography and statesmen, comparing himself to them. Ibid., pp. 22–23.

CHAPTER 3: "SHOW THESE TEXANS WHAT GOOD TASTE REALLY IS"

55 Kennedy image and style. Numerous works have focused on JFK's charisma. See John Hellmann, *The Kennedy Obsession: The American Myth of JFK* (New York: Columbia University Press, 1997); Henry Fairlie, *The Kennedy Promise: The Politics of Expectation* (Garden City, NY: Doubleday, 1973); David M. Lubin, *Shooting Kennedy: JFK and the Culture of Images* (Berkeley: University of California Press, 2003); and Thomas Brown, *JFK: The History of the Image* (Bloomington: Indiana University Press, 1988).

57 Death of Patrick. See O'Donnell, *Johnny*, pp. 377–78.

57 Marina moves to Dallas and lives with Paines. See the testimony of Marina Oswald, WC 1, p. 26, the testimony of Ruth Hyde Paine, WC 3, pp. 4–7, and WR, p. 730.

57 Carrying the rifle. See the testimony of Marina Oswald, WC 1, p. 26.

58 Oswald's trip to Mexico City. With regard to Oswald's activities in Mexico City, see WC, pp. 730–37. The FBI conducted an extensive investigation of Oswald's trip. See CE 2121, FBI Report, May 18, 1964, pp. 570–659. The FBI indentified and interviewed numerous people who interacted with Oswald. Several individuals remembered Oswald on the bus to Mexico and spoke with him. See the affidavit of two British tourists, John Bryan and Meryl McFarland, WC 11, pp. 214–15, and the testimony of Pamela Mumford, a secretary from Los Angeles, WC 11, pp. 215–24.

From Marina Oswald's testimony, as well as the documents she produced, it was confirmed that Oswald had indeed visited Mexico City. See the testimony of Marina Oswald, WC 1, pp. 27–28 and CE 3073, FBI report, Oct. 15, 1964, pp. 667–72. Oswald visited the Cuban Embassy to request a visa. See WC 25, CE 2564, pp. 813–18 (documents that were given to the United States by the Cuban ambassador). Oswald also visited the Soviet Embassy. In a typed note, signed by Oswald on November 9, 1963, he recalled his failed efforts during his visit to Mexico City to get a visa to go to Cuba. See WC 16, CE 15, p. 33. This was drafted with the help of Ruth Paine. See the testimony of Ruth Paine, WC 3, pp. 13–18, 51–52, and WC 9, p. 95.

59 Test ban treaty. See Joint Statement by the Heads of the Delegations to Moscow Nuclear Test Ban Meeting, July 25, 1963; Radio and Television Address to the American People on the Nuclear Test Ban Treaty, July 26, 1963; Special Message to the Senate on Nuclear Test Ban Treaty, August 8, 1963; and Remarks at the Signing of the Nuclear Test Ban Treaty, October 7, 1963, *Public Papers of the Presidents of the United States, John F. Kennedy: January 1, 1963 to November 22, 1963*, pp. 599–601, 601–6, 622–24, 765–66.

59 Oswald hired at the Texas School Book Depository. Ruth Paine was informed by Linnie Mae Randle, the sister of Buell Wesley Frazier, who worked at the Texas School Book Depository, that there might be an opening. Paine called Roy Truly, who then interviewed and hired Oswald. He began work on October 16, 1963. See the testimony of Ruth Paine, WC 3, pp. 33–34; testimony of Marina Oswald, WC 1, p. 29; and the testimony of Roy Truly, WC 3, pp. 213–14, WR, p. 738. It was happenstance that Oswald was assigned to the Book Depository. He easily could have been selected to work at the other building, "the Warehouse," located at 1917 Houston Street, a few blocks away from the Book Depository. Both Oswald and another individual started work that same day, and according to Truly, he chose Oswald to work at the Book Depository. See the testimony of Roy Truly, WC 3, p. 237; Donald Jackson, "Evolution of an Assassin," *Life*, Feb. 21, 1964, p. 78; and Bugliosi, *Reclaiming History*, p. 1455.

59 Oswald moves to boardinghouse at 1026 N. Beckley Avenue. See testimony of housekeeper, Mrs. Earlene Roberts, WC 6, pp. 436–37; testimony of landlord, Mrs. Arnold Carl (Gladys J.) Johnson, WC 10, pp. 293–94; and WR, p. 737.

59 Birth of new daughter. Marina Oswald wrote: "Monday evening Lee visited me in the hospital. He was very happy at the birth of another daughter and even wept a little. He said that two daughters were better for each other—two sisters. He stayed with me about two hours." See WR, p. 738 (this is CE 904, not printed in WC).

59 Oswald attends Walker rally. See testimony of Michael Paine, WC 2, pp. 398–401.

60 Angry note to FBI. Hosty destroyed this note, and it was not known until around the time of the HSCA investigation. See testimony of James Patrick Hosty Jr., WC 4, pp.440–76, and James P. Hosty Jr., *Assignment Oswald* (New York: Arcade Publishing, 1996). Also see U.S. Senate, *The Final Report of the Select Committee to Study Government Operations with Respect to Intelligence Activities* (Church Commit-

tee), book 5, *The Investigation of the Assassination of John F. Kennedy: Performance of the Intelligence Agencies* (Washington, DC: U.S. Government Printing Office, 1976), appendix B, "The FBI and the Destruction of the Oswald Note," pp. 95–97.

60 Decision to visit Texas. JFK's advisers knew that Texas would be a key to his 1964 election. At the time, there were political tensions between the liberal and conservative factions within the Texas Democratic Party. In June 1963, Kennedy convened a private meeting with both LBJ and Governor Connally in El Paso, Texas. Neither Johnson nor Connally wanted the president to visit or meddle in local Texas politics. JFK viewed this visit, among other things, as a fund-raising opportunity. There was an October 4, 1963, meeting at the White House with Connally— Johnson was excluded. A five-city visit was planned, San Antonio, Houston, Fort Worth, Dallas, and Austin, concluding with JFK staying overnight at the LBJ Ranch on the evening of November 22, 1963. See "Why Kennedy Went to Texas," by Governor John Connally Jr., *Life*, Nov. 24, 1967, p. 86. Also see Steven M. Gillon, *The Kennedy Assassination—24 Hours After: Lyndon B. Johnson's Pivotal First Day as President* (New York: Basic Books, 2009), pp. 12–15, and William Manchester, *The Death of a President: November 20–November 25, 1963* (New York: Harper & Row, 1967), pp. 21–25.

61 Kennedy White House parties. See Anne H. Lincoln, *The Kennedy White House Parties* (New York: Viking, 1967).

61 "Remarks of the President at the Dinner Honoring All Living Nobel Prize Winners in the Western Hemisphere," April 29, 1962, ". . . since Thomas Jefferson dined alone." See *Public Papers of the Presidents of the United States, John F. Kennedy: January 1, 1962 to December 31, 1962* (Washington, DC: U.S. Government Printing Office, 1963), p. 347.

61 Planned state dinner and birthday party for John Jr. See Manchester, *Death*, pp. 15–16.

62 Speech in Amherst, October 26, 1963, "one acquainted with the night." See *Public Papers of the Presidents of the United States, John F. Kennedy; January 1, 1963 to November 22, 1963* (Washington, DC: U.S. Government Printing Office, 1964), pp. 815–18.

63 Political climate in Texas—attack on UN Ambassador Adlai Stevenson. See John Barlow Martin, *Adlai Stevenson and the World: The Life of Adlai Stevenson* (New York: Doubleday, 1977), pp. 774–75. For footage of the attack on Stevenson, go to www.youtube.com/watch?v=XWXoAPMcVrE.

63 Warned to stay away from Texas. See Manchester, *Death*, pp. 38–41. Because of right-wing extremism in Dallas, both LBJ and Governor Connally urged Kennedy to make only a quick luncheon speech in Dallas but no public appearances. Connally was especially opposed to a motorcade in Dallas. See John Connally Jr., "Why Kennedy Went to Texas," *Life*, Nov. 24, 1967, p. 86.

63 Fulbright, "dangerous place." See Manchester, *Death*, p. 39.

63 Skelton letter to RFK. Ibid., pp. 33–35.

64 Jackie's choice of clothes and JFK comments. "There are going to be all these rich, Republican women at that lunch. . . . Be simple—show these Texans what good taste really is." See Manchester, *Death*, p. 10.

The dressmaker of Jackie's pink suit also was surrounded with some mystery. Although from the 1950s Jackie was a client of Coco Chanel, the world-renowned French fashion designer, Mrs. Kennedy received some bad press because of her shopping spree while visiting Paris. So although the very expensive pink, double-breasted suit (which was one of JFK's favorites) was first shown by Chanel in her 1961 autumn/winter collection, there is very strong evidence that the suit was actually a knockoff copied by Oleg Cassini in the Chez Ninon dress shop in New York City. See Justine Picardie, *Coco Chanel: The Legend and the Life* (London: HarperCollins, 2010), pp. 304–7, and the comments of Karl Lagerfeld, the current designer of the Chanel label, on the Fashionista website: http://fashionista.com/2012/03/karl-lagerfeld-says-oleg-cassini-knocked-off-that-pink-chanel-suit-jackie-kennedy-wore-the-day-jfk-was-assassinated. There are photographs of Jackie wearing this dress or something very similar several times prior to November 22. See www.pinkpillbox.com/pinksuit.htm.

66 Kennedy's inscribed photo. Currently in a private collection.

67 Oswald's opinion of JFK. According to the testimony of Paul Roderick Gregory (an acquaintance of the Oswalds who visited their home as a student studying Russian in the late summer of 1962), the Oswalds "always" had a copy of *Life* magazine in their living room with a picture of John F. Kennedy on the cover. Gregory stated that Oswald admired Kennedy and made positive comments on JFK as a "nice man" and a "good leader." See testimony of Paul Roderick Gregory, WC 9, p. 148 (complete testimony, pp. 141–60). McMillan places the magazine on the living-room table. See McMillan, *Marina and Lee*, p. 194.

According to Marina, "Lee compared himself to the great men he read about in books and genuinely believed that he was one of them." See McMillan, *Marina and Lee*, p. 321. She was explaining why Lee borrowed so many books about great leaders for the library. In addition, from the records of the New Orleans Public Library, it was determined that in July 1963, Oswald borrowed William Manchester's book on JFK, *Portrait of a President*, and JFK's *Profiles in Courage*. WC 2, CE 2650, pp. 928–31.

67 Publication of the motorcade route from Main to Houston to Elm and then onto the Stemmons Freeway. The complete motorcade route was revealed to the public by the two Dallas newspapers on Tuesday, November 19, 1963. See United States, See WC 22, CE 1362, 1363.

68 Oswald reading day-old papers. There were numerous sources and witnesses who saw Oswald reading newspapers from the previous day left behind at the Texas School Book Depository. Coworker Charles Douglas Givens stated that he never saw Oswald buy a newspaper and that he would read yesterday's papers. However, Givens also stated that two of Oswald's coworkers, Harold Norman and James Jarman, brought in the *Dallas Morning News* every day (WC 6, p. 352), so it is possible that Oswald read a paper or two on the same day.

69 Beginning in 1840 and ending in 1960, every president elected in a year ending in zero died in office. This phenomenon, the "zero-year curse," was finally broken in 1981, when President Ronald Reagan survived an assassin's bullet outside the Washington Hilton Hotel in the District of Columbia.

The first documented assassination attempt on a sitting U.S. president was the attack on Andrew Jackson outside the U.S. Capitol on January 30, 1835, after he attended the funeral for a member of the U.S. House of Representatives. Both pistols of the assassin, Richard Lawrence, misfired, and Jackson began to beat him with a cane. Almost immediately he was restrained by the crowd. The prosecutor at the trial was Francis Scott Key, and Lawrence was quickly found not guilty by reason of insanity. Lawrence was housed at various institutions and eventually was moved to St. Elizabeth's Hospital in Washington, D.C. (the place where another would-be assassin, John Hinckley Jr., the attacker of Ronald Reagan, resides today). For a contemporary account of the Jackson attack, see *Shooting at the President! The Remarkable Trial of Richard Lawrence, Self-Styled "King of the United States," "King of England and of Rome," &c. &c. &c., for an Attempt to Assassinate the President of the United States . . . by a Washington Reporter* (New York: W. Mitchell, 1835), 16 pp. This rare work is the first freestanding pamphlet or book published on the assassination or attempted assassination of a U.S. president. One of the few copies that exist can be found in the rare book collection of the Library of Congress.

For a good general overview of the assassinations and attempts on presidents and other public figures, see James W. Clarke, *American Assassins: The Darker Side of Politics* (Princeton, NJ: Princeton University Press, 1982).

69 Attack on Blair House. See Stephen Hunter and John Bainbridge Jr., *American Gunfight: The Plot to Kill Harry Truman—and the Shoot-out That Stopped It* (New York: Simon & Schuster, 2005).

69 Puerto Rican nationalist shooting up U.S. House of Representatives. See Jim Abrams, "It's Been 50 Years since Worst Attack on Congress," *Puerto Rico Herald*, Feb. 24, 2004, www.puertorico-herald.org/issues/2004/vol8n10/ItsBeen50.html, and Edward F. Ryan, "Terror at the Capitol," *Washington Post*, March 2, 1954, pp. 1, 12–13.

69 Who was Lee Harvey Oswald? For a probing exposé on Oswald, see the compelling 1993 *Frontline* documentary, "Who Was Lee Harvey Oswald?" summarized on the PBS website: www.pbs.org/wgbh/pages/frontline/shows/oswald and the PBS *American Experience* broadcast, "Oswald's Ghost": www.pbs.org/wgbh/amex/oswald.

71 The president's limousine and second car in San Antonio. Because the president was visiting so many cities so quickly, the Secret Service was unable to transport the presidential limousine to all the cities. Instead, the Ford Motor Company provided additional Lincoln town car convertibles. See Gerald Blaine and Lisa McCubbin, *The Kennedy Detail: JFK's Secret Service Agents Break Their Silence* (New York: Gallery Books, 2010), p. 161.

71 Handmade signs, JACKIE, COME WATERSKI IN TEXAS and SEGREGATED CITY. See Patricia Howard Harris, *An Austin Scrapbook of John F. Kennedy* (Austin, TX: Pemberton Press, 1964).

71 Brooks Medical Center, "She really did touch me." See Manchester, *Death*, p. 76.

71 JFK San Antonio speech at Brooks Medical Center, Nov. 21, 1963. See *Public Papers of the Presidents of the United States, John F. Kennedy: January 1, 1963 to November 22, 1963* (Washington, DC: U.S. Government Printing Office, 1964), pp. 882–83.

72 Change in Oswald's routine, breakfast at Dobbs House. See the FBI interview with restaurant waitress, Mrs. Dolores Harrison, WC 26, CE 2009, p. 536, and McMillan, *Marina and Lee*, pp. 414–15.

76 Oswald asks Frazier to drive him on Thursday to get curtain rods. See testimony of Buell Wesley Frazier, Oswald's coworker and acquaintance, WC 2, pp. 222–27 (complete testimony, pp. 210–45) and WC 7, p. 531.

77 Jackie to Rep. Thomas, "Don't leave me!" See Manchester, *Death*.

77 Marina surprised to see Lee on Thursday. See testimony of Marina Oswald, WC 1, p. 65, and McMillan, *Marina and Lee*, p. 414.

78 Oswald tells Marina he will buy her a washing machine. See WC 1, p. 66, and McMillan, *Marina and Lee*, p. 416.

78 Whether Oswald knew president was coming to Dallas. See testimony of Marina Oswald, WC 1, p. 70.
79 Oswald upset at Marina and whether Marina was too hard on him. Ibid., p. 65.
80 LBJ visits JFK room, problems with Yarborough. See Manchester, *Death*, p. 82.
81 JFK's Houston speech to League of United Latin American Citizens (LULAC), Nov. 21, 1963. See *Public Papers of the Presidents of the United States, John F. Kennedy: January 1, 1963 to November 22, 1963* (Washington, DC: U.S. Government Printing Office, 1964), p. 884.
81 Jackie speaks to crowd in Spanish. Because Jackie was able to communicate in a few foreign languages, she endeared herself in Paris with her ability to speak French (her major in college was French literature). She also spoke Spanish. To see her brief remarks in Spanish at the LULAC meeting, go to www.youtube.com/watch?v=c2tQRfyucKY. In Miami in December 1962, she also joined the president and spoke in Spanish at the Orange Bowl to members of Brigade 2506, which participated in unsuccessful Bay of Pigs invasion. See video of her complete Miami speech in Spanish at www.youtube.com/watch?v=72z2025GyBE.
81 Houston Coliseum speech in honor of Rep. Albert Thomas, Nov. 21, 1963. "Your old men shall dream . . ." See *Public Papers . . . 1963*, pp. 884–86.
81 Oswald kicking Marina away when they went to bed. See McMillan, *Marina and Lee*, p. 419.
82 "Not everyone can understand . . ." Marina Oswald's notebook of poems in Russian and translated. See WC 16, CE 106.
83 "You were great today." See Manchester, *Death*, p. 87.

CHAPTER 4: "A BRIGHT PINK SUIT"
84 Morning edition of *Dallas Morning News*, November 22, 1963, p. 14, WELCOME MR. KENNEDY TO DALLAS." See testimony of Bernard William Weismann, who purchased the advertisement, WC 5, pp. 487–535, WC 11, p. 428–34; WC 16, CE 1031, pp. 835, and WC 21, Weismann Exhibit 1, p. 722.
85 Handbill, WANTED FOR TREASON. Robert Alan Surrey, a business partner and political associate of General Walker, was the author of this handbill. About five thousand copies were printed on "cheap colored newspaper print" ("dodger stock") and then were inserted in some morning newspapers or put on car windshields. See testimony of Robert G. Klause, WC 5, pp. 535–46. Also see testimony of Robert Alan Surrey, WC 5, pp. 420–49, and WC 18, CE 996, p. 646.
85 Curry goes on TV, November 20. See Jerry T. Dealey and George Bannerman Dealey, *D in the Heart of Texas* (Dallas: JEDI Management Group, 2002). According to these descendants of the namesake of Dealey Plaza, after the attack on United Nations ambassador Adlai Stevenson, civic leaders held a series of meetings in the office of Stanley Marcus (of Neiman Marcus, the Dallas-based department store). They concluded that there was no graceful way to retract the invitation to JFK and instead had the mayor, Earle Cabell, and the Dallas police chief, Jesse Curry, appear on local TV and radio stations, as well as give interviews in newspapers. They called on the citizens of Dallas to act according to law and warmly welcome the president. They condemned violence and warned that there would be a response to any misbehavior or protests. For instance, see Curry's TV appearance at www.youtube.com/watch?v=n7K2OTw2Ds0. Also see Jesse Curry, *Retired Dallas Police Chief Jesse Curry Reveals His Personal JFK Assassination File* (Dallas: American Poster and Printing Company, 1969.

85 JFK "we're heading into nut country today." See O'Donnell, *Johnny*, p. 23.

85 "If someone wants to shoot me from a window." See Manchester, *Death*.

85 JFK fatalistic attitude. See, O'Donnell, *Johnny*, p. 19.

85 JFK's statement about possible assassination the night before. See Manchester, *Death*.

86 The Palm Beach attempt on JFK. See Blaine, *Kennedy Detail*, pp. 73–74, 136–37, 149–50. The Secret Service expressed fears of JFK riding in a motorcade with an open car. "Unless the president changed his mind about the motorcade, or agreed to ride in a closed-top car—something Blaine knew just wasn't going to happen— the only way to have a chance at protecting the president against a shooter from a tall building would be to have agents posted on the back of the car. Even then, if somebody had the advantage of looking down on the motorcade and was a good enough marksman to hit the president with the first shot, there was little an agent could do. No matter how quickly you reacted, no man was faster than a speeding bullet." (P. 74; note to reader: Blaine wrote his book in third person.)

86 JFK's zero-year curse letter penned when Kennedy was a senator. See "Political Auction," Potomack Company, Nov. 1, 2008, sale 15, lot 100.

In her diary, Evelyn Lincoln, Kennedy's secretary, said that she found a note written by JFK in 1961 (made public in July 1998) that suggested that the president had a premonition of his own death. The note paraphrased a thought that has been attributed to President Lincoln and was sometimes quoted by JFK in his 1960 presidential campaign: "I know that there is a God and I see a storm coming. If he has a place for me, I am ready." *Los Angeles Times* wire service, July 29, 1998, http://articles.latimes.com/1998/jul/29/news/mn-8243.

86 JFK favorite poem, "rendezvous with death." In October 1953, when they returned from their honeymoon, Jack read to Jackie what he said was his favorite poem, by Alan Seeger. See Arthur M. Schlesinger Jr., *A Thousand Days: John F. Kennedy in the White House* (New York/Boston: Houghlin Mifflin, 1965), p. 98.

86 Oswald leaves money and wedding ring on dresser. See testimony of Marina Oswald, WC 1, pp. 72–73.

87 Oswald walks to Frazier's house; Buell's sister remembers about curtain rods and sees him carrying a package. See testimony of Linnie Mae Randle, WC 2, p. 248.

87 Frazier's mother asks "who is that?"—first time Oswald ever walked to Frazier's house. See testimony of Frazier, WC 2, p. 222.

87 "What's the package?" Oswald-Frazier exchange. Ibid., pp. 226–27.

88 Oswald and Frazier did not discuss President Kennedy's visit. Ibid., p. 227.

89 Oswald jumps out of car on arrival at Texas School Book Depository; they usually walked in together. Ibid., pp. 229–30.

89 Frazier's impression of how Oswald carried package. Ibid., pp. 228–29.

90 Oswald begins filling orders. Ibid., p. 231.

91 Weather in Dallas, decision not to install bubble top. See testimony of Secret Service agent Roy H. Kellerman, WC 2, pp. 66–67 (complete testimony pp. 61–112). Also see Clint Hill and Lisa McCubbin, *Mrs. Kennedy and Me: An Intimate Memoir* (New York: Gallery Books, 2012), p. 284; and O'Donnell, *Johnny*, p. 25.

91 Kennedy's speech to crowd outside Texas Hotel. "[N]o faint hearts in Fort Worth." See *Public Papers of the Presidents of the United States, John F. Kennedy: January 1, 1963 to November 22, 1963* (Washington, DC: U.S. Government Printing Office, 1964), p. 887. The JFK Fort Worth tribute website has the complete audio of

this address: www.jfktribute.com/videos. (There is also the complete video of the breakfast address to the Chamber of Congress.)

92 "Where's Jackie?" See O'Donnell, *Johnny*, p. 23.

92 "Mrs. Kennedy is organizing herself." See *Public Papers . . . January 1, 1963 to November 22, 1963*, p. 887.

92 Hill is asked to get Jackie and tells her she is going to breakfast. See testimony of Secret Service Agent Clinton J. ("Clint") Hill, WC 2, p. 133 (complete testimony, pp. 132–44). Also see Hill, *Mrs. Kennedy and Me*, pp. 282–83.

93 JFK comments about Jackie in Paris. See *Public Papers . . . January 1, 1963 to November 22, 1963*, pp. 888–90.

93 Al Olsen, and then Power's comment, "She's not that kind of Bunny." See Manchester, *Death*.

93 Jackie's comment, "I'll go anywhere with you this year." See O'Donnell, *Johnny*, p. 24.

94 Open windows and modern Secret Service protection. The Secret Service had been assigned to protect the president after the McKinley assassination. However, by today's standards, security was woefully inadequate. While the Secret Service agents in 1963 were dedicated, conscientious professionals, it really was not until after the assassination attempt on President Reagan that the protection of the "package" (the president) in the motorcade was really secure. There may now be up to three bulletproof cars that appear to be the president's vehicle (two as decoys and one that actually carries the president, the car being known as the Beast).

    With few exceptions, such as presidential inaugurals and State of the Union addresses, the motorcade route is varied and almost never publicized. All traffic is stopped and all streets are closed. At presidential inaugurals (with the exception of Carter's), when the president gets out of the car to walk on Pennsylvania Avenue, it is usually where governmental buildings, the FBI and Justice Department, are on either side of the street. There are never any open windows. Frequently, the president wears a bulletproof jacket at public functions. The Secret Service is not shy about putting up bulletproof glass to prevent sharpshooter snipers from firing from great distances. (For instance, this was employed when President Obama spoke in Grant Park in Chicago on the night he was first elected. Bulletproof glass was also used in his 2013 speech in Berlin, in contrast to the similar addresses given there by JFK and Reagan.) Oswald shot from a few hundred feet. Today a sharpshooter with a modern sniper rifle accurately can hit a target from a mile away.

    For a very thorough treatment of the tactics and procedures of the modern U.S. Secret Service in protecting the president, see the 2012 three-part series on the Discover-Military Channel: *Secret Service Secrets*—"The Home Front," "Campaign Nightmare," and "On Enemy Soil." Two *National Geographic* DVD releases are also informative: *Inside the Secret Service*, 2004, and the 2013 release, *Secret Service Files: Protecting the President*.

95 No crowd screening. Even up until John Hinckley Jr.'s attempted assassination of Reagan, the "zone-of-protection" was inadequate. Today there are metal detectors at all public functions. Nevertheless, the most difficult protective detail is to guard a president, especially presidential candidates when they are "working the crowd."

    In 1964, the Warren Commission made several recommendations to improve presidential protection. See WR, pp. 425–69, ch. 8, "Protection of the President."

95 "Every possible precaution." Even the early afternoon November 22, 1963, edition of the *Dallas Times Herald* (the first issue, printed before the assassination) contained a first-page story with the headline SECRET SERVICE SURE ALL IS SECURE.

96 Ken O'Donnell, "This trip is turning out to be terrific." See O'Donnell, *Johnny*, p. 26.

96 "Mr. and Mrs. America." Ibid.

96 "There is Mrs. Kennedy . . . bright pink suit." See Cathy Trost and Susan Bennett, *President Kennedy Has Been Shot: Experience the Moment-to Moment Account of the Four Days That Changed America* (Naperville, IL: Sourcebooks, 2003), p. 15.

97 Comments from reporters at Love Field. In 1964, many long-playing 33 rpm records were commercially released that replayed the events of November 22. Some of these include: *The Assassination of a President: The Four Black Days, November 22–25, 1963* (Beverly Hills, CA: American Society of Record Drama); *The Actual Voices of the Four Days That Shocked the World, Nov. 22–25, 1963: The Complete Story* (New York: Colpix Records, "produced in association with United Press International," 1964; rereleased as a CD in the UK by RPM Records, 1998); *The Fateful Hours*, "actual unforgettable news reports on Friday, November 22nd, 1963 by KLIF Dallas" (Hollywood, CA: Capitol Records); and *Four Dark Days in November: A Presentation of WQMR News*, "featuring the actual ON-THE-AIR coverage of the events of November 22, 23, 24, and 25th 1963 as reported and broadcast by Radio Station WQMR, Washington, D.C." (N.p.: Capitol Records).

98 JFK steps up to chain-link fence to greet crowd. See Blaine, *Kennedy Detail*, pp. 200–1.

98 Art Rickerby photos. For a review of the photographs taken the day of the Kennedy assassination, see three books by Richard B. Trask: *Pictures of the Pain: Photography and the Assassination of President Kennedy* (Danvers, MA: Yeoman Press, 1994); *That Day in Dallas: Three Photographers Capture on Film the Day Kennedy Died* (Danvers, MA: Yeoman Press, 1998); and *National Nightmare on Six Feet of Film: Mr. Zapruder's Home Movie and the Murder of President Kennedy* (Danvers, MA: Yeoman Press, 2005).

99 Charles Roberts question to Jackie about campaigning. See Trost, *President*, p. 15.

102 The motorcade. The following listing is reflective of numerous firsthand accounts of journalists and Secret Service agents, as well as information obtained from testimony and exhibits (WC 17, CE 767 and CE 768, and WC, 18, CE 1024 and 1026—Secret Service Memoranda, as well as Commission Document 3—Secret Service Report) to the Warren Commission, interviews, police reports, actual film of the motorcade, obituaries, and many other sources. For the Secret Service code of names of some individuals in the motorcade, see Manchester, *Death*, and Blaine, *Kennedy Detail*. For determining the make and model of some of the cars, the location of some of the police motorcycle riders, and the location of some of the occupants in the vehicles, see: Todd Wayne, Vaughan. *Presidential Motorcade Schematic Listing; November 22, 1963, Dallas, Texas*. Jackson, Michigan: by the author, 1993 (a signed limited edition of 100 copies). Also helpful for compiling this listing are Mary Ferrell's Chronologies, Volume 4, November 22, 1963, Narrative, and Book 1, November 22, 1963 (www.maryferrell.org), and Trask's *Pictures of the Pain*. In the chaos of the events and the "whistle stop" nature of the President's visit to Texas, there are different, and sometimes, conflicting accounts as to the exact location of some of the individuals in this Dallas motorcade. This is especially true of the three buses and the location of some members of JFK's staff.

The Advance car

- Dallas Police Captain Perdue William Lawrence, Assistant in Charge of motorcade detail (Testimony—WC, Vol. VII, pp. 577–589 and WC, Vol. XX, Lawrence Exhibits Nos.1 through 4, pp. 482–98).

The pilot car (a white Ford sedan) (WC, Vol. XVII, CE 768, p. 615; WC, Vol. XXI, Stevenson Exhibit 5053, p. 578—Report to Dallas Police Chief Jesse E. Curry; Curry, *JFK Assassination File*, p. 24; and Mary Ferrell's Chronologies, Volume 4, November 22, 1963, Narrative, p. 369-1b).

- Dallas Police Deputy Chief George L. ("G.L") Lumpkin (driver)
- Dallas Homicide Detective Faye M. ("F.M.") Turner (front passenger seat)
- Dallas Homicide Detective William ("Billy") L. Senkel (left rear seat)
- Lt. Col. George L. Whitmeyer, Jr., East Texas Section Commander of the Army Reserve (middle rear seat)
- Jacob L. ("Jack") Puterbaugh, Democratic National Committee Advance Man (right rear seat)

Three two-wheel advance motorcycles (Harley Davidsons) (viewed from behind motorcade) (WC, Vol. XX, Lawrence Exhibit No. 2, p. 489).

- Dallas Police Sergeant Samuel Q. Bellah (commander of this unit) (middle)
- Dallas Police Officer Glen C. ("G.C.") McBride (left)
- Dallas Police Officer J.B. Garrick (right)

Five two-wheel lead motorcycles (Harley Davidsons) (viewed from behind) (Vaughan, *Presidential Motorcade*, p. 4 and Mary Ferrell's Chronologies, Volume 4, November 22, 1963, Narrative, p. 369-1b)

- Dallas Police Officer Stavis ("Steve") Ellis (commander of this unit) (far left)
- Dallas Police Officer E.D. ("Buddy") Brewer (immediate left of center)
- Dallas Police Officer Leon E. ("L.E.") Grey (center)
- Dallas Police Officer Harold B. ("H.B." or "Harry") Freeman (immediate right of center)
- Dallas Police Officer William G. ("W.G." or "Bill) Lumpkin (far right)

The lead car (a four door Ford Mercury sedan) (WC, Vol. XVII, CE 768, p. 615; numerous sources, including WC, Vol. XVIII, CE 1026—Secret Service Memorandum; and Manchester, *Death*, p. 133)

- Dallas Police Chief Jesse E. Curry (driver seat) (Testimony—WC, Vol. IV, pp. 150–202. Vol. XII, p. 25, Vol. XV, pp. 124–33, 641)
- Secret Service Agent Winston George ("Win") Lawson (front passenger seat) (Testimony—WC, Vol. IV, pp. 317–58)
- Dallas County Sheriff James Eric ("J.E." or "Bill") Decker (left rear seat), (Testimony—WC, Vol. XII, pp. 42–52)
- Secret Service Agent Forrest V. Sorrels, (in charge of the Dallas field office) (right rear seat) (Testimony—WC, Vol. VII, pp. 332–60, 592 and. Vol. XIII, pp. 55–83)

The Presidential limousine (Code name "SS 100 X") (a 1961 Lincoln Continental flown in from Washington, D.C.) (District of Columbia orange License Plate Number

GG-300) (numerous sources, including WC, Vol. XVIII, CE 1024 and WC, Vol. XVIII, CE 1026—Secret Service Memoranda; and Manchester, *Death*, p. 134)

- Secret Service Agent William Robert ("Bill") Greer (driver seat) (Testimony—WC, Vol. II, pp. 112–32 and Statement—WC, Vol. XVIII, CE 1024, p. 723)
- Secret Service Special Agent in Charge of Dallas Detail Roy H. Kellerman (Code name "Digest") (front passenger seat) (Testimony—WC, Vol. II, pp. 61–112)
- Mrs. John H. Connally, Jr. ("Nellie") (left middle "jump" seat) (Testimony—WC, Vol. IV, pp. 146–49)
- Governor John H. Connally, Jr. (right middle "jump" seat) (Testimony—WC, Vol. IV, pp. 129–46).
- First Lady, Mrs. John F. Kennedy (Jacqueline "Jackie") (Code name "Lace") (left rear seat) (Testimony—WC, Vol. V, pp. 178–81)
- President John F. Kennedy ("Jack") (Code name "Lancer") (right rear seat)

Motorcycle escorts for the Presidential limousine (Harley Davidsons viewed from behind) (WC, Vol. XX, Lawrence Exhibit No. 2, p. 489 and Trask, *Pictures of the Pain*, p. 63)

- Dallas Police Officer William Joseph Martin ("Billy Joe") (far left) (Testimony—WC, Vol. VI, pp. 289–93)
- Dallas Police Officer Bobby W. ("B.W.) Hargis (left) (Testimony—WC, Vol. VI, pp. 293–96)
- Dallas Police Officer James M. ("J.M.") Chaney (right)
- Dallas Police Officer Douglas L. ("D.L.") Jackson (far right) (Jackson's recollection's sent to District Attorney Henry Wade, June, 1980, in the Harold Weisberg archives at Hood College)

The Presidential follow-up car (Code name "Half-back") (SS 679X) (a 1956 Cadillac nine-passenger four-door touring sedan with the top down that was flown into Texas from Washington, DC) (WC, Vol. XVIII, CE 1024; WC, Vol. XVIII, CE 1026—Secret Service Memoranda; Manchester, *Death*, p. 134; and Trask, *Pictures of the Pain*, p. 63)

- Secret Service Agent Samuel A. ("Sam") Kinney (driver seat) (Statement—WC, Vol. XVIII, CE 1024, pp. 730–31)
- Secret Service Assistant to the Special Agent in Charge Emory P. Roberts (Code name "Dusty") (right passenger seat) (Statement—WC, Vol. XVIII, CE 1024, pp. 733–38)
- Secret Service Agent Clinton J. ("Clint") Hill (Code name "Dazzle") (left front running board) (Testimony—WC, Vol. II, pp. 132–44 and Statement—WC, Vol. XVIII, CE 1024 pp. 740–45)
- Secret Service Agent William T. ("Tim") McIntyre (left rear running board) (Statement—Warren Commission Document (CD) 3, pp. 99–100)
- Secret Service Agent John D. ("Jack") Ready (right front running board) (Statement—CD 3, p. 102)
- Secret Service Agent Paul E. Landis, Jr. (Code name "Debut") (right rear running board) (Statement—in WC, Vol. XVIII, CE 1024, pp. 751–59)
- Presidential aide Kenneth P. O'Donnell (Code name "Wand") (left middle seat) (Testimony—WC, Vol. VII, p. 440–56)
- Special Assistant to the President David Francis ("Dave") Powers (right middle seat) (Affidavit—WC, Vol. VII, pp. 472–74)

- Secret Service Agent George Warren Hickey, Jr. (left rear seat) (Statement—WC, Vol. XVIII, CE 1024, pp. 730–31)
- Secret Service Agent Glen A. Bennett (right rear seat) (Statement—WC, Vol. XVIII, CE 1024, p. 760)

The Vice Presidential limousine (a 1962 light blue/gray Lincoln 4-door convertible) (numerous sources, including WC, Vol. XVIII, CE 1024 and 1026—Secret Service Memoranda, and Manchester, *Death*, p. 134)
- Texas Highway Patrolman Hurchel D. Jacks (driver) (WC, Vol. XVIII, CE 1024, p. 801)
- Secret Service Agent Rufus Wayne Youngblood (Code name "Dagger") (front passenger seat) (Testimony—WC, Vol. II, pp. 144–55 and Statement—WC, Vol. XVIII, pp. 766–72)
- Senator Ralph Yarborough (left rear seat) (Testimony—WC, Vol. VII, pp. 439–40)
- Mrs. Lyndon Baines ("Lady Bird") Johnson (Code name "Victoria") (middle rear seat) (Testimony—WC, Vol. V, pp. 564–67)
- Vice-President Lyndon Baines Johnson (Code name "Volunteer") (right rear seat) (Testimony—WC, Vol. V, pp. 561–64)

The Vice Presidential follow-up car (Code name "Varsity") (a yellow 1963 Ford Mercury hardtop sedan) (WC, Vol. XVII, CE 768, p. 607; WC, Vol. XVIII, CE 1024 and 1026—Secret Service Memoranda; and Manchester, *Death*, p. 134)
- Texas Highway Patrolman Joe Henry Rich (driver seat) (Statement—WC, Vol. XVIII, CE 1024, p. 800)
- Vice Presidential aide Clifton C. ("Cliff") Carter (front middle seat) (Affidavit—WC, Vol. VII, pp. 474–75)
- Secret Service Agent Jerry D. Kivett (Code Name "Daylight") (right passenger seat) (Statement in WC, Vol. XVIII, CE 1024, pp. 776–77)
- Secret Service Agent Warren W. ("Woody") Taylor (left rear seat) (Statement in WC, Vol. XVIII, CE 1024, pp. 782–84)
- Secret Service Agent Thomas Lemuel ("Lem") Johns (Code Name "Dandy") (right rear seat) (Statement in WC, Vol. XVIII, CE 1024, pp. 773–75)

The car of the Mayor of Dallas (a two-door 1963 white Ford Mercury Comet Caliente convertible) (WC, Vol. XVII, CE 768, p. 615)
- Texas Highway Patrolman Milton T. Wright (driver) (Statement—WC, Vol. XVIII, CE 1024, p. 802—Secret Service Memorandum)
- Dallas Mayor Earle Cabell (front passenger seat) (Testimony—WC, Vol. VII, pp. 467–85)
- Mrs. Earle (Elizabeth "Dearie") Cabell (rear left seat) (Testimony—WC. Vol. VII, p. 485–91)
- Rep. Herbert Ray ("Ray") Roberts (D-4th Congressional District-TX) (read right seat) (In testimony of Earle Cabell)

The White House National Press "pool car" (a 1960 blue-gray or black Chevrolet
- A driver provided by the telephone company (SW Bell Co.)
- Malcolm ("Mac") Kilduff (Code name "Warrior") White House assistant press secretary (front passenger seat)

- Merriman Smith ("Smitty"), White House Correspondent *UPI* (middle front seat)
- Jack Bell, *AP* (left or middle rear seat)
- Robert E. Baskin, Washington Bureau Chief of *The Dallas Morning News* (left or middle rear seat)
- Bob Clark, *ABC News* Washington correspondent (right rear seat)

Press—Camera car No. 1 (a 1964 yellow 2-door Chevrolet Impala convertible) (motion picture cameras) (Trask, *Pictures of the Pain*, pp. 306, 358–90)
- A Texas Ranger (driver)
- John Hoefan, NBC sound engineer (middle front passenger)
- David ("Dave") Wiegman, Jr., *NBC* cameraman (right passenger seat)
- Thomas J. Cravens, Jr., *CBS* cameraman (left rear seat)
- Cleveland ("Cleve") Ryan, a lighting technician (middle rear seat)
- Thomas Maurer ("Ollie") Atkins, Navy White House photographer (right rear seat) ("Remembering JFK," *Ohio Today; For Alumni and Friends of Ohio University*, Fall 2003: www.ohio.edu/ohiotoday/fall03/jfk/)

Press—Camera car No. 2 (a 1964 silver 2-door Chevrolet Impala convertible) (still cameras) (Trask, *Pictures of the Pain*, pp. 306, 391–13)
- A Driver
- Donald C. ("Clint") Grant, *Dallas Morning News* photographer (middle front seat) (*C-Span* broadcast, "Journalists Remember the JFK Assassination," November 20, 1993)
- Frank Cancellare, *UPI* photographer (front passenger seat) (jumped out of motorcade at Dealey Plaza)
- Captain Cecil W. Stoughton, JFK's photographer (left rear seat)
- Arthur ("Art") Rickerby, *Life Magazine* photographer (middle rear seat).
- Henry D. Burroughs, *AP* photographer (right rear seat)

Press—Camera car No. 3 (a gray 2-door 1964 Chevrolet Impala convertible) (Trask, *Pictures of the Pain*, pp. 306, 414–31)
- An individual from the Texas Department of Public Safety (driver)
- James H. Underwood, *KRLD-TV* (Channel 4, the CBS affiliate in Dallas) (middle passenger seat) (Testimony—WC, Vol. VI, pp. 167–71)
- Thomas C. ("Tom") Dillard, chief photographer, *The Dallas Morning News* (right passenger seat) (the flexibility of driving in a car, gave photographers like Dillard an opportunity to pre-arrange with their coworkers a point to stop and drop-off film that he was taking all along the parade route in order to meet newspaper print deadlines) (Testimony—WC, Vol. VI, pp.162–67).
- James ("Jimmy") Darnell, *WBAP-TV* (Channel 5, the NBC affiliate in Fort Worth) (left rear seat) (In Underwood's testimony in WC)
- Malcolm O. ("Mal") Couch, *WFAA-TV* (Channel 8, the ABC affiliate in Dallas) (middle rear seat) (Testimony—WC, Vol. VI, pp. 153–62)
- Robert Hill ("Bob") Jackson, photographer, *The Dallas Times Herald* (right rear seat) (WC, Vol. II, pp. 155–64)

Two motorcycles (Harley Davidsons) (Traveling behind the second camera car on each side of cars) (Trask, *Pictures of the Pain*, p. 306 and Mary Ferrell's Chronologies, Volume 4, November 22, 1963, Narrative p. 372)

- Dallas Police Officer Marrion L. ("M.L.") Baker (right) (Testimony—WC, Vol. III, pp. 242–70)
- Dallas Police Officer Hollis B. ("H.B.") McLain (left) (Testimony HSCA—Vol. V, pp. 617–71)

Members of Congress—Car No. 1 (a white Ford Mercury Comet Caliente two-door convertible) (WC, Vol. XVII, CE 768—Secret Service Memorandum and Manchester, *Death*, p. 132)

- Driver
- Rep. George Herman Mahon (D-19th Congressional District-TX) (right front passenger seat)
- Rep. Walter Edward Rogers (D-18th Congressional District-TX) (left rear seat)
- Rep. W. Homer Thornberry (D-10th Congressional District-TX) (middle rear seat)
- Lawrence F. ("Larry") O'Brien, Special Assistant to the President for Congressional Relations (right rear seat) (Testimony—WC, Vol. VII, pp. 457–72)

Members of Congress—No. 2 (a white Ford Mercury Comet Caliente two-door convertible) (WC, Vol. XVII, CE 768, p. 616—Secret Service Memorandum)

- Driver
- Rep. Albert Richard Thomas (D-8th Congressional District-TX) (middle front passenger seat) (Mary Ferrell's Chronologies, Vol. 4, November 22, 1963, Narrative, p. 371)
- Rep. Jack Brooks (D-2nd Congressional District-TX) (right front passenger seat)
- Rep. Lindley Beckworth, Sr. (D-3rd Congressional District-TX) (left rear seat)
- Rep. Olin Earl ("Tiger") Teague (D-6th Congressional District-TX) (middle rear seat)
- Rep. James Claude ("Jim") Wright, Jr. (D-12th Congressional District-TX (right rear seat)

Members of Congress—Car No. 3 (a gray 1964 Lincoln sedan) (WC, Vol. XVII, CE 768, p. 616)

- Driver
- Rep. John Andrew Young (D-14th Congressional District-TX) (front passenger seat)
- Rep. Henry Barbosa Gonzalez (D-20th Congressional District-TX) (left rear seat)
- State Senator William Neff ("Bill") Patman (D-14th Texas District) (middle rear seat)
- Rep. Graham Boynton Purcell, Jr. (D-13th Congressional District-TX) (right rear seat)

White House Staff Car (a 1964 Ford Mercury Colony Park Station Wagon) (Manchester, *Death*, p. 131 and Trask, *Pictures of the Pain*, p. 210) (some sources, place other senior White House Staff in this vehicle, contrary to Manchester, who places them in the VIP bus)

- Major General Chester V. ("Ted") Clifton (Code name "Watchman"), U.S. Army Presidential Aide (driver)
- Major General Godfrey T. ("God") (Code Name "Wing") McHugh, U.S. Air Force Presidential Aide (front passenger seat)

Two motorcycles (Harley Davidsons) (Vaughan, *Presidential Motorcade*, p. 24 and Mary Ferrell's Chronologies, Volume 4, November 22, 1963, Narrative p. 372)

- Dallas Police Officer James W. ("J.W.") Courson (left)
- Dallas Police Officer Clyde. A. Haygood (right) (Testimony—WC, Vol. VI, pp. 296–301)

The Staff and Dignitary Bus (a Continental Trailways Bus) (Curry, *JFK Assassination File*, p. 16 and WC, Vol. XVII, CE 768, p. 616—Secret Service Memorandum, list this as the first bus. But WC, Vol. XVII, CE 767, p. 596—Secret Service Memorandum, list this as the third bus)

- Driver
- Evelyn Lincoln (Code Name "Willow"), personal secretary to the President (Manchester, *Death*, p. 132)
- Rear Admiral George G. Burkley, MD, (Code Name "Market") the President's physician (Manchester, *Death*, p. 132)
- Mary Barelli Gallagher, personal secretary to Jacqueline Kennedy (Manchester, *Death*, p. 174)
- Pamela Turnure, press secretary to Jacqueline Kennedy (Manchester, *Death*, p. 174)
- Marie Fehmer, secretary to Vice President Lyndon Johnson (Manchester, *Death*, p. 132)
- Elizabeth ("Liz") Carpenter, Executive Assistant to the Vice President Manchester, *Death*, p. 132)
- Jack Valenti, political consultant (Manchester, *Death*, p. 132)
- Sergeant Paul Glynn, U.S. Air Force, personal aid to the Vice President (Vaughan, *Presidential Motorcade*, p. 34)
- Love Field—Airport Reception Committee (WC, Vol. XX, CE 767, p. 599) (though on the Secret Service list, some of these individuals did not participate in the motorcade)
  - Robert B. ("Bob") Cullum, President, Chamber of Commerce
  - Rev. Luther Holcomb, Head of Council of Churches (In an oral history interview Holcomb's believed that he was traveling in one of the cars—www.eeoc.gov/eeoc/history/35th/voices/oral_history-luther_holcomb-dana_whitaker.wpd.html) (About 1:14 p.m. (C.S.T.) Holcomb, who was supposed to give the invocation at the Trade Mart, offered a prayer for the president from the podium, after having learned that JFK and Governor Connally had been shot)
  - Clifton ("Cliff") Cassidy, Vice President, Democratic Committee
  - John Grey, Committeeman

- ○ Eugene M. Locks, State Chairman (Not in motorcade—Flew to Austin to be Master of Ceremonies for the planned Austin Dinner)
- ○ Mr. J. Erik Jonsson, Head, Citizen's Council (met party at Love Field, but went to Trade Mart and announced to the audience that the President had been shot)
- ○ Mrs. Erik Jonsson (met party at Love Field, but probably went to Trade Mart with her husband)
- ○ W. Dawson Sterling, Head, Dallas Assembly
- ○ Charlie King, President AFL-CIO
- ○ Allen Maley, Executive Secretary, AFL-CIO
- ○ James E. Smith, President, Negro Chamber of Commerce
- ○ Dave Moss, Head, Democratic Clubs
- ○ Lew Sterrett, County Judge
- ○ Harold Barefoot Sanders, Jr., U.S. Attorney for the Northern District of Texas (appointed by JFK) (Manchester, *Death*, p. 174)
- ○ Dallas Women's Club Council President
- ○ George Miner, Vice-President, State AFL-CIO
- ○ David Keeler, Vice-President, State AFL-CIO

Press—Bus No. 1 (a Continental Trailways Bus) (Three buses were filled and as the motorcade briefly stopped after the assassination a few reporters, such as Robert MacNeil, chose to leave the first bus and cover the story from Dealey Plaza. The first press bus stopped; the second, did not). Some of the individuals included:

- • Joe Savage, Continental Trailways bus driver (Oral Interview, Sixth Floor Museum—stated he drove the first press bus)
- • Bo Byers, *Houston Chronicle*, reporter (*C-Span* broadcast, "Journalists Remember the JFK Assassination," November 20, 1993. But the *Houston Chronicle* puts him in the second bus ("Houston, Nation Shaken by the Kennedy Assassination," *Houston Chronicle*, September 23, 2001: www.chron.com/life/article/Houston-nation-shaken-by-Kennedy-assassination-2032469.php)
- • Harry Cabluck, *Fort Worth Star-Telegram* reporter (Trask, *Pictures of the Pain*, p. 331)
- • Richard Beebe Dudman, *St. Louis Post Dispatch* (Vaughan, *Presidential Motorcade*, p. 25)
- • Douglas Kiker, *New York Herald Tribune* (obituary, *New York Times*, August 15, 1991)
- • Robert MacNeil, *NBC* White House correspondent (MacNeil, *Right Place*, p. 207)
- • James ("Jim") Mathis, *The Advance Syndicate (Newhouse)* reporter ("The Assassination; The Reporter's Story; How Journalists broke the news of JFK's Death," *Columbia Journalism Review*, Winter 1964)
- • Marianne Means, *Hearst Newspapers* (her appearance on the *C-Span* broadcast, "Covering the Kennedy Assassination," sponsored by Close Up Foundation, October 30, 2003: www.c-spanvideo.org/program/KennedyAss)
- • Robert Charles Pierpoint, *CBS*, White House correspondent (Roberts, *Truth*, p. 13)

- Charles ("Chuck") Roberts, *Newsweek* (Roberts, Charles. *The Truth about the Assassination*, New York: Grosset & Dunlap, 1967, p. 11)
- Hugh Sidey, *Time Magazine*, White House correspondent ("That Infamous Day in Dallas" by Hugh Sidey, *Time Magazine*, March 31, 2003)
- Tom Wicker, *New York Times*, White House correspondent *Columbia Journalism Review* article)

Local Press Pool Car (*Dallas Morning News*) (Vaughan, *Presidential Motorcade*, p. 26, citing several *Dallas Morning News* articles and "The Reporter's Notes on JFK Assassination are Opened" by Will Pry, *Dallas Morning News*, June 5, 2013, which includes notes by Harris and Quinn) (though not on any manifest, this car found a place between the first and second press buses) www.dallasnews.com/news/jfk50/explore/20130 605-the-reporters-notes-on-jfk-assassination-are-opened.ece
- Lewis Harris (driver)
- Mike Quinn (right passenger seat)
- Kent Biffle (left rear seat)
- Larry Groves (right rear seat) (Mary Ferrell's Chronologies, Vol. 4, November 22, 1963, Narrative, p. 397)

Press—Bus No. 2 (a Continental Trailways bus)
- Driver
- David Broder, *Washington Star.* (Donovan, *Boxing*, p. 114)
- Sidney ("Sid") Davis, *Westinghouse Broadcasting* ("We Heard the Shots . . ." by Sid Davis *American Heritage Magazine*, November–December 2003, Vol. 54, Issue 7)
- Robert J. Donovan, Washington bureau chief, *Los Angeles Times* (Donovan, Robert J. *Boxing the Kangaroo; A Reporter's Memoir.* Columbia, Missouri: University of Missouri Press, 2000, p. 113)
- Ronnie Dugger, *Texas Observer* (*Columbia Journalism Review* article)
- Seth Kantor, *Scripps Howard* (Testimony—WC, Vol. XV, pp. 71–96)
- Jerry terHorst, *Detroit News* ("The Assassination; The Reporter's Story; How Journalists broke the news of JFK's Death," *Columbia Journalism Review* article)
- Robert ("Bob")Young, *Chicago Tribune* (his obituary, *Chicago Tribune*, June 11, 2000 (probably in this bus)

Some of the others on one of the two press buses
- Julian Read, press secretary to Governor Connally (Oral history interview Sixth Floor Museum) (Vaughan identifies him as possibly traveling in White House Staff Car based on the one of the Willis photographs, Vaughan, *Presidential Motorcade*, p. 23)

From the Washington Press Corps
- Oscar Henry ("Henry") Brandon, Washington correspondent, *London Sunday Times* (Zelizer, *Covering the Body,* p. 78 and Manchester, *Death*, p. 661)
- Frank Cormier, White House correspondent for the *Associated Press* (obituary, *New York Times*, February 11, 1994) (www.nytimes

.com/1994/02/11/obituaries/frank-cormier-66-covered-5-presidents-for
-associated-press.html)

- Allan ("Al") Cromley, Washington bureau chief, *Daily Oklahoman*
  (obituary in the *Daily Oklahoman*, August 9, 2011: http://newsok.com/
  al-cromley-former-washington-bureau-chief-for-the-oklahoman-dies/
  article/3593006)
- Thomas H. Flaherty, Jr., *Life Magazine* (www.zoominfo.com/p/Thomas
  -Flaherty/1197548983)
- Edward T. Folliard, staff writer for the *Washington Post* ("President
  Kennedy Shot Dead; Lyndon B. Johnson is Sworn In," by Edward T.
  Folliard, *Washington Post*, www.washingtonpost.com/wp-dyn/content/
  article/2009/07/01/AR2009070102229.html)
- Jack Gertz, *AT&T* (Columbia Journalism Review article)
- Robert Hollingsworth, Washington bureau chief, *Dallas Times Herald*
  (Oral history interview, Sixth Floor Museum)
- Carlton Kent, reporter Chicago Sun-Times (his coverage of the assassi-
  nation in articles published in his newspaper)
- Peter Lisagor, Washington bureau chief, *Chicago Daily News* (numerous
  sources, including *Columbia Journalism Review* article)
- William R. MacKaye, Washington bureau chief, *Houston Chronicle* (his
  coverage of the assassination in articles published in his newspaper)
- William May, reporter for the *Newark News* (Sachsman, David B. and
  Sloat, Warren. *The Press and the Suburbs; the Daily Newspapers of New
  Jersey*. New Brunswick, New Jersey: Center for Urban Policy Research,
  1985, p. 29)
- Alan L. Otten, White House reporter, *Wall Street Journal* (Obituary, *Bos-
  ton Globe*, August 22, 2009) www.boston.com/bostonglobe/obituaries/
  articles/2009/08/22/alan_l_otten_political_reporter_for_wall_street_
  journal_at_88/)
- Robert Roth, reporter for the *Philadelphia Bulletin* ("The Bulletin
  Fondly Recalled in Philadelphia . . ." *Philadelphia Inquirer*, January 30,
  2007, http://articles.philly.com/2007-01-30/news/25220399_1_bulletin
  -building-audit-bureau-circulation)
- Robert L. Riggs, Washington Bureau Chief for the *Louisville
  Courier-Journal* his coverage of the assassination in articles published
  in his newspaper)
- Alvin Silverman, Washington Bureau Chief for the *Cleveland Plain
  Dealer* ("Kennedy Assassinated; Johnson Takes Oath; Reporter's Account of the Horror," by Alvin Silverman, *Cleveland Plain Dealer*, November 22, 2011: http://blog.cleveland.com/pdextra/2011/11/kennedy_
  assassinated_johnson_t.html)
- Felton West, Washington bureau chief, *Houston Post* (his obituary,
  Houston Chronicle, June 25, 2005 and his coverage of the assassination
  in articles published in his newspaper)

In a list reporter Seth Kantor provided to Todd Wayne Vaughan (Vaughan, *Presidential Motorcade* p. 28) additional White House Press Corp on the buses included:

- Robert Cahn, *United States Information Agency*
- William Costello, *Mutual Broadcasting System*
- Jack Doyle, *AT&T*
- Paul Healey, *New York Daily News*
- William Hercher, *US News and World Report*
- Robert Hilburn, *Fort Worth Star-Telegram*
- Fred Lawrence, *UPI-News*
- Carrol S. Linkins, *Western Union*
- James M. Perry, political writer, *National Observer*
- Philip ("Phil") Potter, reporter for the *Baltimore Sun*
- Jack Schultz, *UPI-News*
- Jack Williams, *Kansas City Star*
- Andrew Willoner, soundman, *CBS-TV*

From the local press

- Bruce Neal, *Wendall Mays Radio Stations* (Oral history interview with Sixth Floor Museum)
- Keith Shelton, *Dallas Times Herald* ("The Story through the Eyes of a Storyteller," *Denton Record-Chronicle*, June 29, 2013: www.dentonrc .com/local-news/local-news-headlines/20130629-the-story-through -the-eyes-of-a-storyteller.ece)

In the same Seth Kantor listing, additional local reporters on the buses included:

- Mary Rice Brogan, *Houston Chronicle*
- Sam Kinch, Jr. *Fort Worth Star-Telegram*
- Gene Kraft, (Gene Ashcraft) *KFJZ- Radio, Fort Worth*
- Ted Rozumalski, photographer, *Houston Chronicle*
- Johnny Tackett, *Scripps-Howard*
- Arthur Uhlmann, *Houston Chronicle.*

Two Motorcycles (Harley Davidsons) (WC, Vol. XX, Lawrence Exhibit No. 2, p. 489)

- Dallas Police Officer R. Smart (left)
- Dallas Police Officer Bobby Joe ("B.J.") Dale (right)

The Western Union car (a 1957 black Ford) (WC, Vol. XXI, Semingsen Exhibit No. 3001, Attachment No. 7, p. 420)

- R.C. Johnson, sales manager (driver)
- Mr. Yates, Sales Representative

The White House "Communications" Car carrying the nuclear code "football" car (a 1964 white Chevrolet Impala) (WC, Vol. XVII, CE 768, p. 616 and Manchester, *Death*, pp. 62–63 and 129)

- Chief Warrant Officer Arthur W. Bales, Jr., (Code Name "Sturdy")U.S. Army, White House Signal Corps officer
- Chief Warrant Officer Ira Gearhart (Several different Code names associated with him "Bagman" "Shadow") (the one who held the codes to launch a nuclear attack—"the football")

A Dallas Police car (1963 black and white Ford) (Vaughan, *Presidential Motorcade,* p. 35)

- Dallas Police Officer J.M. Phillips (driver)
- J.S. Davenport (Mary Ferrell's Chronologies, Volume 4, November 22, 1963, Narrative, p. 373)

One three wheel police motorcycle (Vaughan, *Presidential Motorcade,* p. 36)

103 Secret Service waved away from standing on president's car. Secret Service Assistant to the Special Agent in Charge Emory Roberts waved off Secret Service Agent Don Lawton. This was in some way a poor attempt at an inside joke, since Lawton knew that JFK did not want agents to block the view of the crowds, and he also knew that his assignment was actually to stay behind to keep Love Field and Air Force One secure. He was not assigned to the motorcade detail. Instead, Secret Service Agent Clint Hill was crouched on the running board behind the president's car for much of the motorcade. See Blaine, *Kennedy Detail,* pp. 200–206. Also, see footage of presidential limousine departing from Love Field at www.youtube.com/watch?v=x7WrBVm9ALM.

103 JFK speech not delivered at the Dallas Trade Mart. See *Public Papers of the Presidents of the United States, John F. Kennedy: January 1, 1963 to November 22, 1963,* pp. 890–94.

104 Austin dinner speech not delivered at the Austin Civic Auditorium. See *Public Papers . . . January 1, 1963 to November 22, 1963* pp. 894–98.

Although there have been unconfirmed reports that most of the copies of the twenty-page dinner program for the Texas Democrats Welcome Dinner were picked up from the printer and then destroyed that afternoon, many of these programs continue to find their way into the collector marketplace. In his eerie introductory remarks in this program, Governor Connally mentioned: "This is a day long to be remembered in Texas." See *Texas Welcome: Texas Welcomes the President of the United States and the Vice President of the United States* (Austin: State Democratic Committee, November 22, 1963).

According to Stanley Marcus, one of the civic leaders in Dallas whose last name adorned the department store Nieman Marcus, Lyndon Johnson planned to deliver a closing line at this fund-raiser dinner to elicit laughter: "And thank God, Mr. President" then with a silent pause, "that you came out of Dallas alive." Stanley Marcus, *Minding the Story: A Memoir* (Boston: Little, Brown, 1974), p. 255. However, as journalist Max Holland noted, this retelling may not be a true account. Max Holland, *The Kennedy Assassination Tapes: The White House Conversations of Lyndon B. Johnson Regarding the Assassination, the Warren Commission, and the Aftermath* (New York: Knopf, 2004), p. 6.

105 Frazier's comment about viewing the motorcade. See testimony of Buell Wesley Frazier, WC 2, p. 232. The Warren Commission testimony of some of Oswald's coworkers demonstrated that they all knew that the presidential motorcade was going to pass by their building, and that they planned to watch it.

105 Oswald asked if he planned to watch motorcade. See the testimony of Buell Wesley Frazier, WC 2, p. 232–33.

107 Speed of limousine throughout the motorcade. Secret Service Agent Rufus Youngblood stated that the motorcade began with a speed of 20 to 25 miles an hour, and Secret Service Agent Gerald Blaine believed that the motorcade began at a

speed of 25 to 30 miles per hour. But the presidential car slowed considerably as the crowds began to form and especially when the motorcade proceeded west on Main Street. As the crowds thinned, the motorcade was going about 10 miles per hour when it began to turn onto Houston Street. Manchester said that the president's car was going 11.2 miles per hour. See the testimony of Rufus Youngblood, WC 2, pp. 47–48; Blaine, *Kennedy Detail*, p. 204; and Manchester, *Death*, p. 154.

108 How to look and wave at crowd. See O'Donnell, *Johnny*, p. 26.

108 Jackie, "The sun was so strong." See Manchester, *Death*, and Theodore White Camelot interview.

108 "No sign of hostility." See O'Donnell, *Johnny*, p. 26.

108 "[W]ild welcome." Ibid., p. 27.

108 Home videos. See the 8mm film *President Kennedy's Final Hour, Dallas, Texas, November 22, 1963* (Dallas Film Associates, 1964). Special Assistant to the President Dave Powers also was filming the motorcade while he was traveling in the Secret Service follow-up car directly behind JFK. He ran out of film at 12:17 (EST). For this account, go to the CNN website at www.cnn.com/US/9611/21/kennedy.lost.film.

108 Motorcade stops when JFK sees sign from children, and also for nuns. See the testimony of Governor John D. Connally Jr., WC 4, p. 132.

109 Nixon attacked by mob in Venezuela. His wife was in the car with him in 1958. See Richard M. Nixon, *Six Crises* (Garden City, NY: Doubleday, 1962), pp. 197–233.

From Love Field to Dealey Plaza. For a firsthand account by Governor Connally's wife, who traveled in the presidential limousine, from the time of the arrival at Love Field, see Nellie Connally and Mickey Herskowitz, *From Love Field: Our Final Hours with President John F. Kennedy* (New York: Rugged Land Books, 2003). For a DVD with rare video footage, see *Lost JFK Tapes: Assassination* (National Geographic, 2009), excerpted at http://channel.nationalgeographic.com/channel/videos/jfks-assassination.

112 "You can't say that Dallas doesn't love you." Through the years, Nellie Connally often quoted herself, and slightly varied these remarks. For instance, in her statement to the press at Parkland Hospital on November 24, 1963, she said, "I had just turned around and said to him, you can't say Dallas doesn't love you, Mr. President. That was it." See www.youtube.com/watch?v=P5QkgDXzQ0M. In her Warren Commission testimony in the spring of 1964, she changed this slightly, and remarked that "after we got off Main Street . . . the receptions had been so good every place that I showed much restraint by not mentioning something about it before. I could resist no longer. When we got past this area I did turn to the President, and said 'Mr. President, you can't say Dallas doesn't love you.' " See the testimony of Nellie Connally, WC 4, p. 147. Years later, in her book, she remembered, "Mr. President, you certainly cannot say that Dallas does not love you." See Connally, *From Love Field*, p. 7. When she was promoting her book on ABC's *The View* in 2003, she stated: "So, I turned to the President and said, 'Mr. President, you can't say Dallas doesn't love you.' " See www.youtube.com/watch?v=YfR09lJ4H6U&NR=1&feature=endscreen.

113 The three black men on fifth floor. These three individuals, all coworkers of Oswald, were in the closest proximity to him when the shots were fired. They all heard the shots coming from above them on the sixth floor, heard the sound of cartridge shells dropping on the floor, and experienced some cement falling from the ceiling. See the testimony of Bonnie Ray Williams, WC 3, pp. 161–84; Harold Norman, WC 3, pp. 186–98; and James Jarman Jr. ("Junior"), WC 3, pp. 198–211.

113 Williams comments on why he didn't decide to watch from sixth floor, and near encounter with Oswald. See the testimony of Bonnie Ray Williams, WC 3, pp. 165–69.

113 Position of Williams, Jarman, and Norman. See WC 3, pp. 173–74.

114 Arnold Louis Rowland and his wife came to Dealey Plaza to watch motorcade before his going to work at pizza restaurant. See the testimony of Arnold Louis Rowland, WC 2, pp. 165–90.

114 Rowland commenting on security presence and attack on Ambassador Adlai Stevenson. Ibid., p. 169.

115 Rowland recognized that there were individuals in the Texas School Book Depository windows. Ibid., pp. 169, 174–79.

115 Rowland. "He was standing there and holding a rifle." Ibid.

115 Rowland's description of man. Thought he might be "a security agent" or from "Secret Service." Ibid., pp. 169, 171–72, 174.

116 Distraction of epileptic, man in sixth-floor window disappears, Rowland continues to look at window constantly. Ibid., p. 174.

116 Brennan also notes epileptic. See the testimony of Howard Leslie Brennan, a steamfitter who was working on a job nearby and immediately after lunch went to view the motorcade, WC 3, p. 141 (complete testimony WC 3, pp. 140–61, 184–86, 211). Years later, Brennan would also write about his observations before and after the assassination. See Howard L. Brennan and J. Edward Cherryholmes, *Eyewitness to History: The Kennedy Assassination, as Seen by Howard Brennan* (Waco, TX: Texian Press, 1987).

117 Photograph capturing image of Brennan. See the testimony of Howard Brennan, WC 3, p. 479, and WC 17, CE 479, p. 178 (Zapruder frame used for Brennan to identify his location).

117 Brennan observing crowd. See the testimony of Howard Brennan, WC 3, p. 143.

117 Brennan sees man on sixth floor. Ibid., p. 144.

117 Brennan sees JFK car turn Main, Houston, Elm. Ibid.

118 Rowland, "maybe we should tell someone." See the testimony of Arnold Rowland, WC 2, p. 189.

119 Rowland commenting on pink suit. "My wife likes clothes." Ibid., p. 175.

119 Abraham Zapruder. Initially Abraham Zapruder, manufacturer of ladies' dresses, did not bring his camera to work. At the urging of his secretary, Lillian Rogers, he returned home and retrieved his "high end" 8mm Bell & Howell 414PD Director Series camera. Around 11:30 A.M., he began to consider the best location from which he could film the motorcade. This included his own office window at Jennifer Jones Inc. of Dallas at 501 Elm Street, just across the street and to the east of the Texas School Book Depository. Finally Zapruder and his receptionist, Marilyn Sitzman, decided that the best location would be to stand on top of a four-foot-high concrete pedestal, which extended to the left of the pergola on the north side of Elm Street. See the testimony of Abraham Zapruder, WC 7, p. 570 (complete testimony pp. 569–76), and the obituary of Marilyn Sitzman, *Dallas Morning News*, Aug. 14, 1993, p. 40A.

Today, there is always a "pool" camera covering the president's public movements. So any future attempt or attack on a president will always be filmed. Some refer to this camera coverage as "the death watch." This was not the case in November 1963. Nevertheless, there were many amateur and a few professional photographers and filmmakers in Dealey Plaza. Zapruder's film was the most important and famous. Other important films include those taken from the other side of the Grassy Knoll by Orville Nix and Marie Muchmore. Also, important still photographs were taken by Phil Willis, James "Ike" Altgens, and Mary Moorman. For an illustration of where these individuals stood in Dealey Plaza, see Josiah Thompson, *Six Seconds in Dallas: A Micro-Study of the Kennedy Assassination* (New York: Bernard Geis Associates, 1967), p. 12. Copies of the Zapruder film were first "bootlegged" by Penn Jones Jr., publisher of the *Midlothian Mirror*, after Jim Garrison subpoenaed it as a part of his New Orleans investigation. The film was widely distributed and sold to the public. For black-and-white stills (along with Nix and Muchmore), see WC 18, pp. 1–95; several color stills of the Zapruder film were published in *Life* magazine a few years later. Dan Rather of CBS News gave an audio narration of the film on TV in 1963. The first TV viewing of the film occurs on Geraldo Rivera's *Good Night America* on ABC in 1975. To view the first commercially released version of this film in VHS and later in DVD, see *Images of the Assassination: A New Look at the Zapruder Film* (Orland Park, IL: MPI Home Videos, 1998). This comprehensive 1½-hour documentary, which also includes the original televised interview with Abraham Zapruder, can be seen at www.youtube.com/watch?v=vVafiaW4AdY. Also see David R. Wrone, *The Zapruder Film: Reframing the Kennedy Assassination* (Lawrence: University Press of Kansas, 2003).

CHAPTER 5: "SOMEONE IS SHOOTING AT THE PRESIDENT"

122 Motorcade running late. See O'Donnell, *Johnny*, p. 27, and Manchester, *Death*, p. 154.

123 Jackie "cool in the tunnel." See the testimony of Mrs. John F. Kennedy, WC 5, p. 179 (complete testimony, pp. 178–81).

123 Amos Euins, a ninth grader, gets excused from school. See the testimony of Amos Euins, WC 2, p. 202 (complete testimony, pp. 201–10).

123 Euins, "waved back at me." Ibid., p. 203.

125 Jackie says, "They were gunning the motorcycles." See Theodore White Camelot interview.

125 Jack Bell, "they are shooting off firecrackers." See Trost, *President*, p. 22.

126 Brennan, there was a sound . . . "backfire" . . . "firecracker." See the testimony of Howard Brennan, WC 3, pp.143–44, 154.

126 Brennan immediately sees man holding rifle. Ibid., p. 144–45.

126 Zapruder hears shots and jostles camera. Many who viewed the Zapruder film believed that the camera shakes at the point where Zapruder is startled by the first shot.

126 Rowland, "backfire." See the testimony of Arnold Rowland, WC 2, p. 179.

127 Worrell hears first shot. See the testimony of high-school student, James Richard Worrell Jr., who skipped school to watch motorcade, WC 2, p. 193 (complete testimony pp. 190–201).

127 A "motorcycle backfire." See the testimony of Buell Wesley Frazier, WC 2, p. 234.

127 Euins thinking it was a "backfire." See the testimony of Amos Euins, WC 2, p. 205.

127 Williams, president pushing hair back. See the testimony of Bonnie Ray Williams, WC 3, p. 175.

128 Fragment hits bystander. See the testimony of car salesman James Thomas Tague, who went downtown to take his wife to lunch and "accidentally came upon the motorcade," WC 7, pp. 552–58.

129 Williams (see above).

130 Worrell sees rifle. See the testimony of Worrell, WC 2, p. 194.

130 Williams hears shots. See the testimony of Bonnie Ray Williams, WC 3, p. 175.

132 Dillard's comments. See Trost, *President*, p. 23.

132 Rowland knew it was rifle fire. See the testimony of Arnold Rowland, WC 2, p. 179.

132 Brennan comments on shots. See the testimony of Howard Brennan, WC 3, pp. 144, 154.

132 Frazier's comments on shooting. See the testimony of Buell Wesley Frazier, WC 2, p. 234.

133 AP Reporter, James William "Ike" Altgens photographs. See the testimony of Altgens, WC 7, pp. 515–25, and FBI interview, WC 22, CE 1407, pp. 790–92. Also see WC 16, CE 203, p. 594 (photo is cropped). For the entire image, see *Saturday Evening Post*, Dec. 14, 1963, pp. 24–25.

133 Dave Powers, "President's been shot." See O'Donnell, *Johnny*, p. 27.

133 Connally, "They're going to kill us all!" See the testimony of Governor John D. Connally Jr., WC 4, p. 133. Also see John Connally and Mickey Herskowitz, *In History's Shadow: An American Odyssey* (New York: Putnam's, 1967), pp. 175–92.

134 Hill runs toward back of car. See the testimony of Clint Hill, WC 2, pp. 138–40, and Hill, *Mrs. Kennedy and Me*, pp. 290–91.

135 Billy and Gayle Newman and their two children witness the assassination. See the Sheriff's Department "Voluntary Statement" of Gayle Newman, WC 19, Decker Exhibit 5323, p. 488 (Dallas County Sheriff's Office records of the events surrounding the assassination).

135 Brennan's seeing rifle. See the testimony of Howard Brennan, WC 3, pp. 144–45.

135 Amos Euins looks at man in sixth-floor window. See the testimony of Amos Euins, WC 2, pp. 203–5.

136 Dillard and Jackson's comments. See Trost, *President*, pp. 23–24.

136 Euins, third shot. See the testimony of Amos Euins, WC 2, p. 207.

137 Dillard photos, after assassination. See WC 19, Tom C. Dillard Exhibit, A, B, and C, pp. 563–65, as well as his testimony, WC 6, p. 166.

137 Merriman Smith, Bob Clark, and Pierce Allman comment on number of shots. See Trost, *President*, pp. 24, 27.

138 Mrs. Robert Reid on Elm Street, three shots, black men on fifth floor. See testimony of Mrs. Robert Reid, WC 3, 273.

139 Jackie's "Oh no!" See the testimony of Jacqueline Kennedy, WC 5, p 180. Also see Associated Press, ed., *The Torch Is Passed: The Associated Press Story of the Death of a President* (New York: Associated Press, 1964), p. 14.

139 O'Donnell and Powers, seeing third shot, know he's dead. See O'Donnell, *Johnny*, pp. 27–28.

140 Brennan's concern that "there were going to be bullets flying from every direction." See the testimony of Howard Brennan, WC 3, p. 145.

141 Walker note. See the testimony of Marina Oswald, WC 1, p. 17. Also see WC 16, CE 1, pp. 1–2. Marina discussed the note in a Secret Service interview on December 5, 1963, WC 23, CE 1785, pp. 392–94. For the translation, see WC 23, CE 1786, pp. 395–97, where Marina confirmed the note.

142 "They have shot his head off." See Hill, *Mrs. Kennedy and Me*, p. 291.

In addition to governmental activities, there have been numerous independent studies as well as reenactments, confirming the number of shots and trajectory. See *Life-Itek Kennedy Assassination Film Analysis* (Lexington, MA: Itek Corp., 1967) and *John Kennedy Assassination Film Analysis* (Lexington, MA: Itek Corp., 1976). Also see the DVD *JFK: Beyond the Magic Bullet* (Discovery Channel, 2010) and the analysis in the CBS television documentary and companion book: Stephen White, *Should We Now Believe the Warren Report?* (New York: Macmillan, 1968).

I have attempted to capture the excitement at the moment when shots were fired in Dealey Plaza, as well as to convey the impressions of those witnesses closest to the sixth-floor window of the Texas School Book Depository. At the time of the assassination, there were over two hundred witnesses in Dealey Plaza. The majority of these witnesses thought that they heard no more than three shots and that they came from behind JFK's car. Some of these witnesses and others believed that they initially heard the sounds of fireworks or motorcycles that backfired before they realized what had happened. Those closest to the sixth-floor window all testified that they heard and felt three shots.

While some of the witness testimony changed over time and not all testimony was consistent, credible investigations both governmental and independent that reviewed the totality of the evidence concluded that the shots came from behind JFK and were fired from the sixth-floor window. Viewed by itself and without corroboration from other evidentiary sources, eyewitness testimony can be unreliable in many cases. See James M. Doyle and Elizabeth Loftus, *Eyewitness Testimony: Civil and Criminal*, 4th ed. (Los Angeles: LexisNexis, 2007), and Elizabeth Loftus, *Eyewitness Testimony* (Cambridge, MA: Harvard University Press, 1979). Both the Warren Commission and the House Select Committee on Assassinations were charged with weighing the credibility of these witnesses, and both concluded that there were three shots fired from the sixth-floor window of the Texas School Book Depository.

143 Rufus Youngblood, "Get down." See the testimony of Rufus Wayne Youngblood, WC 2, p. 149; Rufus W. Youngblood, *20 Years in U.S. Secret Service: My Life with Five Presidents* (New York: Simon & Schuster, 1973); and Merle Miler, *Lyndon: An Oral Biography* (New York: Putnam's, 1980), p. 313.

143 Reporter, "something has happened," Dallas radio broadcast of reporter Ronald Lee Jenkins of KBOX. While this recording was widely perceived to be contemporaneous, this audio was actually a re-creation made by the station a few days after the assassination. Apparently the original tapes were lost. The complete dramatization can be found on a long-playing phonographic recording, *The Actual Voices and Events of the Four Days That Shocked the World, Nov. 22–25, 1963: The Complete Story* (New York: Colpix Records, "produced in association with United Press International," 1964; rereleased as a CD in the UK by RPM Records, 1998). A coworker of Jenkins, Sam Pate, who began this broadcast, told the FBI in 1964 that it was a re-creation. See Warren Commission Document 1245 and Warren Commission Document 1070). In an interview, Pate, Jenkins, and the sound board engineer all recalled interrupting the original KBOX broadcast on November 22, 1963. See www.youtube.com/watch?v=lIMMXcJbtfg. Also see interview with Sam Pate, oral history, Sixth Floor Museum.

CHAPTER 6: "THEY'VE SHOT HIS HEAD OFF"

144 Truly's encounter with Oswald. See the testimony of Roy Truly, WC 3, pp. 224–26, 239.

145 Mrs. Reid, Oswald's coworker who ran into building and encountered Oswald. See the testimony of Mrs. Robert A. Reid, WC 3, pp. 274–75 (complete testimony, pp. 270–81).

146 MacNeil's encounter with Oswald. See Robert MacNeil, *The Right Place at the Right Time* (Boston: Little, Brown, 1982), pp. 198–216, and Trost, *President*, p. 35. Also see his recollections on the NBC News website at www.nbcnews.com/id/3476061/ ns/msnbc-jfk_the_day_that_changed_americ a/t/covering-jfk-assassination.

    One other reporter, Pierce Allman of WFAA-TV and radio, also believed that he ran into Oswald just as he was leaving out of the front door of the Texas School Book Depository. See Trost, *President*, p. 35.

146 Smith and Bell struggle for first national press notice, UPI "Three Shots Were Fired." See Manchester, *Death*, p. 168.

147 "[E]verybody was running. See the testimony of Bonnie Ray Williams, WC 3, 175, 177, 211.

148 Brennan, officers search. See the testimony of Howard L. Brennan, WC 3, p. 145–46.

148 Jenkins outside Trade Mart, car rushes by, "Something is wrong." Jenkins, KBOX broadcast, *Actual Voices and Events.*

149 Rowland, silence of man in window. See the testimony of Arnold Rowland, WC 2, p. 174.

150 First CBS broadcast. Walter Cronkite interrupts the CBS daily broadcast of the soap opera *As the World Turns*; go to the CBS website at www.cbsnews.com/2100 -500202_162-584646.html. Comprehensive coverage can be found on the YouTube .com website in ten parts. For the complete coverage, beginning with the first interruption, go to www.youtube.com/watch?v=qtXfZs0-Bn0 and also for an additional two hours of uninterrupted coverage go to www.youtube.com/watch?v=t_Ry9 -bpixM.

151 Marina Oswald's first awareness of assassination. See the testimony of Marina Oswald, WC 1, pp. 73–74.

153 Jackie, "Dave, he's dead." See O'Donnell, *Johnny*, p. 29.

154 Hill attempts to move JFK from the presidential limousine. See Hill, *Mrs. Kennedy and Me*, pp. 293–94.

154 Merriman Smith ("Smitty") viewed president at Parkland. See Trost, *President*, p. 39. Merriman Smith received a Pulitzer Prize for his coverage of the assassination. Go to the UPI website at www.upi.com/Top_News/Special/2011/11/22/UPI -Archives-Merriman-Smiths-account-of-JFKs-assassination/PC-9391321983592.

155 Bob Clark's comments on wounds. See Trost, *President*, pp. 39–40.

155 Clint Hill's jacket to cover head. See Hill, *Mrs. Kennedy and Me*, p. 293. Subsequent to the assassination, Hill would destroy his jacket.

155 Jenkins broadcast. Jenkins, KBOX broadcast, *Actual Voices and Events.*

156 Oswald's escape compared to that of Booth. See James L. Swanson, *Manhunt: The 12-Day Chase for Lincoln's Killer* (New York: Morrow, 2006).

157 Bus driver's account. See the testimony of Cecil J. McWatters, WC 2, pp. 262–92. This bus driver gave Oswald a bus transfer that was found on his person at the time of arrest. See WC 16, CE 381-A, p. 974. Oswald's former landlady happened to be on the bus as well, and she identified him. See the testimony of Mary E. Bledsoe, WC 6, pp. 408–27 (complete testimony pp. 400–27).

157 Cab ride. See the testimony of William Wayne Whaley, WC 2, pp. 253–62, 294, and WC 6, pp. 428–33. Also see his taxicab manifest for November 22, 1963, WC 16, CE 370, p. 966.

157 Housekeeper's observations. Mrs. Earlene Roberts, worked for Mrs. Arthur Carl (Gladys) Johnson, the owner of the property. Roberts was the person who rented Oswald the room and also encountered him shortly after the assassination at the

rooming house at 1026 North Beckley. See the testimony of Earlene Roberts, WC 6, pp. 434–44. Gladys Johnson also testified, WC 10, pp. 292–301.

157 Possible escape to Mexico. Warren Commission Assistant Counsel David W. Belin speculated that Oswald may have been on his way to catching a bus to Mexico before his encounter with Officer Tippit. This hypothesis was in draft chapter 6 of the Warren Report but was excluded in the final version. See Gerald Posner, *Case Closed: Lee Harvey Oswald and the Assassination of JFK* (New York: Random House, 1993), p. 273.

158 Oswald not accounted for at work. See the testimony of Roy Truly, WC 3, p. 230, 239.

158 Wicker, speculation and rumor at Trade Mart. See Trost, *President*, pp. 52–53. Eddie Barker at KRLD-TV, provided about 32 minutes of live reports from the Trade Mart, beginning at approximately one P.M. (CST). Some of this coverage was broadcast on CBS. Barker mentioned that three shots were fired in Dealey Plaza but in error reported that a Secret Service agent was killed. Go to www.youtube .com/watch?v=utBnHGEBvQk.

158 Kilduff's comments on Jackie's hair. See Trost, *President*, pp. 40–41.

158 O'Donnell, "I did not have the heart." See O'Donnell. *Johnny*, p. 30.

158 Merriman Smith, wire report "seriously wounded." See Trost, *President*, p. 42. At this time, almost everyone knew that JFK was dead. Clint Hill told Smith of Kennedy's death, but the reporter would need official confirmation before he communicated this over the UPI wire service. See Manchester, *Death*, p. 168, and also the small pamphlet, *The Murder of the Young President*, by Merriman Smith, which United Press International inserted into the phonographic recording, *The Actual Voices and Events*, Colpix Records, 1964.

159 Local radio station WQMR announcement. *Four Dark Days in November: A Presentation of WQMR News, Featuring the Actual On-the-Air Coverage of the Events of November 22,23, 24 and 25th, 1963, as Reported and Broadcast by Radio Station WQMR, Washington, D.C.*, 331/3 rpm phonographic recording (Capitol Records, 1964).

160 Hugh Sidey sees Secret Service washing the car. See Trost, *President*, p. 59.

160 Jackie's "red roses" and blood comment. See Theodore White, Camelot Interview.

160 Jackie, "I want to be with him when he dies," to Dave Powers. See O'Donnell, *Johnny*, p. 30.

160 Navy admiral, "It is her right." See Manchester, *Death*.

161 Jackie holding a piece of JFK's brain. Ibid.

161 Doctor's comment, "a fatal wound," and Jackie, "I know." See Manchester, *Death*, p. 188.

161 Minutes after the assassination, there was another car racing to Parkland Hospital. After the broadcast on television that the president had been shot and had been taken to Parkland Hospital, a decision quickly had been made at Holy Trinity Catholic Church. Independently, Jackie Kennedy requested hospital officials to get a priest. In fact, the church's pastor, the Very Reverend Oscar L. Huber, C.M., was already on his way there in a car driven by his assistant pastor, Rev. James N. Thompson. C.M. Parkland Hospital was regularly served by this parish. They were concerned about the traffic, but Thompson had a "secret way" to get there quickly. As they approached the hospital, a policeman stopped them. Thompson said, "Look, officer, the President is in there and he is either dying or dead, one of us has to go." As Thompson began to shut his car down, another officer came running, saw they were priests, and stated, "OK, let them through." Since Parkland Hospital

was surrounded in the chaos of onlookers, vehicles, and police, Rev. Thompson let Father Huber out of the car while he parked it, and Huber was escorted by police to the emergency room.

It was Father Huber who gave JFK last rites close to one P.M. (CST). Mrs. Kennedy was very concerned that since her husband was already dead, it was too late to perform this religious ritual. Huber assured her that the president's soul had not yet left his body, as the pastor removed the sheet already covering the president's limp body. See the self-published limited-edition signed pamphlet, Reverend James Thompson, C.M., *Around One O'Clock* (Dallas: Holy Trinity Catholic Church, 1964).

Father Oscar Huber also personally signed typed copies of his recollections of this day. In addition, see Gillon, *Kennedy Assassination*, pp. 75–80, and Manchester, *Death*, pp. 213–19.

161 Placing the ring on finger. See Manchester, *Death*, pp. 293–94.

162 Notifying LBJ about death, calling him Mr. President. LBJ is first notified of JFK's death by Kenny O'Donnell ("He's gone"). See the testimony of Lyndon Baines Johnson, WC 5, p. 693. The assistant press secretary, Mac Kildruff, addressing LBJ, as Mr. President, requested that he be able to make the official announcement to the press. See Robert A. Caro, *The Years of Lyndon Johnson: The Passage to Power* (New York: Knopf, 2012), pp. 320–21.

162 Delay in official notification of the press until LBJ leaves. At this time, private concerns were first discussed about a possible conspiracy to attack other officials, including the vice president. See Caro, *Years*, p. 320.

At the time of the assassination, the Speaker of the House was next in line to succeed President Lyndon Johnson. See the Presidential Succession Act of 1947, 3 U.S.C. §19. But there was no constitutional directive for replacing the vacancy of the vice presidency due to JFK's death. Also, there was no guidance for addressing the problematic situation of a disabled president, should Kennedy have lived. This situation would result in the 1967 ratification of the 25th Amendment to the U.S. Constitution.

162 Johnson's unmarked car travels to Love Field. With a few motorcycles leading the way, Dallas Police Chief Jesse Curry drove LBJ back to Love Field in an unmarked police car. In this vehicle, Rep. Thomas (a longtime friend of Johnson who served with him in the U.S. House of Representatives) sat in the passenger seat, and in the backseat, the new president crouched in the middle between Rep. Thornberry (who followed LBJ as the congressman representing the 10th Congressional District of Texas) and Secret Service agent Rufus Youngblood. A few cars raced behind them, one carrying Lady Bird Johnson and several Secret Service agents. Behind them was a car with more agents, along with Chief Warrant Officer Ira Gearhart, who held the "football," the codes to launch a nuclear attack, and Liz Carpenter, executive assistant to the vice president, who began to draft Johnson's statement, which he would deliver to the American people that evening. The last car in this group included Jack Valenti, a political consultant to Johnson, Cliff Carter, LBJ's aide, and Captain Cecil W. Stoughton, JFK's photographer. Ibid., pp. 247–48, and see Manchester, *Death*, pp. 237–39.

162 Cronkite announcement that priests administered last rites and that Dan Rather says JFK is dead. See Trost, *President*, p. 80.

163 Kilduff's official announcement. See Manchester, *Death*, p. 221, and the brief announcement and news conference at www.youtube.com/watch?v=fHQ4EoDZKdg.

163 Merriman Smith UPI eyewitness account. See United Press International, ed., *Four Days: The Historical Record of the Death of President Kennedy* (Rockville, MD: American Heritage, 1964), pp. 32–33.

163 KLIF announcement. *Four Dark Days in November* (Capitol Records, 1964).

163 CBS, "the flash, apparently official . . ." See Trost, *President*, p. 85.

164 Every afternoon paper in the country issued extra editions, and some subsequently published commemorative editions. The *Dallas Times Herald* issued four editions that afternoon, three on the assassination as further information became available. See the recollection of reporter Keith Shelton, Trost, *President*, p. 103. See also Charles Cameron, C-Span broadcast, "Journalists Remember the JFK Assassination," November 20, 1993. www.c-spanvideo.org/program/FKASS.

The very first edition of the *Dallas Times Herald* was printed before the assassination and is the scarcest of all the newspapers from that day in Dallas, but not as desired by collectors. Fewer copies of it were saved than the assassination editions. This edition was the final ("1 star") edition with the normal-size above-the-fold headline "Texas Progress, U.S. Projects Linked, JFK Tells Forth Worth." This edition included the front-page story "Secret Service Sure All Secure." Subsequent afternoon editions were: final edition ("2 stars") with the large above-the-fold headline, "President Dead"; and the two final editions ("3 stars") one with the lead-in headline, "JFK Ambushed in Dallas" and the large above-the-fold headline, "President Dead; Connally Shot"; and the other with the lead-in headline, "Suspect Arrested" and the large above-the-fold headline, "President Dead; Connally Shot." Inasmuch as the *Dallas Morning News* was a morning newspaper, there is the November 22 morning edition with no coverage of the assassination, and the Saturday, November 23, 1963, edition with its up-to-date reporting of the shooting.

164 *Chicago American.* This newspaper box is in the author's own collection.

166 Ordering of the casket over the telephone. Hill ordered the most expensive casket from the O'Neal funeral home. See the testimony of Clint Hill, WC 2, pp. 142; Hill, *Mrs. Kennedy and Me*, pp. 297–98; and Manchester, *Death*, p. 291–92.

166 The tensions between J. Edgar Hoover and Bobby Kennedy were evident. Hoover used to say that presidents were going through his administration. Because of the "secret files" he kept, no president would dare terminate his employment as director of the FBI. For his relationship, or lack of one, with Robert Kennedy, see Burton Hersch, *Bobby and J. Edgar: The Historic Face-Off between the Kennedys and J. Edgar Hoover That Transformed America* (New York: Carroll & Graf, 2007).

167 RFK office call from Hoover, "president has been shot." See Manchester, *Death*, pp. 195–96.

167 Ted Kennedy is notified in the Senate. See Trost, *President*, p. 46 (recollection of Richard Reidel, press liaison aide, U.S. Senate).

CHAPTER 7: "I HAVEN'T SHOT ANYBODY"

169 Confrontation over removing the body without an autopsy performed in Texas. See Manchester, *Death*, p. 289–329.

171 Broadcast of the description of Oswald. See Trost, *President*, p. 57 (Dallas Police Department radio transmission, channel 1, at about 12:45 CST).

171 Oswald's encounter with J. D Tippit. Over twelve witnesses were present during the shooting of J. D. Tippit. See WR, pp. 156–76. Several identified Oswald in a subsequent lineup, including: Mrs. Helen Markham, William W. Scoggins (he did not see the shooting, but identified him as the man running away), Mrs. Barbara Jea-

nette Davis, and her sister-in-law, Mrs. Charlie Virginia Davis. See the testimony
of Mrs. Helen Markham, WC 3, p. 318 (complete testimony pp. 305–22, 340–42);
the testimony of William W. Scoggins, WC 3, pp. 322–27 (complete testimony
pp. 322–40); the testimony of Mrs. Barbara Jeanette Davis, WC 3, pp. 349–50 (complete testimony pp. 342–50); and the testimony of her sister-in-law, Mrs. Charlie
Virginia Davis, WC 6, pp. 461–63 (complete testimony, pp. 454–68).

    The cartridges and bullets from the revolver were traced to Oswald's gun, which
was in his possession when he was captured. See WR, appendix 10, pp. 558–60.
Years later, the House Select Committee on Assassinations would again conclude
that Oswald shot Officer Tippit. See HSCA *Final Report*, pp. 59–60. For a comprehensive review of the evidence implicating Oswald in the Tippit murder, see
Dale K. Myers, *With Malice: Lee Harvey Oswald and the Murder of Officer J. D. Tippit* (Milford, MI: Oak Cliff Press, 1998).

174 LBJ aboard Air Force One. See Holland, *Kennedy Assassination Tapes*, pp. 19–51;
Caro, *Years*, pp. 323–64; Gillon, *Kennedy Assassination*, pp. 87–166; and Manchester, *Death*, pp. 339–51, 386–88.

176 Breaking the coffin handle to get it aboard. See Manchester, *Death*, p. 308.

176 WQMR gets word of LBJ oath. *Four Dark Days in November* (Capitol Records,
1964).

177 Getting Jackie to stand near LBJ. See Manchester, *Death*, p. 322.

177 Lady Bird's interaction with Jackie, "I want them to see . . ." See Lady Bird Johnson,
A White House Diary (New York: Holt, Rinehart & Winston, 1970), p. 6.

178 Jackie's comments about keeping on the same clothes, but washing the blood
off—"I should have kept the blood on." See Theodore White, Camelot Interview.

178 O'Donnell interaction with Jackie, her decision to participate in swearing in.
O'Donnell, *Johnny*, p. 36.

179 Merriman Smith observation, "Mrs. Kennedy walked down the narrow corridor."
The Actual Voices and Events (Colpix Records, 1964).

179 Stoughton's photos. To view the photographs taken by Cecil Stoughton on Air
Force One, go to the "November 22, 1963 and Beyond" Web page of the Lyndon
Baines Johnson Library and Museum at www.lbjlib.utexas.edu/johnson/kennedy/
index.htm. See "Historic Notes: the Full Record," *Time*, Feb. 24, 1967, pp. 17–19.
Also see Richard B. Trask, "The Day Kennedy Was Shot," *American Heritage* 39,
no. 7 (Nov. 1988), www.americanheritage.com/content/day-kennedy-was-shot.

180 The oath, the dictabelt, and the photographs. To demonstrate to the nation that a
new president was now the commander-in-chief, Stoughton immediately raced off
the plane with the recording and his camera. By the time Air Force One landed at
Andrews Air Force Base, one of the iconic photographs had already been broadcast
on NBC. See Gillon, *Kennedy Assassination*, pp. 141–42.

180 Shoe salesman follows Oswald to theater. See the testimony of Johnny Calvin
Brewer, manager of Hardy's Shoe Store, WC 7, pp. 1–8.

181 Arrest of Oswald. See the testimony of M. N. ("Nick") McDonald, WC 3,
pp. 295–305. Also see Dale K. Myers, *With Malice: Lee Harvey Oswald and the
Murder of Officer J. D. Tippit* (Milford, MI: Oak Cliff Press, 1998), pp. 139–87.

181 Oswald's comments upon arrest, "Well it's all over now," police brutality. Ibid.,
pp. 301, 303. See also WC 24, CE 2003, p. 240 (Dallas Police Department file on the
investigation of the assassination of President Kennedy, pp. 195–404). Testimony
of theater patron John Gibson, WC 7, p. 73. "Civil rights." Bugliosi, *Reclaiming
History*, p. 106.

182 Oswald's comments in police car, "You fry for that" and "I hear they burn for murder." See the testimony of Dallas Police Sergeant Gerald Lynn Hill, WC 7, p. 58 (complete testimony, pp. 43–66) and the testimony of C. T. Walker, with Accident Investigations for the Dallas Police Department, WC 7, pp. 40–41 (complete testimony, pp. 34–43).

182 First identification as A. J. Hidell. See the testimony of Dallas Police Sergeant Gerald Lynn Hill, WC 7, p. 58.

182 "Why should I hide my face?" Ibid., p. 59.

183 Shelley's identification. "Well, that is Oswald. He works for us. He is one of my boys." See the testimony of Dallas Police Officer C. W. Brown, WC 7, p. 248 (complete testimony, pp. 246–51).

183 "You find out." See testimony of Dallas Police Detective Guy F. Rose, WC 7, p. 228.

183 Fritz first interview. See generally the testimony of Captain J. W. Fritz, WC 4, pp. 202–47; the testimony of James Patrick Hosty Jr., WC 4, pp. 440–76; James P. Hosty and Thomas Hosty, *Assignment Oswald* (New York: Arcade Publishing, 1996); WC 29, CE 2003, pp. 195–404 (Dallas Police Department file on the investigation of the assassination of the president, especially pp. 264–79, interview with Oswald); and WC 17, CE 832, pp. 785–86 (FBI report of interview with Lee Harvey Oswald). For the most comprehensive attempt to capture the conversation and the chronology of all Oswald's comments while in custody during interviews and in comments to the press, see Bugliosi, *Reclaiming History*, pp. 123–273.

183 "Well I struck an officer." See the testimony of Dallas Police Detective Elmer L. Boyd, WC 7, p. 135 (complete testimony, pp. 119–35).

In the almost forty-eight hours Oswald was in custody, he was interviewed for more than twelve hours by several individuals from the Dallas police, FBI, Secret Service, and the U.S. Post Office—often by more than one at the same time. While there were no recordings of these encounters, there were several reports, summaries, and testimonies that captured what was said during these sessions. The reports of some of these interviews can be found in appendix 11 of the Warren Report, pp. 599–636, and additional interaction can be found in witness testimony, commission exhibits, and subsequent books written by some of the principal players. It is not surprising that there would be no recordings. Even Dallas Police Chief Jesse Curry would admit in his testimony before the Warren Commission that there should not have been so many people in a room at once when questioning Oswald. Curry said "we were violating every principle of interrogation." See the testimony of Jesse Edward Curry, WC 4, p. 152 (complete testimony, pp. 150–202).

Many judge Oswald's interrogation by the Miranda warning rule adopted by the Supreme Court in 1966 and other subsequent protections of a suspect's rights. This was definitely not the case in 1963, and the custodial interrogation of criminal suspects was very different. When Oswald mentions a few times, including to the press, that he was hit by an officer, it is difficult if not impossible to know whether this was only during the struggle in the Texas Theater. Some police officers who testified stated that Oswald was not struck in the theater, but there was a scuffle. See WR, p. 200. Recognizing the concept of "the thin blue line," it is difficult to assess Oswald's treatment.

The tactics used in 1963 would not meet the rigorous standards of today, and it is doubtful that any official would have admitted to some of these approaches if they had occurred. For standards of acceptable police conduct during this time period, see Fred E. Inbau and John Reid, *Criminal Interrogation and Confessions* (Baltimore: Williams & Wilkens, 1962).

In 1963, the Supreme Court decided in the case of *Gideon v. Wainright*, 372 U.S. 335 (1963), that state courts were required to provide counsel in criminal cases to represent defendants who were unable to afford to pay their own attorneys. However, it was not until the Supreme Court ruled in *Miranda v. Arizona*, 384 U.S. 436 (1966), that police also were mandated to notify defendants of their rights. In 1963, under Texas state law, the police were required to inform the suspect in custodial interrogation that there was the right to remain silent and that anything said could be used against him in a court of law. Oswald was informed of his rights, as required by Texas law. See generally Bugliosi, *Reclaiming History*, p. 115–16.

184 "Oh, so *you're* Hosty, the agent who's been harassing my wife." See Hosty, *Assignment*, p. 20.

184 "My wife is a Russian citizen." Ibid., p. 20.

185 "Accosted" your wife. See the testimony of Captain J. W. Fritz, WC 4, p. 210.

185 Adjustment of handcuffs by Fritz and "thank you, thank you." See the FBI report on interview with Oswald, Nov. 23, 1963, WC 28, CE 832, pp. 785–86; the testimony of Captain J. W. Fritz, WC 4, p. 238; and Hosty, *Assignment Oswald*, p. 22.

185 "I'm sorry for blowing up at you . . ." See Hosty, *Assigment Oswald*, p. 22.

185 "Do you own a rifle?" "Did you ever own a rifle?" See the testimony of J. W. Fritz, WC 4, p. 214.

185 "Yes, sir. He told me he had seen a rifle at the building 2 or 3 days before that Truly and some men were looking at it." Ibid.

In fact, Roy Truly, Oswald's supervisor, later testified before the Warren Commission that Warren Caster brought two weapons to work at the Texas School Book Depository on November 20, 1963. See the testimony of Roy Truly, WC 7, pp. 380–86. Warren Caster, the assistant manager of Southwestern Publishing Co., who had offices on the second floor of the Texas School Book Depository, identified these weapons as a Mauser and a .22-caliber rifle. See the testimony of Warren Caster, WC 7, pp. 386–88. Caster was unable to identify all the workers who might have seen his weapons that day, since they were removed from a carton in the hallway in front of the entrance to the warehouse. It is not in his testimony, but it is possible that Oswald may have seen these weapons as well. Castor stated that he showed them to them to Truly and William Shelley, another employee in the building. William Shelley remembers the seeing the two weapons, and mentions actually holding the .22 but not the Mauser. See the testimony of William Shelley, WC 7, pp. 390–93.

186 "I had one a good many years ago. It was a small rifle or something, but I have not owned one for a long time." Ibid.

186 "Sure. Sure I've been to Mexico." See Hosty, *Assignment Oswald*, p. 25.

186 Marina at Ruth Paine's home, discovery of rifle missing. See the testimony of Marina Oswald, WC 1, p. 74.

187 "I knew that it was Lee." Ibid.

188 The animosity between the Kennedy and LBJ camps was legendary. In early historical works, authors seemed to side with one group or the other. Today, modern historians have provided a more balanced approach, weighing the arrogance of Kennedy's Eastern intellectual establishment against the manipulative political sense of LBJ. For a recent thoughtful appraisal of this rivalry in the hours and days

immediately before and after the assassination, see Robert A. Caro, *The Years of Lyndon Johnson: The Passage to Power* (New York: Knopf, 2012), pp. 308–436.

188 LBJ labeled "cornpone" by Kennedy's aides. Ibid., p. 344.

188 First thoughts of planning of funeral. For a comprehensive overview of the planning of the funeral, see Billy C. Mossman and B. C. Stark, *The Last Salute: Civil and Military Funerals 1921–1969* (Washington, DC: Department of the Army, 1971), ch. 23, "President John F. Kennedy State Funeral, 22–25 November 1963," pp. 188–215. See also U.S. Army Center for Military History, www.history.army .mil/books/Last_Salute/ch23.htm.

189 "How often do you go out there?" Ibid., 210.

189 "I don't want to stay there . . ." Ibid.

189 Medals for marksmanship. See WC 24, CE 2003, p. 265.

189 "Why were you registered at the boarding house as O.H. Lee?" See the testimony of Captain J. W. Fritz, WC 4, p. 211.

189 Whether Oswald worked at Book Depository. See the testimony of Captain J. W. Fritz, WC 4, p. 213, and WC 17, CE 832, p. 785.

189 When Oswald went to lunch. See WC 17, CE 832, p. 786.

190 Getting a Coca-Cola, Oswald's location at time of shooting, and going home. Ibid., and WC 24, CE 2003 (Dallas Police Department file), p. 265.

190 "I felt like it." "You know how boys do when they have a gun, they just carry it." See WC 24, CE 2003, p. 265 (Dallas Police Department file); WC 17, CE 832 (FBI report) p. 786; and Hosty, *Assignment*, p. 23.

190 "Did you shoot officer Tippit?" See the testimony of Captain J. W. Fritz, WC 3, p. 214.

190 "I hit the officer in the show; he hit me in the eye and I guess I deserved it. That is the only law I violated. That is the only thing I have done wrong." Ibid., p. 214.

191 "Common laborer." See Hosty, *Assignment Oswald,* p. 22.

191 Oswald denies shooting the president. See WC 17, CE 832, p. 786.

192 Questioning of Oswald by Secret Service Agent Forrest V. Sorrels. See the testimony of Forrest V. Sorrels, WC 7, pp. 332–60.

192 "I don't know who you fellows are, a bunch of cops." . . . "Aren't you supposed to get me an attorney?" See the testimony of Forrest V. Sorrels, WC 7, p. 353.

192 "I just want to ask you some questions." . . . "I don't care to answer any more questions." Ibid.

192 Calls from Air Force One. For a comprehensive treatment of the calls LBJ made from Air Force One, see Max Holland, *The Kennedy Assassination Tapes: The White House Conversations of Lyndon B. Johnson Regarding the Assassination, the Warren Commission, and the Aftermath* (New York: Knopf, 2004).

192 LBJ call to Rose Kennedy. In this brief conversation, the first LBJ made aboard Air Force One, Mrs. Kennedy addressed Johnson as Mr. President. Johnson was apparently "choked up" in this brief exchange and handed the telephone over to Lady Bird to finish the call. Ibid., pp. 45–47. Because Joseph Kennedy had suffered a stroke, it was the decision of the family not to immediately tell the patriarch about JFK's death.

193 Lineup. See the testimony of Dallas Police Officers W. E. Perry and Richard L. Clark, who participated posing as suspects in this first "show-up." WC 7, pp. 232–39.

193 Reporter's question and Oswald's response, "No, sir. Nobody charged me with that . . ." Bugliosi, *Reclaiming History*, p. 132 (from newsreel footage in the library of the Sixth Floor Museum in Dallas).

194 The search of Oswald's pockets. "What are these doing in there?" "I just had them in my pockets." See the testimony of Dallas Police Detective Elmer L. Boyd, WC 7, p. 136.

194 Decision to have the normal press coverage for arrival at Andrews Air Force Base. See Holland, *Kennedy Assassination Tapes*, pp. 32–33.

194 RFK rushing on plane and ignoring LBJ, "Jackie, I'm here." See Manchester, *Death*, p. 387.

194 Unloading of coffin. Ibid., pp. 387–92.

195 LBJ's brief comments to the nation. See *Public Papers of the Presidents of the United States, Lyndon B. Johnson*, bk. 1, Nov. 22, 1963–June 30, 1964 (Washington, DC: U.S. Government Printing Office, 1965), p. 1. For the audio recording of the president's remarks, including his handwritten notes on his statement, go to the Lyndon Baines Johnson Library and Museum Web page at www.lbjlib.utexas.edu/johnson/kennedy/Remarks%20at%20Andrews/remarks.htm.

195 LBJ's handwritten letters to the Kennedy children. See Manchester, *Death*, pp. 405–6, and Caro, *Years*, p. 369.

195 Johnson's decision to return home and give Jackie as much time as needed to remain in White House. See Manchester, *Death*, pp. 403–4, 547.

196 Jackie at Bethesda Naval Hospital. See Manchester, *Death*, pp. 398–99, 406–7, 414–19, 427–30, 434–35, 439–40.

196 "They think they found the man who did it." "He says he's a Communist." See Manchester, *Death*, p. 407.

196 Oswald's statements during his arraignment. These quotes come from interviews with William Alexander, assistant Dallas district attorney, by Gerald Posner and Vincent Bugliosi. See Posner, *Case Closed*, p. 348, and Bugliosi, *Reclaiming History*, p. 153. For a general description of the arraignment, see the testimony of David L. Johnston, elected justice of the peace, who presided over this judicial procedure, WC 15, pp. 503–13.

197 Oswald's comments in the hallway. See Bugliosi, *Reclaiming History*, p. 156 (from news video in the library of the Sixth Floor Museum in Dallas).

197 Questioning to Oswald about rifle. See the testimony of Captain J. W. Fritz, WC 4, p. 217.

197 "The people out at the Paine residence say you did have a rifle, and that you kept it out there wrapped in a blanket." See the testimony of Captain J. W. Fritz, WC 4, p. 217.

197 "You know you've killed the president." Ibid., p. 225.

198 Kennedy autopsy. Several factors inhibited the ability to perform a thorough autopsy and gather all evidence that should have been available to make the determination as to the direction of the bullets fired and ultimately the cause of JFK's death. These included:

- The desire of Jackie Kennedy and JFK's advisers to immediately take the body of JFK back to Washington, D.C. (contrary to Texas law—at the time, killing a president was not a federal offense)
- The quick wrapping of JFK's body in Dallas (so as not to damage the expensive coffin that had been purchased). This initial coffin would later be weighted and buried at sea.
- The lack of preservation of crime-scene evidence (such as the Secret Service wiping off some of the blood in the presidential limousine) and the benign careless handling of other forensic evidence

- Other problems with the chain of evidence (with the focus on ballistic evidence, such as the cartridges on the sixth floor of the Texas School Book Depository and CE 399, the near "so called pristine bullet," which the Warren Commission determined caused seven wounds in JFK and Governor Connally)
- The decision by one doctor, Captain James J. Humes, MD, to destroy his original notes
- Eyewitness statements concerning JFK's wounds that were not accurate
- Family wishes at Bethesda as to how the very rushed autopsy would proceed that resulted in the lack of communications between Parkland Hospital and Bethesda Naval Center doctors
- The Kennedy family's "control" over the autopsy evidence
- The mystery surrounding what happened to the metal container containing JFK's brain

Nevertheless, over a decade later, the medical panel of the House Select Committee on Assassinations (HSCA) confirmed the Warren Commission's findings. The chairman of this medical panel, nationally known forensic pathologist Michael M. Baden, MD, chief medical examiner for the City of New York, summed up and shared the views of the other forensic pathologists who reviewed the medical evidence. Baden stated: ". . . we did agree as a group with the basic bottom-line conclusions of the original autopsy doctors: *Two bullets from behind struck the President and only two bullets* [emphasis added]. However, we had a great deal of concern on many levels as to how the autopsy was performed, beginning immediately with the assumption of jurisdiction by what appears to be the Federal Government and the family of the President, intruding into what was at that time a State crime, homicide. The effect of that was to remove the body from Dallas, the jurisdiction which had a very competent forensic pathologist in charge, Dr. Earl Rose, who happens to be a member of our panel presently, to Bethesda at, apparently, the request of the family." See the testimony of Michael M. Baden, HSCA 1, p. 310 (complete testimony, pp. 180–324), and HSCA 7, *Medical Evidence and Related Issues Pertaining to the Assassination of President John F. Kennedy* (the Medical Panel Report).

198 "I don't know what you're talking about. That's the deal, is it?" See the testimony of Jesse E. Curry, WC 4, p. 156.

198 Oswald's arraignment for murder of JFK, "Well, I guess this is the trial." See the testimony of Justice of the Peace David L. Johnston, WC 15, p. 507, and Bugliosi, *Reclaiming History*, p. 194.

198 "I want Mr. John Abt of New York." See the testimony of David L. Johnston, WC 15, p. 508 (complete testimony, pp. 503–13).

199 Preparation of body for an open casket. At this time, a decision had not yet been made whether there would be an open casket. So after the autopsy, the body was prepared for a public viewing. All the previous assassinated presidents, Lincoln, Garfield, and McKinley, had state funerals with open caskets. In fact, especially in the nineteenth century, it was the practice to have open caskets with a public viewing. Prior to JFK's death, every president who lay in repose in the U.S. Capitol, and many others who did not, had open caskets. This all changed with JFK, and it is now the more common practice for closed caskets, especially for coffins that are in repose in the U.S. Capitol Rotunda. After Kennedy's death, no president has had an open casket, and only two individuals who lay in repose in the Capitol have

had their caskets open: General Douglas MacArthur and Representative Claude Pepper (D-FL). For a complete summary of all individuals who have received this honor of lying in repose in the U.S. Capitol, see the website of the Architect of the Capitol www.aoc.gov/nations-stage/lying-state; also the U.S. Senate Historical Office website, www.senate.gov/artandhistory/history/resources/pdf/LyingState.pdf.

Jackie wanted the casket to remain closed. But because of past precedents, Robert McNamara thought otherwise. At around four A.M. (EST) in the East Room, Bobby Kennedy opened the casket and viewed his brother. It did not look like him, and the embalmers had not done a good job. He said, "Close it." See Manchester, *Death*, pp. 442–43, and Arthur M. Schlesinger Jr., *Robert Kennedy and His Times* (Boston: Houghton Mifflin, 1978), p. 610.

199 Oswald's midnight press conference. See WR, pp. 200–1, WC 24, CE 2166, p. 617, and Bugliosi, *Reclaiming History*, p. 189 (taken from a newsreel at the Sixth Floor museum).

200 Oswald paraffin test. See the testimony of Sergeant W. E. Barnes, WC 7, pp. 297, 282–83 (complete testimony, pp. 270–86).

CHAPTER 8: "WE HAD A HERO FOR A FRIEND"

201 Jackie's return to the White House. See Manchester, *Death*.

203 Barlett note, "We had a hero for a friend." November 29, 1963, *Washington Star* article, Charlie Bartlett "hero for a friend." See Manchester, *President*, p. 446.

204 McGrory's tribute to JFK was also published as a small pamphlet: Mary McGrory, *In Memoriam* (Washington, DC: Evening Star Newspaper Company [1964]).

204 Maud Shaw bearing the bad news to JFK's children. Although reluctant to do so, Maud Shaw, at the request of Jackie's mother, Janet Auchincloss, told the children. There may have been a miscommunication and/or misunderstanding, since Jackie wanted to be the one to convey this tragic news. This perhaps was the beginning of Shaw's ultimate termination of employment. See C. David Heyman, *American Legacy: The Story of John and Caroline Kennedy* (New York: Atria Books, 2007), pp. 110–14. Also see generally Maud Shaw, *White House Nannie: My Years with Caroline and John Kennedy, Jr.* (New York: New American Library, 1966), pp. 12, 20–21.

204 Private mass in the East Room for the family and close aides. See Manchester, *Death*, pp. 460–64.

205 The misunderstanding about when LBJ would move into the Oval Office. The relationship, or lack of one, between Robert Kennedy and LBJ was well known. Similar to the strained interaction between the Kennedy and Johnson camps, historians shortly after JFK's death tended to side with one group or the other. More recently, historian Robert A. Caro has assessed fairly the misunderstandings in the immediate few hours and days after the Kennedy assassination. In addition to RFK running past LBJ on Air Force One immediately after it landed, and the miscommunication on occupying the Oval Office, there were tensions about Johnson's decision to hold a cabinet meeting on Saturday and to give an address to a Joint Session of Congress so soon after Kennedy's death. See Caro, *Years*, pp. 373–77. Also see generally Jeff Shesol, *Mutual Contempt: Lyndon Johnson, Robert Kennedy, and the Feud That Defined a Decade* (New York: Norton, 1997).

205 The decision to bury JFK at Arlington. For a discussion on the decision by Jackie to bury JFK at Arlington National Cemetery and for other information concerning the burial, go to the official website of the cemetery at www.arlingtoncemetery .mil/visitorinformation/monumentmemorials/jfk.aspx.

206 Jackie's visit to Arlington and choosing the site. See Manchester, *Death*, pp. 491–97.

207 Fritz questioning of Oswald. See generally WC 20, Kelley Exhibit A, pp. 440–46 (memorandum of interview with Lee Harvey Oswald); WC 24, CE 1988, pp. 18–20 (FBI report on interrogation of Lee Harvey Oswald); WC 29, CE 2003, pp. 195–404 (Dallas Police Department file on the investigation of the assassination of the president, especially pp. 264–79, interview with Oswald); and the testimony of Captain J. W. Fritz, WC 4, pp. 202–47.

207 Questioning about curtain rods. See the testimony of Captain J. W. Fritz, WC 4, pp. 218–19, and WC 24, CE 1988, p. 18.

207 Questioning about eating lunch. See generally, the testimony of Captain J. W. Fritz, WC 4, p. 223; WC 20, Kelley Exhibit A, p. 440; WC 24, CE 1988, pp. 18–19; and WC 24, CE 2003, p. 267.

208 Oswald's belongings at Paine house. See WC 20, Kelley Exhibit A, p. 440; WC 24, CE 2003, p. 267; and WC 24, p. 19.

208 Oswald's purchase of guns through the mail. See WC 20, CE 1988, p. 19.

208 Finding pistol at time of arrest. See WC 20, Kelley Exhibit A, p. 440, and WC 24, CE 1988, pp. 18–19.

209 Oswald's anger about questioning by FBI. See WC 20, Kelley Exhibit A, pp. 440–41, and WC 24, p. 18.

209 "I am not a malcontent; nothing irritated me about the president." See WC 20, Kelley Exhibit A, p. 441.

209 Refusal to take a polygraph examination. See WC 20, Kelley Exhibit A, p. 441, and WC 24, CE 1988, p. 19.

209 Questions about ID Card. See WC 20, Kelley Exhibit A, p. 443, and CE 1988, p. 19.

210 Viewing the parade and shooting the president. See WC 20, Kelley Exhibit A, p. 441, and WC 24, CE 1988, p. 20.

211 Marina's conversation with Oswald. See McMillan, *Marina and Lee*, p. 435–36; testimony of Marina Oswald, WC 1, pp. 78–79; and Bugliosi, *Reclaiming History*, pp. 223–24.

212 Wade interview and death penalty comments. For a transcript of these comments made in the corridor of the jail, see WC 24, CE 2172, pp. 843–45 (Nov. 23, 1963, interview with District Attorney Henry Wade); and Bugliosi, *Reclaiming History*, pp. 222–23 (quoting from newsreel footage in the Sixth Floor Museum Library).

213 Captain Fritz comments to the press, "case is cinched." See the transcript of Fritz's remarks, WC 24, CE 2153, pp. 787–88 (Nov. 23, 1963, interview with J. W. Fritz), and Bugliosi, *Reclaiming History*, p. 226 (quoting from newsreels in the Sixth Floor Museum Library).

214 Meeting of Lee with his brother. This conversation is recounted by Oswald's sibling. See Robert Oswald, Myrick Land, and Barbara Land, *Lee: A Portrait of Lee Harvey Oswald by His Brother Robert Oswald* (New York: Coward-McCann, 1967), pp. 142–46.

215 There is no complete transcript of all the statements Oswald made in the hallways during the time he was moved from room to room while he was in custody. I have listened to several newsreels and TV and radio recordings of these encounters, and the quotations from Oswald and the reporters are verbatim from these recordings.

217 Sale of Zapruder film. On November 25, 1963, *Life* magazine purchased the rights to the film plus royalties for $150,000, of which Abraham Zapruder donated $25,000 to Officer Tippit's widow. In 1975, the rights to the film, which was stored at the National Archives, were transferred back to the Zapruder family for the price of one

dollar. In 1999, a special arbitration panel awarded the Zapruder family $16 million in compensation for the government's possession of the film. Later that year, the family donated all the copyrights of the film to the Sixth Floor Museum at Dealey Plaza in Dallas. For the timeline of the activities concerning this historical film, see the Web page of the museum at www.jfk.org/go/collections/about/zapruder -film-chronology. See also Richard B. Trask, *National Nightmare on Six Feet of Film: Mr. Zapruder's Home Movie* (Danvers, MA: Yeoman Press, 2005); Oyvind Vagnes, *Zaprudered: The Kennedy Assassination Film in the Visual Culture* (Austin: University of Texas Press, 2011); and David R. Wrone, *The Zapruder Film: Reframing the Kennedy Assassination* (Lawrence: University Press of Kansas, 2003).

218 LBJ, national day of mourning. See *Public Papers of the Presidents of the United States, Lyndon B. Johnson*, bk. 1, Nov. 22, 1963–June 30, 1964 (Washington, DC: U.S. Government Printing Office, 1965), p. 2.

218 Curry tells reporters about transfer. See Bugliosi, *Reclaiming History*, pp. 240–41 (citing newsreel footage in the Sixth Floor Museum in Dallas). For a three-part series on the interviews of Jesse Curry with the press on November 22 and 23, 1963, go to www.youtube.com/watch?v=htgn-ZH1_oQ&list=PLAA712DECA2E71AC9, www.youtube.com/watch?v=4goKpDybUmY&list=PLAA712DECA2E71AC9, and www.youtube.com/watch?v=AVnw5lnf8AI&list=PLAA712DECA2E71AC9.

219 Oswald death threats and Newsom reporting threats. See FBI Report of November 24, 1963, WC 19, CE 2013, p. 429; FBI report of Milton L. Newsom, Nov. 25, 1963, telephone call to Dallas sheriff's office, WC 24, CE 2018, p 434; WC 19, Frazier Exhibit 5086, pp. 770–71; and WC 19, Frazier Exhibit 4086, p. 772. Also see the testimony of Sheriff J. E. ("Bill") Decker, WC 12, pp. 48–51, 110 (complete testimony, pp. 42–52); the testimony of Captain Cecil E. Talbert of the Dallas Police Department, WC 2, p. 110 (complete testimony, pp. 108–28); and the testimony of Captain W. B. Frazier, WC 2, pp. 52–58, 69–78.

221 Last interview with Oswald. See generally WC 20, Kelley Exhibit A, pp. 440–46 (memorandum of interview with Lee Harvey Oswald); testimony of Postal Inspector Harry D. Holmes, WC 7, pp. 289–308, 525–30; and WC 24, CE 2064, pp. 488–94 (FBI report on Postal Memorandum from Harry D. Holmes).

222 Questioning about P.O. boxes and guns. See WC 24, CE 2064, p. 489, and the testimony of Harry Holmes, WC 7, pp. 298–99.

223 Questioning about use of name A. J. Hidell. See WC 24, p. 489; the testimony of Harry Holmes, WC 7, pp. 299, 303; and the testimony of Dallas Police Detective James R. Leavelle, WC 7, pp. 267–68.

223 Curry's last statement to press. For the transcript of this interview, see WC 24, CE 2147, pp. 771–79 (KRLD-TV broadcast of Nov. 24, 1964, interview with Jesse Curry).

224 Oswald's comments about Cuba. See WC 24, CE 2064, pp. 489–90, and the testimony of Harry Holmes, WC 7, p. 301.

224 Oswald's involvement with the FPCC. See WC 20, Kelley Exhibit A, p. 443.

225 Oswald's views on Johnson and Kennedy. See testimony of Dallas Police Detective James R. Leavelle, WC 7, p. 267; WC 20, Kelley Exhibit A, p. 443; and WC 24, CE 2003, p. 269.

225 Questioning on whether Oswald is a communist or a Marxist. See WC 24, p. 490, and testimony of Postal Inspector Harry Holmes, WC 7, p. 298.

225 Questioning about Oswald's subscriptions. See WC 24, Kelley Exhibit A, p. 443.

226 Questioning about Oswald's discharge from the Marines. See WC 24, p. 490.

226 Questioning Oswald about his map. See WC 20, Kelley Exhibit A, p. 444, and WC 24, CE 2064, pp. 490–91.

227 Questioning about the sack Oswald carried into the Book Depository and his activities immediately prior to and following the assassination. See WC 24, pp. 490–91.

227 Questioning about Oswald's movement at lunch and immediately after the assassination. See WC 24, CE 2064, 24 H pp. 490–91.

229 "We'll be through in a few minutes." "I never used the name of Hidell." See the testimony of Secret Service Agent Forrest Sorrels. WC 7, p. 357.

229 Oswald's use of the Selective Service card and the name A. J. Hidell with Oswald's photograph attached. See WC 24, p. 491.

CHAPTER 9: "LEE OSWALD HAS BEEN SHOT!"

231 The transfer of Oswald from city to county jail. See generally the testimony of Captain J. W. Fritz, WC 15, pp. 144–50, and the testimony of Jesse E. Curry, WC 12, p. 38.

231 Leavelle's suggestion on moving Oswald. See the testimony of James Robert Leavelle, WC 13, p. 17 (complete testimony, pp. 13–21).

232 Oswald changes clothes. See generally WC 24, CE 2003, pp. 270–71; the testimony of Harry Holmes, WC 7, p. 300; and the testimony of Detective L. D. Montgomery, WC 13, p. 27.

232 "I would not move . . ." See the testimony of Forrest V. Sorrels, WC 13, p. 63 (complete testimony, pp. 55–83).

233 Oswald's interaction with Leavelle, joking about being shot. These quotes are taken from Bugliosi, *Reclaiming History*, p. 269 (citing interviews of James. R. Leavelle by Dale K. Myers, April 7, 1983, and Nov. 13, 1999, and Leavelle's statement on *Peter Jennings Reporting: The Kennedy Assassination—Beyond Conspiracy*, ABC News special, Nov. 20, 2003.

233 Oswald's last words, "I'd like to contact . . . representative of the American Civil Liberties Union." See WC 21, Pappas Exhibit 4, pp. 23–24.

234 Reporter comments, Pappas, "Now the prisoner . . . ," "Do you have anything to say in your defense?" See Trask, *President*, p. 207.

234 "Jack, you son of a bitch, don't!" See WC 19, Combest Exhibit 5101, p. 350 (FBI report of interview with Dallas police detective Billy H. Combest).

Jack Ruby was well known by the Dallas Police Department and frequently visited the station. He was seen standing with the press during Oswald's brief news conference. However, the actual shooting resulted from a series of coincidences that placed Ruby in the building when he made the impulsive decision to shoot Oswald. Oswald's transfer was delayed by the last interrogation and the decision to change his clothes. Ruby's contact with a stripper and his desire to send her money placed him in the nearby Western Union office late that morning. He waited in line just a few minutes before his encounter with Oswald. So a longer line would have delayed him, and he would have missed the transfer of Oswald. In addition, the decision to move Oswald in an unmarked car gave Ruby the opportunity to jump in front of this vehicle just as it was backing up, actually blocking others and giving him a better opportunity to lunge forward. Finally and perhaps most important, Ruby left his beloved dog, Sheba, in his parked car. Given Ruby's strange obsessive love for this dachshund, he would have never left his dog in a car, if he intended not to return. Some of the dogs Jack Ruby owned he called his children, and sometimes he referred to Sheba as his "child" or his "wife."

234 There were several cameras in the basement of the Dallas Jail when Jack Ruby shot Oswald (still, film, and videotape—NBC was the only network that broadcast the incident live). Robert Jackson of the *Dallas Times Herald* won a Pulitzer Prize for his photo capturing Oswald the moment he was shot. For a very good documentary covering this shooting, see *JFK: The Ruby Connection*, Discovery Channel, 2009.

For background material on the life of Jack Ruby, see Melvin M. Belli and Maurice C. Carroll, *Dallas Justice: The Real Story of Jack Ruby and His Trial* (New York: David McKay, 1964); Elmer Gertz, *Moment of Madness: The People vs. Jack Ruby* (Chicago: Follett, 1968); Garry Wills and Ovid DeMaris, *Jack Ruby: The Man Who Killed the Man Who Killed Kennedy* (New York: New American Library, 1968); and John Kaplan and Jon R. Waltz, *The Trial of Jack Ruby* (New York: Macmillan, 1965). For reminiscences by two strippers who worked for Jack Ruby, see Diane Hunter and Alice Anderson, *Jack Ruby's Girls* (Atlanta: Hallux, 1970).

235 "I hope I killed the son of a bitch." See the testimony of Dallas police officer Don Ray Archer, WC 12, p. 400 ("I hope I killed the son-of-bitch"); and the testimony of Dallas police detective Lewis D. Miller, WC 12, p. 308 ("I hope the son of a bitch dies." "It will save you guys a lot of trouble," or "It will save everybody a lot of trouble").

235 "Oh hell! You guys know me, I'm Jack Ruby." See the testimony of Dallas police detective Lewis D. Miller, WC 12, p. 308 (complete testimony, pp. 297–314). For similar remarks, see the testimony of Dallas police captain Cecil E. Talbert, WC 12, p. 120 ("I'm Jack Ruby. Everybody knows me. I'm Jack Ruby"; complete testimony, pp. 108–28); testimony of Dallas police reserve captain Charles Oliver Arnett (occupation truck driver), WC 12, p. 153 ("I am Jack Ruby. All of you know me"; complete testimony, pp. 128–58); and Dallas police officer Don Ray Archer, WC 12, p. 400 ("You all know me, I'm Jack Ruby").

235 Pettit, "He's been shot! He's been shot! . . ." Ibid., p. 209.

237 "Do you think I'm going to let the man who shot our president get away with it?" See WC 24, CE 2002, p. 105 ((interview of W. J. Harrison, Dallas Police Department file on the Investigation of the Operational Security Involving the Transfer of Lee Harvey Oswald, November 24, 1963; complete file, pp. 48–194).

237 Ruby shooting Oswald, "Somebody had to do it." See the testimony of Dallas Police Detective Barnard S. Clardy WC 12, pp. 412–13 (complete testimony pp. 403–14); the testimony of Dallas police officer Don Ray Archer, WC 12, p. 401 (complete testimony, pp. 395–403); and the testimony of District Attorney Wade, WC 5, p. 245 (complete testimony, pp. 213–54).

237 Pappas, "Now the ambulance is being rushed in here . . ." Ibid., p. 214.

237 "Ladies and gentlemen, Lee Harvey Oswald has been shot and is on his way to Parkland Hospital." See Bugliosi, *Reclaiming History*, p. 278.

240 "He cry." See the testimony of Marguerite Oswald, WC 1, p. 162.

240 "I think someday you'll hang your heads in shame." Ibid.

240 Dallas guilt. Some individuals viewed Dallas as a city dominated by right-wing extremists and Kennedy-haters. The climate in preassassination Dallas, which made it a likely place for JFK's assassination, was captured in the early work by Warren Leslie, *Dallas: Public and Private* (New York: Grossman, 1964). Most recent, see Bill Minutaglio and Steven L. Davis, *Dallas 1963* (New York: Twelve, 2013).

Many civic leaders were faced with the challenge of repairing the national perception of somehow the entire city being responsible for JFK's death. On New

Year's Day 1964, Stanley Marcus, whose father founded the Dallas-based department store Neiman-Marcus, purchased a full-page ad appearing in national newspapers with the store's logo. The message, with the headline WHAT'S RIGHT WITH DALLAS, was signed by Marcus. See Stanley Marcus, *Minding the Store: A Memoir* (Boston: Little, Brown, 1974), pp. 257–58.

241 Chief Justice Warren, climate of hate. *John Fitzgerald Kennedy: Eulogies to the Late President Delivered in the Rotunda of the United States Capitol,* November 24, 1963. (Washington, DC: U.S. Government Printing Office, 1964).

 For Earl Warren's reaction to the assassination and his experience of being appointed to head the presidential commission to investigate JFK's death, see Earl Warren, *The Memoirs of Chief Justice Earl Warren* (Garden City, NY: Doubleday, 1977), pp. 351–72.

242 Opening of casket for last time. See Manchester, *Death,* pp. 516–17.

243 Trip to Capitol. Ibid., pp. 529–43.

244 Mansfield's comments. For Senator Mansfield's "took a ring" eulogy and other speeches in the U.S. Congress, see U.S. Senate, *Memorial Addresses in the Congress of the United States and Tributes in Eulogy of John Fitzgerald Kennedy, Late a President of the United States* (Washington, DC: U.S. Government Printing Office, 1964).

245 Crowds in line in front of Capitol. For an account of the long lines, forty blocks long and four abreast, of people waiting to pay their respects to the fallen president, see the comments of CBS News reporter Roger Mudd in Trost, *President,* p. 238.

246 The comments of Assistant Secretary of Labor Daniel Patrick Moynihan, who one day would be senator. See Trost, *President,* p. 236.

248 Dignitaries walking and concern about LBJ's safety. See Melville Bell Grosvenor, *The Last Full Measure: The World Pays Tribute to President Kennedy* (Washington, DC: National Geographic, 1964), p. 346.

249 St. Matthew's. "Where's my daddy?" See Manchester, *Death,* p. 584.

249 Cushing's comments and St. Matthew's ceremony. Ibid., pp. 584–90.

249 John Jr. salute. See the account, "A Little Soldier's Salute," by Robert M. Andrews, a UPI correspondent, in United Press International, ed., *Four Days: The Historical Record of the Death of President Kennedy* (Rockville, MD: American Heritage, 1964), pp. 114–15.

251 Eternal Flame. Robert McNamara was Jackie Kennedy's personal representative at Arlington National Cemetery when the site of JFK's final resting place was chosen. Lieutenant General Walter K. Wilson Jr., as chief of the Army Corps of Engineers, was responsible for carrying out her wishes for the creation of the eternal flame, and Colonel Clayton B. Lyle was actually assigned this task. For an account of how the corps responded to Jackie's wishes, see Lieutenant General Walter K. Wilson Jr., *Engineer Memoirs* (Washington, DC: U.S. Army Corps of Engineers, Office of the Chief of Engineers, May 1984), EP 870-1-8, http://140.194.76.129/publications/eng-pamphlets/EP_870-1-8/toc.htm), pp. 194–96. See also recollections at both Arlington National Cemetery and Army Corps of Engineers websites, www.arlingtoncemetery.net/eternalflame.htm and www.usace.army.mil/About/History/HistoricalVignettes/CivilEngineering/060JFKsEternalFlame.aspx.

252 Jackie's comments about meeting visitors (reception of foreign guests). See Manchester, *Death,* pp. 605–15.

252 Tippit's funeral. The Tippit family maintains a website with comprehensive information on Officer Tippit's life and death at www.jdtippit.com. For a short video clip on his funeral go to www.youtube.com/watch?v=eiDXCktjvAw.

Also, for video coverage of Tippit's funeral, Oswald's funeral, and President Johnson's telephone call to Tippit's widow, Marie Frances Gasway Tippit, go to www.youtube.com/watch?v=tFP9a8LeuvA. A special fund was established for Marie Tippit and his three minor children, which received over $600,000 in contributions.

253 Oswald funeral: For AP reporter Mike Cochran's retelling of his volunteering to be a pallbearer and other reminiscences of the funeral, see Trost, *President*, pp. 254–57. For a more in-depth interview that weaves in the timeline of the three funerals, of John F. Kennedy, Officer J. D. Tippit, and Lee Harvey Oswald, which occurred on Monday, November 25, 1963, see *Nov. 22, Twenty-Five Years Later*, pullout section (*Dallas Morning News*, Nov. 20, 1988), www.writespirit .net/wp-content/cache/supercache/www.writespirit.net/soulful-tributes/political -figures/president-kennedy/funeral-timel ine-kennedy-john.

In addition to the few photographs published in T. Thompson, "In Texas, a Policeman and an Assassin Are Laid to Rest," *Life*, Dec. 6, 1963, pp. 52B–52E, the Warren Commission hearings volumes contain several photographs of the funeral. See WC 16, CE 165–168, 170–179, pp. 522–24, 526–29. Marguerite Oswald also published a magazine. See Marguerite Oswald, *Aftermath of an Execution: The Burial and Final Rites of Lee Harvey Oswald as Told by His Mother* (Dallas: Challenge Press, 1965).

253 Oswald's exhumation. This exhumation was assented to by Oswald's wife, remarried as Marina Oswald Porter, after the persistent urging of author Michael Eddowes, author of *The Oswald File* (New York: Clarkson N. Potter, 1997). Lee Harvey Oswald's brother, Robert, had opposed this action, and after a more-than-two-year legal battle, the exhumation occurred early in the morning of October 4, 1981. Shortly after this event, a press conference was held, which indeed confirmed that the corpse was Lee Harvey Oswald. His identification was based largely on dental records and other forensic evidence. See Linda E. Norton, James A. Cottone, Irvin M. Sopher, and Vincent J. Dimario, "The Exhumation and Identification of Lee Harvey Oswald," *Journal of Forensic Sciences* 29, no. l (Jan. 1984): 19–38 (the Norton Report).

Several years later, Oswald's deteriorated casket from this exhumation was sold in 2010 for $87,468 by Nate D. Sanders Auctions in Santa Monica, CA. See http:// natedsanders.com/viewuserdefinedpage.aspx?pn=LeeHarveyOswaldCasket-Consignment.

Today, the Shannon Rose Hill Cemetery in Fort Worth still discourages visitors to Oswald's burial site. Since the cemetery was not providing any help with directions to Oswald's grave, a living comedian purchased the plot adjacent to the plot where Lee and his mother are buried. This adjacent tombstone has the inscription NICK BEEF to aid individuals trying to find Oswald's grave. See Jack Douglas Jr., "A JFK Mystery: Who Is Nick Beef?" *Fort Worth Star-Telegram*, March 13, 2005, www.chron.com/news/houston-texas/article/A-JFK-mystery-Who-is-Nick -Beef-1475555.php. While today the cemetery refuses to provides directions to this tombstone, for those so inclined there are several written and video sources of directions.

The original headstone: This original, more ornate headstone contained Oswald's complete name and years of birth and death. After it was stolen and returned shortly thereafter, Marguerite Oswald removed it and replaced it with the current headstone, a slab with only the word "Oswald." She stored the original in her basement, where it was forgotten until found by a new owner of her house. It subsequently was sold to a museum in Illinois, but the ownership of this item was being challenged. See, Steven Yaccino, "A Macabre Tussle over a Historic Slab," *New York Times*, Apr. 11, 2012, p. A16, www.nytimes.com/2012/04/11/us/a-dispute -over-lee-harvey-oswalds-tombstone.html?_r=0.

254 "[V]isit our friend." See Manchester, *Death of a President*, pp. 619–20.

263 Souvenirs. An entire book can be written on all of hundreds, perhaps thousands, of souvenirs and other "relics" regarding the JFK assassination. Especially in the age of the Internet and eBay, most of these items, even those that were once thought to be scarce, are quite common and appear frequently on the marketplace. Even wood and bricks from the Texas Book Depository, pieces of the grassy-knoll fence, and Oswald's furniture come on the marketplace.

Desk ornament of Dealey Plaza. This infamous desk ornament was not well received in Dallas. In William Manchester's book, *The Death of a President*, he labels this item to be one of "continuing embarrassment to the civic leaders" and a "grotesque sideshow" (p. 634). There are several different versions of this item, with and without the paper clip well. It almost always is missing the original pen.

Night-light. This is very hard to find in the original packaging, and it is likely that few were made. Also, some entrepreneur produced the large "flame of hope" candles and to boost sales was able to get press coverage by having various Kennedy family members pose as they received their boxed sets.

The first memorial pamphlet was probably the forty-page *John Fitzgerald Kennedy and the Federal City He Loved*, published in 1963 by Tatler Publishing Company just a few days after the assassination and funeral (with a few photos of the funeral).

Figurines. Inarco produced three-piece figurine dolls with a large Jackie head in a mourning veil accompanying smaller figurines of the two children, Caroline and John Jr. saluting.

Oswald radio interviews in New Orleans. Similar to the reissuing and repackaging of identical Beatles long-play (33⅓ rpm) phonographic recordings by Vee Jay Records prior to and during the Capitol Records 1964 release of *Introducing the Beatles*, so too, did entrepreneurs cash in on the radio interviews and other statements of Lee Harvey Oswald. There were several different phonographic records, but they essentially provided the same material. See *Oswald: Self-Portrait in Red*, Information Council of America, Eyewitness Records, EW-1001, 1967 (probably the most common), and *Lee Harvey Oswald Speaks*, Information Council of America, Eyewitness Records, EW-1002, 1967 (essentially the same record with a different cover); *Lee Harvey Oswald Speaks*, Truth Records, 1966; *The President's Assassin Speaks*, Key Records, KLP 880; and *Hear Kennedy's Killer: An Interview with Lee Harvey Oswald*, S.S. Records, 1964 (this independent label is slightly more scarce).

Needless to say, there are more than one hundred phonographic memorial recordings in tribute to JFK, including foreign-language releases, as well as many long-playing documentary records on the assassination itself. No one yet has published a complete discography of all of the long-playing records, 45 rpm singles, reel-to-reel tapes, and CDs that have been released.

Marguerite Oswald's financial enterprises. After her son's death, Marguerite Oswald engaged in some commercial efforts. See her self-published booklet on

Lee's funeral: Marguerite Oswald, *Aftermath of an Execution: The Burial and Final Rites of Lee Harvey Oswald as Told by His Mother* (Dallas: Challenge Press, 1965). She also participated in a phonographic recording, *The Oswald Case: Mrs. Marguerite Oswald Reads Lee Harvey Oswald's Letters from Russia* (New York: Broadside Records, 1964). This was part of two-record series from Broadside Records. The other recording was by Mark Lane, the attorney she retained to represent her son before the Warren Commission, *The Oswald Case: Mark Lane's Testimony to the Warren Commission* (New York: Broadside Records, 1964). For Marguerite Oswald's own biography, see Jean Stafford, *A Mother in History* (New York: Farrar, Straus & Giroux, 1966). Bob Schieffer, the current moderator of *Face the Nation* on CBS, was then a young police reporter for the *Fort Worth Star-Telegram* and had the serendipitous opportunity to drive Marguerite Oswald to the police station just a few hours after her son was arrested. His observation about their conversation was that "she was obsessed about money." See Trost, *President*, p. 129.

263 LBJ address to the joint session of Congress, Nov. 27, 1963. See *Public Papers of the Presidents of the United States, Lyndon B. Johnson*, bk. 1, Nov. 22, 1963–June 30, 1964 (Washington, DC: U.S. Government Printing Office, 1965), pp. 8–10.

264 Creation of Warren Commission. Ibid., p. 13. For Executive Order 11130, see p. 14.

264 November 29, 1963, *Washington Star* article, Charlie Barlett, "hero for a friend." See Manchester, *President*, p. 446.

CHAPTER 10: "ONE BRIEF SHINING MOMENT"

266 White interview, *Life* magazine. See *Life*, December 6, 1963, pp. 158–59.

269 See Benjamin Bradlee, *That Special Grace* (Philadelphia: Lippincott, 1964).

270 Magazine memorial issues. During the last fifty years, there have been hundreds of serial magazines with images of the Kennedys printed on the cover. *Life* magazine, perhaps more than any other, was especially smitten with placing the Kennedys on its covers. This is perhaps because of the relationship Henry Luce, the publisher of *Life*, and his wife had with Joseph and Rose Kennedy. JFK, his wife, other family members, and even their children, appeared on the covers of *Life* nineteen times up until November 22, 1963. Including memorial and other special issues, the Kennedys have appeared on the cover of *Life* over seventy times. Scores of specialty individual magazines ("one-shots) were issued to cover the Kennedy presidency, his life and death, his funeral, various aspects of the JFK assassination, Oswald's life, and the Ruby trial.

Some special items: Stanley Marcus of the Dallas-based department store Nieman Marcus commissioned a very special limited edition of five hundred copies of JFK's undelivered Dallas Trade Mart speech. This publication was created by book designer Carl Hertzog.

The State Democratic Executive Committee Mailing. All of the more than two thousand individuals who bought the $100 gold tickets to the Austin dinner fund-raiser for the evening of November 22 received in the mail a commemorative packet, which included in a black cardboard mailer: a short paper message acknowledging the contribution, a long-play phonographic recording, *His Last 24 Hours*, a reprint of the two JFK speeches never delivered, and a specially produced one-page dinner program.

The Funeral Cortege of President John F. Kennedy (toy soldiers). This boxed set contains twenty-six pieces, including gun carriage, six-horse team, riderless horse, flag-draped coffin, military escorts and honor guard, Ted, Robert, Jackie, Caroline, and John Jr., as well as instructions for their positioning. It was produced in 1997 by S.T.E. Ltd., in a limited edition of five hundred sets.

Stamps and coins: In the years immediately following the assassination, many countries issued commemorative JFK stamps, perhaps the most unusual being a series of eight stamps issued by Umm Al Qiwain of the United Arab Emirates depicting scenes from Kennedy's funeral. See Helen Emery, *Stamps Tell the Story of John F. Kennedy* (New York: Meredith Press, 1968). One of the more unusual commemorative coins was issued in Germany by Deutsche Numismatikin in various sizes in gold and silver with JFK being shot on the obverse and Ruby shooting Oswald on the reverse. See Aubrey Mayhew, *The World's Tribute to John F. Kennedy in Medallic Art: Metals, Coins and Tokens—an Illustrated Standard Reference* (New York: Morrow, 1966).

270 CBS employee brochure, dated Dec. 18, 1963, with a cover letter from Frank Stanton. Other networks created similar items for their employees. For instance, ABC produced a two-disc long-playing record, *November 22nd, 1963*, for internal use by their staff. In addition, networks and radio stations issued items for commercial sale, for instance, the phonographic recording *The Fateful Hours: Actual Unforgettable News Reports of Friday, November 22nd, 1963, by KLIF Dallas, a McLendon Station* (Capitol Records RB-2278, n.d.). NBC issued two books on its broadcast: *Seventy Hours and Thirty Minutes: As Broadcast on the NBC Television Network by NBC News* (New York: Random House, 1966), and an illustrated version, *There Was a President* (New York: Ridge Press, 1966). Even *TV Guide* published a special issue on the television coverage of Kennedy's assassination. Both AP and UPI issued books on their own coverage: *The Torch Is Passed: The Associated Press Story of the Death of a President* (New York: Associated Press, 1964), and *Four Days: the Historical Record of the Death of President Kennedy* (American Heritage, 1964). Local newspapers imprinted their own names on these wire-service editions, and some entrepreneurial newspapers, such as the *Kansas City Star*, actually printed their own dust jackets to cover the AP edition. Many newspapers also issued special pamphlets touting their own coverage, and of course just about every newspaper had a special pullout section. There were scores of these types of publications, some of which were reprinted for anniversary editions. See, for instance, *Good Night, Brave Spirit: John F. Kennedy, 1917–1963* ("as covered by the *Boston Globe*," forty-page magazine format); *Oh, No! Oh No! Four Tragic Days as Reported by the Milwaukee Journal* (large stiff-paper format); or *Assassination of a President: A Chronicle of the Six Days from November 23 to November 28, 1963, Reprinted from the Pages of the New York Times* (New York: Viking, 1964; newspaper size in both boxed and envelope editions).

272 "I wanted my old house back." Pottker, Jan, *Janet & Jackie: The Story of a Mother and Her Daughter, Jacqueline Kennedy Onassis* (New York: St. Martin's, 2001), p. 232.

272 Honoring Clint Hill. See Hill, *Mrs. Kennedy and Me*, pp. 326–27.

273 LBJ Medal of Freedom Ceremony, Dec. 6, 1963, State Dining Room of the White House. See *Public Papers of the Presidents of the United States, Lyndon B. Johnson*, bk. 1, Nov. 22, 1963–June 30, 1964 (Washington, DC: U.S. Government Printing Office, 1965), pp. 29–34.

275 McNamara note. See "The White House Years of Robert S. McNamara," Sotheby's Auctions, Oct. 23, 2012.

275 LBJ candlelight memorial service for President Kennedy at the Lincoln Memorial, Dec. 22, 1963. See *Public Papers of the Presidents of the United States, Lyndon B. Johnson*, bk. 1, Nov. 22, 1963–June 30, 1964 (Washington, DC: U.S. Government Printing Office, 1965), pp. 79–80.

276 Presentation of inaugural-address books to McNamara and Powers. See "The White House Years of Robert S. McNamara," Sotheby's Auctions, Oct. 23, 2012.

277 Short thank-you film in movie theaters. On January 14, 1964, Jacqueline Kennedy appeared in a newsreel, which preceded the featured film in movie theaters throughout the country, to thank the nation for its outpouring of sympathy. For a video of her brief remarks, go to www.youtube.com/watch?v=oJhAkD8LGwg. For some of the letters of the nation's outpouring of grief, see Ellen Fitzpatrick, ed., *Letters to Jackie: Condolences from a Grieving Nation* (New York: Ecco Press, 2010), and Jay Mulvaney and Paul De Angelis, *Dear Mrs. Kennedy: The World Shares Its Grief—Letters, November 1963* (New York: St. Martin's Press, 2010).

278 JFK Library. In 1964, the trustees, hoping to raise funds for the John Fitzgerald Kennedy Library, launched a traveling exhibit. A fund-raising pamphlet was produced for this purpose. See *The John F. Kennedy Library Exhibit* (Rockville, MD: Haynes Lithograph Company, 1964).

279 Kennedy Funeral Tribute. See Melville Bell Grosvenor, *The Last Full Measure: The World Pays Tribute to President Kennedy* (Washington, DC: National Geographic, March 1964), pp. 307–55. This article was subsequently reprinted as a freestanding pamphlet.

280 Move to New York. Though she avoided the press and public comments, hundreds of articles and numerous books were written after Jackie's White House years. See Bill Adler, *The Eloquent Jacqueline Kennedy Onassis: A Portrait in Her Own Words* (New York: William Morrow, 2004); Christopher Anderson, *Jackie after Jack* (New York: William Morrow, 1998); Michael Beschloss, ed., *Jacqueline Kennedy: Historic Conversations on Life with John F. Kennedy* (New York: Hyperion, 2011); Sarah Bradford, *America's Queen: The Life of Jacqueline Kennedy Onassis* (New York: Viking, 2000); C. David Heymann, *A Woman Named Jackie* (New York: Carol Communications, 1989); William Kuhn, *Reading Jackie: Her Autobiography in Books* (New York: Nan A. Talese, 2010); and Gregg Lawrence, *Jackie as Editor: The Literary Life of Jacqueline Kennedy Onassis* (New York: Thomas Dunne Books, 2011).

281 In 1972, Jacqueline Kennedy Onassis, exasperated by her unwelcomed obsessive photographer stalker, Ron Galella, successfully obtained a restraining order. Gallella was enjoined from "(1) keeping the defendant and her children under surveillance or following any of them; (2) approaching within 100 yards of the home of defendant or her children, or within 100 yards of either child's school or within 75 yards of either child or 50 yards of defendant; (3) using the name, portrait or picture of defendant or her children for advertising; (4) attempting to communicate with defendant or her children except through her attorney." *Galella v. Onassis*, 353 F. Supp. 196 (1972). This judgment was affirmed in part and modified in part by the appellate court, 487 F.2d 986 (1973). Galella's claim against the Secret Service agents' aggressive response to his presence when he was near Jackie's family was dismissed. In 2010, the cable television company HBO produced a documentary, *Smash His Camera*, on this paparazzo photographer—the titled borrowed from Jackie own request to her Secret Service detail as to how to respond to Galella's intrusive behavior.

283 LBJ Choice for VP. Almost immediately after the assassination, Senator Hubert H. Humphrey perceived a political opportunity to ingratiate himself with Johnson—especially on the civil-rights issue—perhaps with the hope of securing the number-two spot on the Democratic ticket in 1964. His first meeting with LBJ, along with other congressional leaders, occurred almost immediately after Johnson returned to his office in the Old Executive Office Building. See Gillon, *Kennedy Assassination*, pp. 174–77. Also see Hubert H. Humphrey, *Education of a Public Man: My Life in Politics* (Garden City, NY: Doubleday, 1976), p. 260 ("desire to be of all possible assistance").

283 RFK speech at convention. Go to the PBS website at www.pbs.org/wgbh/amex/rfk/filmmore/pt.html.

283 LBJ election mandate. See U.S. House of Representatives, *Statistics of the Presidential and Congressional Election of November 3, 1964*, corrected to Aug. 15, 1965 (Washington, DC: U.S. Government Printing Office, 1965), p. 53.

284 *Life* interview. See Manchester, *Death*, p. 625.

285 LBJ announcement that he would not seek reelection, March 31, 1968. See *Public Papers of the Presidents of the United States, Lyndon B. Johnson*, bk. 1, Jan. 1–June 30, 1968 (Washington, DC: U.S. Government Printing Office, 1970), pp. 469–76, and Horace W. Busby, *The Thirty-First of March: An Intimate Portrait of Lyndon Johnson's Final Days in Office* (New York: Farrar, Straus & Giroux, 2005).

286 Jackie's note to McNamara about Harvard and library. See "The White House Years of Robert S. McNamara," Sotheby's Auctions, Oct. 23, 2012.

287 Jackie's burial at Arlington. As the widow of an American serviceman, Jackie Kennedy Onassis was entitled to be buried next to the president. While ill with non-Hodgkin's lymphoma, she decided that this would be her final resting place. Paul F. Horvitz, "Jacqueline Kennedy Onassis Laid to Rest Near Eternal Flame," *New York Times*, May 24, 1994, www.nytimes.com/1994/05/24/news/24iht-subjackie .html.

At this site in the cemetery, she is buried next to JFK, her unnamed stillborn daughter, and their infant Patrick.

EPILOGUE

288 Sixth Floor Museum: See Conover Hunt, *A Visitor's Guide to Dealey Plaza National Historic Landmark, Dallas, Texas* (Dallas: Sixth Floor Museum, 1995), and Conover Hunt, *The Sixth Floor: John F. Kennedy and the Memory of a Nation* (Dallas: Dallas County Historical Society, 1989). The Sixth Floor Museum's website is www.jfk.org.

289 Touring Dallas sites. In Texas, the first commercially released self-published stitch-stapled vest-pocket guide that contained numerous illustrations and maps locating the many sites associated with the assassination was by John Wesley Tackett, *Nov. 22—Where It Happened: The Historical Guide to the Assassination Site, Oswald's Grave, Ruby's Trial, Marina Oswald's Home, and Others* ([Fort Worth]: Author, 1964). In subsequent, years, others have also published similar works.

290 Conspiracy street bazaar: Much like Ford's Theatre, where President Lincoln was shot, the Sixth Floor Museum provides a somber, thoughtful, and historical response to the JFK assassination. However, the visit to Dealey Plaza itself is a very different experience. It has become a major Dallas tourist attraction. Over the years, a tourist could be greeted by a person trying to sell a souvenir newspaper, such as *JFK Assassination: Historical Journal* (reprinted numerous times, as a "a

Dealey Plaza special edition" or a special anniversary edition), or one might spot a well-known conspiracy theorist lecturing on the Grassy Knoll. One such individual and perennial vendor of books in Dealey Plaza is Robert Groden. He has waged a legal battle against the city after being ticketed repeatedly next to the pedestal on the Grassy Knoll where Abraham Zapruder filmed. Groden was arrested in 2010 for selling his assassination literature at this location. This feud with the city is ongoing. See Rudolph Bush, "Attorney for JFK Author Frustrated by Sixth Floor Museum's Dealey Plaza Permit," *Dallas Morning News*, Nov. 14, 2011, http://city hallblog.dallasnews.com/2011/11/attorney-for-jfk-author-frustr.html.

Professional tour guides provide guided visits to the various JFK sites of interest in the city, and at one time, there was even a conspiracy museum, in a storefront just a few blocks away from the Texas School Book Depository.

291 Common relics. Everyone saved newspapers and magazines that were printed in the hundreds of thousands, sometimes the millions. Many were reprinted numerous times over the last fifty years.

292 The following is a listing of some of the most interesting artifacts and relics.

Air Force One, Boeing VC-137C (SAM 26000). The plane is on display in the Presidential Gallery at the National Museum of the U.S. Air Force in Dayton, Ohio, www.nationalmuseum.af.mil/index.asp.

Parkland Hospital's trauma room one. Since 1963, Parkland Hospital has been both expanded and remodeled. Today the wall tiles and contents of trauma room one are the property of the National Archives and remain in its underground storage facility in Lenexa, Kansas.

Cars associated with the assassination. JFK's presidential limousine was rebuilt and is now on display at the Henry Ford Museum in Dearborn, Michigan, www.hfmgv.org/museum/limousines.aspx. The ambulance that drove Oswald to Parkland Hospital after he was shot can be found at Historic Auto Attractions, Roscoe, Illinois, http://historicautoattractions.com/Exhibits.html. In 2012, the Cadillac hearse owned by O'Neal Funeral Home that carried JFK's body accompanied by Jackie from Parkland Hospital to Air Force One was sold to a private collector for $176,000 at Barrett-Jackson's annual auction in Scottsdale, Arizona, www.barrett-jackson.com/application/onlinesubmission/lotdetails.aspx?ln= 1293&aid=443. In 2011, this auction house also sold to a private collector the "alleged" ambulance that carried JFK's coffin from Air Force One to the Bethesda Naval Hospital. However, the John F. Kennedy Library and Museum stated that the car was a fake, the real vehicle having been donated to the institution in 1980 and then destroyed in 1986. The car in which Wesley Frazer drove Oswald to work is on display at Ripley's Believe It or Not! Odditorium in San Antonio, Texas, www .ripleysnewsroom.com/worlds-largest-and-most-interactive-ripleys-believe-it-or -not-odditorium-opens-in-san-antonio.

Jack Ruby's revolver. In 1991, this gun, which was then in the possession of his brother, Earl Ruby, was sold for $220,000 by Herman Darvick Autograph Auctions in New York to a Florida real estate developer. It was offered by him in Las Vegas in 2008 through Guernsey's Auctioneers and Brokers, but it was not sold, failing to meet the reserve price of more than $1 million, although at the same auction, the hat that Ruby wore when he shot Oswald and some other artifacts were sold. "The Pugliese Pop Culture Collection," Guernsey's Auction Catalog, March 15, 2008.

The flags on the JFK limousine. These flags were originally given to Evelyn Lincoln and were acquired in 2005 by the Zaricor Flag Collection, for $450,000 at the sale by Guernsey Auctioneers in New York City. "John F. Kennedy; The Robert L. White Collection," Guernsey's Auction Catalog, Dec. 15, 2005, www.flagcollection .com/itemdetails-print.php?CollectionItem_ID=2546.

Signed 1963 official White House Christmas cards. Before they left for Dallas, Jack and Jackie began to inscribe Christmas cards to be mailed upon their return from Dallas. Many of these cards, which have a picture of a crèche nativity display in the East Room of the White House, were in the possession of JFK secretary Evelyn Lincoln and were later given to Kennedy collector Robert White. Several of these cards have been sold individually at auction and continue to emerge in the marketplace.

JFK's last signature. It is believed that Kennedy inscribed his last documented signature when he signed the November 22, 1963, edition of the *Dallas Morning News* for a maid at the Hotel Texas. This newspaper was sold at Heritage Auctions in Dallas in 2009 for $38,837. See "20th Century Icons Autographs," Heritage Auction Catalog, Nov. 6–7, 2009.

Signed first-day covers (JFK commemorative stamps). Many of the principals involved in the Kennedy administration, especially participants in various events (such as Judge Sarah Hughes and the Secret Services agents), signed envelopes with the first-day issue November 22, 1964, postmark of the JFK commemorative stamp.

Controversial commercially released items considered to be beyond bad taste. In 1999, a board game, Conspiracyland (a takeoff of the children's game Candyland) was sold by the "Patsy Brothers Company" in Dallas. This game's box had cartoon drawings of Jack and Jackie in her pink suit dancing before a Book Depository gingerbread house and the promotional description, "For kooks of all ages. 2 to 4 skeptical players. Average playing time: 6 seconds." In 2004, a computer game, JFK Reloaded, produced by Traffic Software in Scotland, allowed participants to play the role of Oswald and shoot at Kennedy from the sixth-floor window in Dealey Plaza. The participant was then able to keep score on the accuracy and number of shots fired.

292 Jacqueline Kennedy's pink suit is stored at the National Archives and is unavailable to be viewed, sealed until 2103. For autopsy photographs, X-rays, and other similar material, permission must be obtained by JFK's family through their representative, Paul Kirk. Most of the photographs, documents, and other major Warren Commission exhibits and artifacts that are at the National Archives and Records Administration can be viewed online at www.archives.gov/research/jfk.

293 JFK's brain. See Attorney General Ramsey Clark, *Panel Review of the Photographs, X-Ray Films, Documents and Other Evidence Pertaining to the Fatal Wounding of President John F. Kennedy on November 22, 1963, in Dallas, Texas,* 1968.

295 To date, the most thorough review of the numerous conspiracy theories can be found in Bugliosi's work, *Reclaiming History,* wherein he devotes almost five hundred pages of text to this subject. Since the time the JFK assassination occurred, the basic conspiratorial motifs largely have remained the same. Recent books further amplify these theories. Some works focus on single theories, while others weave or provide multiple conspiracy views of the assassination. For a summary of the various conspiracy motifs, see Harold Hayes, ed., *Smiling through the Apocalypse: Esquire's History of the Sixties* (New York: McCall Publishing Company, 1969), which includes a reprinting of the article "Sixty Versions of the Kennedy Assassination," Edward J. Epstein, pp. 467–505; Tom Miller, *The Assassination Please Almanac* (Chicago: Henry Regnery, 1977); or Carl Oglesby, *The JFK Assassination:*

The Facts and the Theories (New York: Signet, 1992). Entire books have focused on certain aspects of these recurring themes. A small sampling of this vast conspiratorial literature includes:

Russians/Communists/KGB. Almost immediately after the assassination, U.S. government officials were concerned that there was a Soviet plot to kill the president. Oswald's activities in the Soviet Union have been the primary subject of several works.

Right wing. Because of the full page WELCOME MR. KENNEDY TO DALLAS ad in the *Dallas Morning News*, as well as the WANTED FOR TREASON handbills, the right wing was a major suspect in having a hand in the assassination. For an early work, see Morris A. Bealle, *Guns of the Regressive Right, or How to Kill a President* (Washington, DC: Columbia Publishing Company, 1965).

Mafia/organized crime. See David E. Scheim, *Contract on America: The Mafia Murders of John and Robert Kennedy* (New York: Shapolsky Publishers, 1988); G. Robert Blakey and Richard N. Billings, *The Plot to Kill the President* (New York: Times Books, 1981); and David E. Kaiser, *The Road to Dallas* (Cambridge, MA: Belknap Press of Harvard University Press, 2008).

CIA: Because many perceive the CIA to be the least cooperative agency with regard to the assassination, and with many files still being withheld, this has been fertile ground for conspiracy theories. See John Newman, *Oswald and the CIA* (New York: Carroll & Graf, 1995); Michael Canfield and Alan J. Weberman, *Coup d'État in America: The CIA and the Assassination of John F. Kennedy* (New York: Third Press, 1975); and Mark Lane, *Plausible Denial: Was the CIA Involved in the Assassination of JFK?* (New York: Thunder's Mouth Press, 1991).

Cuba. See Gus Russo, *Live by the Sword: The Secret War Against Castro and the Death of JFK* (Baltimore: Bancroft Press, 1998); Lamar Waldron, *Ultimate Sacrifice: John and Robert Kennedy, the Plan for a Coup in Cuba, and the Murder of JFK* (New York: Carroll & Graf, 2005); and Warren Hinkle and William W. Turner, *Deadly Secrets: The CIA-Mafia War against Castro and the Assassination of J.F.K.* (New York: Thunder's Mouth Press, 1992).

FBI/Hoover. See Mark North, *Act of Treason: The Role of J. Edgar Hoover in the Assassination of President Kennedy* (New York: Carroll & Graf, 1991).

Business/Wall Street/Texas oilmen. For an early treatment of this subject, see Alvin H. Gershenson, *Kennedy and Big Business* (Beverly Hills, CA: Book Company of America, 1964).

LBJ. See early works, such as Bernard M. Bane, *The Bane in Kennedy's Existence* (Boston: BMB Publishing Company, 1967); Joachim Joesten, *The Dark Side of Lyndon Baines Johnson* (London: Peter Dawnay, 1968), and his self-published mimeographic work, *The Case Against Lyndon B. Johnson in the Assassination of President Kennedy* (Munich, 1967). More recent books include: Barr McClellan, *Blood, Money and Power: How LBJ Killed JFK* (New York: Hanover House, 2003); and Phillip F. Nelson, *LBJ: The Mastermind of the JFK Assassination* (New York: Skyhorse Publishing, 2011).

Multiple/fake/second Oswalds. See books such as Richard H. Popkin, *The Second Oswald* (New York: Avon, 1966); Michael Eddowes, *The Oswald File* (New York: Clarkson Potter, 1977); and John Armstrong, *Harvey & Lee: How the CIA Framed Oswald* (Arlington, TX: Quasar, 2003); and the video release of Jack White, *The Many Faces of Lee Harvey Oswald: Unmasking the Secret Agent Who Was Framed as Kennedy's Killer* (JFK Videos, 1991).

Multiple gunmen. For one of the early works that focused on the number of shots, see Josiah Thompson, *Six Seconds in Dallas: A Micro-Study of the Kennedy Assassination* (New York: Bernard Geis Associates, 1967). One of the early theorists was Robert B. Cutler, who self-published numerous works regarding multiple assassins, including: *The Flight of CE 399: Evidence of Conspiracy* (Beverly, MA: Omni-Print, 1969), and *Two Flightpaths: Evidence of Conspiracy* (Manchester, MA: Cutler Designs, 1971). So too is Raymond Marcus, *The Bastard Bullet: A Search for Legitimacy for Commission Exhibit 399* (Los Angeles: Rendell Publications, 1966). Though originally self-published in very rare signed and numbered limited spiral-bound editions, these works have been reprinted numerous times.

Secret Service. These theories have been alluded to in several conspiracy works. Perhaps one of the more unusual theories appeared in Bonar Menninger, *Mortal Error: The Shot That Killed JFK* (New York: St. Martin's Press, 1992), accusing a Secret Service agent of accidentally firing the shot that killed Kennedy. Needless to say, that Secret Service agent sued the publishers. While the suit was dismissed on technical grounds, the retired Secret Service agent settled with St. Martin's Press for an undisclosed sum.

296 From November 22, 1963, until today, there actually have been over ten governmental, congressional, and presidential commission investigations that have partially or completely covered aspects of the Kennedy assassination. In addition, there have been numerous Freedom of Information Act lawsuits and requests, producing thousands of additional documents. While there have been some inconsistencies and new documents have emerged, the essential findings and conclusions of the Warren Commission have stood the test of time and have not been contradicted. There is no credible evidence of any conspiracy, and that Lee Harvey Oswald alone did not fire his rifle from the sixth floor of the Book Depository on November 22, 1963. The major government investigations include:

The initial FBI Investigation and multivolume report, December 9, 1963.

Warren Commission. *Report of the President's Commission on the Assassination of President John F. Kennedy*, and the subsequent *Hearings Before the President's Commission on the Assassination on President John F. Kennedy*, 26 vols. (Washington, DC: U.S. Government Printing Office, 1964).

National Commission on the Causes and Prevention of Violence, vol. 8, *Assassination and Political Violence* (Washington, DC: U.S. Government Printing Office, 1970).

Several congressional hearings and reports in the 1960s and 1970s on the preservation of evidence of the Warren Commission and on FBI oversight.

U.S. Rockefeller Commission, *Report to the President by the Commission on CIA Activities within the United States* (Washington, DC: U.S. Government Printing Office, 1975).

U.S. Senate, 94th Congress, 2nd Session, *Final Report of the Select Committee to Study Government Operations with Respect to Intelligence Activities in the United States* (Sen. Rpt. 94-755, Church Committee Report), bk. 5, *The Investigation of the Assassination of President John F. Kennedy: Performance of the Intelligence Agencies* (Washington, DC: U.S. Government Printing Office, 1976). This multivolume report also includes several hearing volumes.

U.S. House of Representatives, 95th Congress, 2nd Session, *Final Report of the Select Commission on Assassinations*, and the accompanying twelve hearing and appendix volumes on the JFK Assassination (Washington, DC: U.S. Government Printing Office, 1979).

Attorney General Ramsey Clark, *Panel Review of the Photographs, X-Ray Films, Documents and Other Evidence Pertaining to the Fatal Wounding of President John F. Kennedy on November 22, 1963, in Dallas, Texas* (1968).

National Research Council, Commission on Physical Sciences, Mathematics, and Resources, *Report of the Committee on Ballistics Acoustics* (Washington, DC: National Academies Press, 1982).

Assassination Records Review Board, *Final Report* (Washington, DC: U.S. Government Printing Office, 1998), along with the House and Senate hearing volumes and committee reports that created the board and were published by the U.S. Government Printing Office in 1992.

297 There has yet to be published a comprehensive scholarly historiography of the JFK assassination conspiracy literature. Although there have been more than thirty freestanding bibliographies on JFK assassination books, pamphlets, and articles, no one source is complete, and many of the early works contained numerous errors. Also, unlike the Lincoln assassination, while there are several JFK assassination specialty collections being amassed, especially the Sixth Floor Museum in Dallas, as well as a few university libraries, there is no institutional library whose collection comes even close to completeness.

Within months after the assassination and before the release of the Warren Commission Report, there were three major hardcover works, all published outside the United States: Nerin Gun, *Red Roses from Texas* (London: Frederick Muller, 1964; also published in Italian, French, and Chilean Spanish); Thomas Buchanan, *Who Killed Kennedy?* (London: Secker & Warburg, 1964; also published in a few foreign languages as well as in America in hardcover and paperback); and Joachim Joesten, *Oswald: Assassin or Fall Guy?* (London: Merlin Press, 1964; also published in the United States in hardcover and paperback, and revised after the release of the Warren Report). Joesten became a prolific writer on the JFK assassination, producing several additional books in German, French, and English, all published outside the United States, as well as a self-published newsletter and mimeographed essays printed in English from his home in Germany.

Immediately after the Warren Commission Report was published, thousands of conspiracy-inclined authors produced works, and bestsellers emerged in every decade and every generation. In 1965, Harold Weisberg self-published in Hyattstown, Maryland, his *Whitewash: The Report on the Warren Report* (reprinted by Dell the following year), and the floodgates opened. Weisberg also turned out to be a prolific writer on the JFK and Martin Luther King Jr. assassinations, self-publishing, publishing revised editions, and commercially publishing more than ten works on the assassination and creating many more unpublished manuscripts. Upon his death, the Weisberg archives were donated to Hood College, and many of the holdings are available online: http://hood.edu/library/special-collections/weisberg-archive.html.

The other early writer who should be mentioned is Mark Lane. His article "Oswald Innocent? A Legal Brief" was published in the December 19, 1963, edition of the *National Guardian*, and since the subsequent request by Marguerite Oswald to represent her son before the Warren Commission, Lane has been immersed in writing about the JFK assassination and similar subjects. Lane wrote the runaway bestseller, *Rush to Judgment: A Critique of the Warren Commission's Inquiry into the Murders of President John F. Kennedy, Officer J.D. Tippit, and Lee Harvey Oswald* (New York: Holt, Rinehart & Winston, 1966). He also produced a documentary movie based on the book, along with a long-playing phonographic record. This was followed by his book *A Citizen's Dissent: Mark Lane Replies* (New York: Holt, Rinehart & Winston, 1968). He wrote a work of fiction with Donald Freed, *Executive Action: Assassination of a Head of State* (New York: Dell, 1973) and cowrote the screenplay that was made into the movie, *Executive Action*. In the following years, he penned two more books on the JFK assassination: *Plausible Denial: Was the CIA Involved in the Assassination of JFK?* (New York: Thunder's Mouth Press, 1991), and *Last Word: My Indictment of the CIA in the Murder of JFK* (New York: Skyhorse Publishing, 2011). Lane created "The Citizens Commission on Inquiry" and was on the lecture circuit for decades. Before the House Select Committee on Assassination, Lane also happened to represent James Earl Ray, the assassin of Martin Luther King Jr. This resulted in the publication of his book with Dick Gregory, *Code Name Zorro: The Murder of Martin Luther King, Jr.* (Englewood Cliffs, NJ: Prentice-Hall, 1977). Perhaps even more coincidental, Lane was the legal counsel to Jim Jones in Guyana when the People's Temple mass suicide occurred in 1978. Somehow Lane was able to escape the carnage of this horrific event. At the same time, Representative Leo Ryan, visiting Jonestown, became the first congressman killed while conducting official business while in office. This misadventure resulted in one more Lane book, *The Strongest Poison* (New York: Hawthorn Books, 1980). Mark Lane is only one of the many conspiracy theorists that have written several books on the assassination of JFK, joining the ranks of individuals such as Robert B. Cutler, Penn Jones Jr., Robert J. Groden, and others.

299 Jackie's quotes. White Camelot interview.

299 For Jacqueline Kennedy's commentary, see the Theodore White Camelot interview notes; the December 6, 1963 issue of *Life* magazine, and the November 17, 1964 issue of *Look* magazine.

INDEX

Abt, John, 196, 200, 215
Aftermath of an Execution (Marguerite Oswald), 263
Air Force One
  appearance of, 65
  in Dallas, 94–95, 96, 98, 107–8
and details of JFK trip to Dallas, 67
  and flight to Texas, 65–66
  in Fort Worth, 81, 82
  in Houston, 77
  and JFK's burial at Arlington Cemetery, 250–51
  JFK's views about, 65–66
  and LBJ's return to Washington, 162, 174–80, 281–82
  and LBJ's swearing in as president, 175–80
  and return of JFK body to Washington, 169, 174–80, 192–93, 194–95, 248
  in San Antonio, 70, 72
  as symbol of presidency, 65
Air Force Two, 174
airplane hijacking: Oswald's fantasy about, 53
Allman, Pierce, 137–38, 156
Altgens, James, 133
American Civil Liberties Union (ACLU), 224, 233
American exceptionalism: JFK speech about, 103–4
American University: JFK speech about Soviet-U.S. relations at, 58–59
Amherst College: JFK speech about Frost at, 62
Anderson, Marian, 273
Arlington National Cemetery

and eternal flame at JFK grave site, 251–52, 257, 259, 263, 287, 295
  Jackie's grave in, 287
  Jackie's visits to JFK grave at, 254, 256–57
  JFK burial at, 205–6, 250–52
  Oswald proposed burial in, 240
  picture of, 254
Army, U.S., 30
arraignment
  of Oswald for JFK assassination, 200
  of Oswald for Tippet murder, 196–97
arts
  Jackie's interest in, 62, 262
  JFK's interest in, 61, 62
Arvad, Inga, 166
assassination, JFK
  anniversaries of, 284–85, 287, 290
  and attempted assassination of JFK in 1960, 86
  blame for, 240–42
  conspiracy theories about, 162, 170, 223, 283, 290, 295–97
  and death of JFK, 151–52, 153–54, 155, 158, 161, 162–63
  first shot in, 123–28, 136, 150
  fourth shot in, 140
  as great American tragedy, 298
  "ifs" concerning, 149–50, 298
  impact on American history of, 298
  initial reactions to, 133–43
  JFK's views about possibility of, 85
  and manhunt for Oswald, 174
  Oswald as suspect in, 183

assassination, JFK (*cont.*)
  Oswald's arraignment for, 200
  Oswald's choice of weapon for, 69, 75–79
  and Oswald's decision about when to shoot
    JFK, 121–24
  and Oswald's decision to assassinate JFK, 67,
    68–69, 70, 72, 76
  and Oswald's denial about shooting JFK,
    191, 193, 197, 199, 206–7, 210, 215, 216,
    222
  Oswald's escape after, 72, 73–74, 129, 140,
    141, 144–45, 146, 147, 156–58, 173–74,
    225, 289
  Oswald's feelings about, 172, 211–12
  Oswald's motivations for, 221, 297–98
  Oswald's plans for, 66–67, 72–79, 226
  and Oswald's wait for the motorcade, 97,
    105–6, 110–20, 121–28
  and public announcement of death of JFK,
    162
  public reactions to, 164–66, 262–65
  second shot in, 129–33, 135, 136, 292
  as shared event, 291
  souvenirs/relics of, 261, 262, 263, 291–94
  Texas reaction to, 241
  third shot in, 134, 135–37, 138, 139, 141, 150
  and visitors to Dallas, 289–90
  witnesses to, 113–17, 118–20, 123–28,
    129–30, 132, 135–37, 147–48, 156
  See also Dallas police; Dallas, Texas—JFK's
    trip to; Parkland Hospital; Secret Service;
    Warren Commission; *specific person*
assassinations
  historical curiosity about presidential, 86
  Oswald's choice of weapons for, 4
  Oswald's feelings about, 172
  and Oswald's use of rifle for presidential
    assassination, 141
  pistols as weapons for, 69
  See also specific person
Associated Press, 125, 146, 253
Auchincloss, Janet, 196, 202–3, 204, 292
Austin, Texas: JFK campaign trip to, 60, 83,
  104–5
autopsy
  of JFK, 169–70, 196, 198, 293
  of Oswald, 253

Baker, Officer (Dallas policeman), 145
banners, JFK memorial, 256, 257
Bartlett, Charlie, 23, 203–4, 264–65, 285
Bartlett, Martha, 23
Batchelor, Charles, 230
Bay of Pigs invasion, 30–31, 32, 43–44, 49, 53

Beers, Jack, 234
Bell, Jack, 125, 146
Berlin, Germany: JFK trip to, 34–37, 91
Bethesda Naval Hospital
  and conspiracy theories about JFK
    assassination, 296
  JFK autopsy at, 196, 198, 201
Bird, Sam, 201
Booth, John Wilkes, 157, 213, 288, 291–92, 298
Boyd, Elmer, 183
Bradlee, Ben, 269–70
brain, JFK
  and Bethesda autopsy of JFK, 198
  as missing from National Archives, 293
  at Parkland Hospital, 238
  in presidential limousine and on Jackie, 138,
    139, 152, 160, 161, 162, 172
  and shot to brain as cause of JFK death, 163
Brennan, Howard Leslie, 116–17, 125–26, 132,
  135, 136, 137, 140, 147–48, 171
Brewer, Calvin, 180–81
Bringuier, Carlos, 44, 45, 46, 49
Brooks Medical Center (San Antonio): JFK
  speech at, 71–72
Brown v. Board of Education, 33
Burkley, George, 170, 175–76
Butler, Ed, 45, 47, 48, 50

Camelot
  "death" of, 287
  enduring myth of, 294–95
  and LBJ's first year as president, 282
  and White-Jackie interview, 268, 269, 284
Campbell, Mr. (Texas School Book Depository
  employee), 138
Capitol, U.S.
  funeral procession to White House from,
    247–48
  Jackie's viewing of JFK coffin in Rotunda of,
    246–47
  memorial service for JFK at, 239, 241–45
  public viewing of JFK coffin at, 245–47
Caroline (JFK airplane), 66
Casals, Pablo, 62, 273
Castro, Fidel
  and Batista overthrow, 49–50
  CIA attempted assassination of, 31
  and conspiracy theories about JFK
    assassination, 296
  FPCC and, 39, 45
  and JFK as "ruffian and thief", 49
  JFK's interest in, 30
  and Oswald's motivations for assassinating
    JFK, 297

and Oswald's pro-Cuban activities, 40, 41, 42, 43, 44, 49, 51, 52
Soviet relations with, 30, 45
Walker (Edwin) views about, 12
See also Cuba
Catholicism
of JFK, 26, 110, 204–5
and mass for JFK, 204–5
CBS News, 150, 152, 270–71, 291
Central Intelligence Agency (CIA), 30–31, 44, 49, 50, 296
Cermak, Anton, 69
Chicago American, 164
Christmas (1963): Jackie's gifts to friends for, 276–77
Churchill, Winston, 11, 26
"cinched" interview, Fritz's, 213–14, 242
citizenship: Oswald's attempted renunciation of American, 45, 47, 48, 193, 226
civil rights
as concern in early years of JFK presidency, 33–34
elections of 1960 and, 28
and "ifs" of JFK assassination, 298
JFK's views about, 33–34, 35
LBJ's views about, 34, 35
and warnings about JFK's trip to Texas, 63
Civil War, 217, 251
Clark, Bob, 137, 155
Clark, Ramsey, 293
Cochran, Mike, 253
Cold War, 26
Combest, Billy H., 234, 236
Communism
and conspiracy theories about JFK assassination, 296, 297
and Dallas ad attacking JFK, 84–85
and early years of JFK presidency, 29–30
elections of 1960 and, 24, 25, 27–28
Jackie's comment about, 196, 241
JFK's views about, 24, 27–28, 31
Marxism distinguished from, 47
Nixon's views about, 24, 27–28
and Oswald's arrest in New Orleans, 40
and Oswald's assassination of JFK, 196
Oswald's interest in, 37, 38, 42–43, 44–48, 49, 174
and Oswald's plans to assassinate Walker, 11–12
questioning of Oswald about, 225
in Russia, 25–26
in Southeast Asia, 31
Walker (Edwin A.) and, 11–12
and warnings about JFK's trip to Texas, 63

Congress, U.S.
LBJ address to joint session of, 263–64
learns of JFK assassination, 167–68
and memorial service for JFK, 239, 241–45
military budget and, 31
Connally, John
"day to be remembered" comment of, 105
initial reactions to shooting by, 133
Jackie's views about, 80–81
and JFK's Dallas motorcade, 100, 102, 121
and JFK's Houston campaign trip, 80–81
JFK's views about, 80–81
Oswald's denial of shooting, 210
and Oswald's dishonorable discharge from the Marines, 40, 226
and Oswald's feelings about shootings, 172
and Oswald's first shot, 125
and Oswald's marksmanship, 121
Parkland Hospital and, 152, 153, 238, 294
picture of, 89, 100, 101
questioning of Oswald about shooting of, 210
secrecy about injuries to, 163
shooting of, 131, 134, 135
as witness in Oswald trial, 212
Connally, Nellie, 100, 101, 102, 112, 135, 152, 153, 294
Connor, Eugene "Bull", 34
Conversation Carte Blanche (WDSU radio show): Oswald as guest on, 44–50
Couch, Malcolm, 137
Cronkite, Walter, 150, 151, 152, 162–64, 271
Cuba
and Batista overthrow, 49–50
CIA secret plans concerning, 30–31
and conspiracy theories about JFK assassination, 296
and Dallas ad attacking JFK, 84
elections of 1960 and, 28
and "ifs" concerning JFK assassination, 149
JFK interest in, 30, 31, 41
Oswald's attempt to visit, 58
Oswald's interest in, 40–51, 53, 54, 174, 224–25, 295
and Oswald's motivations for assassinating JFK, 297
and Oswald's plans for Walker assassination, 12
questioning of Oswald about, 224–25
Soviet relations with, 30, 42
Walker (Edwin) views about, 12
See also Bay of Pigs invasion; Castro, Fidel; Cuban Missile Crisis
Cuban Missile Crisis, 31–32, 49, 53

Curry, Jesse, 102, 121, 218–19, 220, 221,
  223–24, 229, 230, 231, 232, 238–39, 241
Cushing, Richard Cardinal, 249

Dallas Citizens Council: JFK speech for, 103–4
Dallas Morning News, 17, 67, 84–85, 96–97, 132
Dallas Police
  and blame for Ruby shooting of Oswald, 242
  and death threats against Oswald, 219–21, 223,
    242
  and initial reactions to shots at JFK, 143,
    147, 156
  and JFK arrival at Love Field, 99
  JFK motorcade and, 95
  and JFK ride to Parkland Hospital, 152
  and manhunt for Oswald, 173–74
  media relations with, 217, 221
  Oswald in custody of, 182–94, 197, 198–200,
    206–12, 213–14, 215–17, 218–29, 230–35,
    242
  Oswald's arrest by, 181–82
  and Oswald's decision about when to shoot
    JFK, 122
  and Oswald's escape after JFK assassination,
    144–45, 228
  and Oswald's motives for assassinating JFK,
    297
  at Paine home, 187
  at Parkland Hospital, 155, 170
  questioning of Oswald by, 182–84, 185–86,
    188–92, 193, 196–97, 206–10, 213–14,
    221–29, 297, 2362
  reputation of, 216–17, 239, 288
  and Ruby's shooting of Oswald, 234–37,
    239, 288
  search of Oswald by, 193–94
  and transfer of Oswald from City Hall jail to
    County jail, 218–19, 220–21, 223–24, 229,
    230–35, 242
  See also specific policeman
Dallas, Texas
  LBJ takes oath of office in, 174–80
  and Oswald escape after JFK assassination,
    144
  Oswald's move to, 59
  Oswald's rooming house in, 75, 78, 157, 171,
    184, 189, 190, 192, 208, 226–27, 232, 289
  reputation of, 216–17, 240–42, 288
  Stevenson visit to, 63, 115, 116, 240
  as victim of Oswald's actions, 240–42
Dallas, Texas—JFK's trip to
  and ad attacking JFK, 84–85
  anticipation of, 82–83
  and flight from Fort Worth to Dallas, 93–95

JFK motorcade in, 67–69, 70, 79, 94, 95,
    97–98, 99, 102–3, 106–20, 121–22
  and JFK security at Love Field, 95
  JFK's arrival in, 94–102, 109
  JFK's reactions to, 96, 108, 241
  plans for, 60
  warnings about JFK's trip to, 63–64, 85
  See also assassination, JFK; Texas School
    Book Depository
Dallas Times Herald, 70
Daniel, Damon, 165
Dealey Plaza (Dallas)
  and anniversaries of JFK assassination, 290
  desk set, 258, 263
  initial reactions to assassination shots in,
    146, 149
  JFK motorcade and, 68, 79, 95, 97, 105–6,
    107, 111–20, 121–32
  map of, 111
  and Oswald's shots at JFK, 123–43
  photographs of, 289
  post-assassination reactions in, 156
  visitors to, 289–90
death threats against Oswald, 219–21, 223, 242
Decker, Bill, 219, 220, 237
democracy: Oswald's definition of, 43
Dillard, Tom, 132, 136, 137
dying declaration by Oswald, 236

Eisenhower, Dwight D., 12, 24, 30
elections of 1960, 1, 24, 26–27, 60, 188, 259
elections of 1964
  campaign for, 281–82, 283–84
  and "ifs" of JFK assassination, 298
  Jackie-LBJ relationship and, 281–82
  JFK plans for, 60, 62, 83, 98
  Kennedy (Bobby) and, 281–82, 283, 284, 285
  LBJ and, 281–82, 284–85
  Warren Commission and, 264
elections of 1968, 285
Erhard, Ludwig, 61, 249–50
eternal flame: at JFK grave site, 251–52, 257,
    259, 263, 287, 295
Euins, Amos Lee, 123–24, 127, 135–37
eulogies: for JFK, 203–4, 244–45, 285
evidence, JFK assassination
  against Oswald, 221
  and conspiracies theories about JFK
    assassination, 296
experience issue: elections of 1960 and, 26–27

Fair Play for Cuba Committee (FPCC), 39,
    40, 41–44, 45, 46–47, 49, 50, 51, 53, 58,
    224–25

Federal Bureau of Investigation (FBI)
   and conspiracy theories about JFK
      assassination, 296
   and death threats against Oswald, 219, 220
   early interest in Oswald by, 10, 59–60
   and manhunt for Oswald, 174
   Oswald's letter about Hosty to, 185
   and Oswald's motives for assassinating JFK,
      297
   Oswald's polygraph examination for, 209
   and Oswald's purchase of rifle, 206
   Oswald's views about, 184–85
   and questioning of Oswald, 183, 184–85,
      186, 191, 206, 208, 209, 297
   visit to Oswald (Marina) by, 59–60, 184–85,
      209
   See also specific agent
film: of Jackie thanking public for condolence
   letters, 277–78
"For President Kennedy: An Epilogue" (White's
   Life article), 266–69
Ford Motor Company, 71
Ford's Theatre (Washington, D.C.), 288, 290,
   291–92
foreign affairs
   as dominant in early years of JFK presidency,
      29–30
   See also specific person, nation, or event
Fort Worth, Texas
   JFK campaign trip to, 60, 81, 82–83, 84–86,
      90–94
   and JFK comment about an assassination
      attempt, 85
   Oswald return from Soviet Union to, 39
Foster, Bob, 66
The Four Dark Days: From Dallas to Arlington
   (CBS documentary), 270
Fox, Charles: JFK painting by, 274–75
Frankfurter, Felix, 273
Frazier, Wesley Buell
   law enforcement suspicions about, 133
   Oswald first meets, 59
   and Oswald's retrieval of rifle from Paine
      home, 76, 77, 87–90, 146, 227
   and Oswald's shots at JFK, 127, 132–33
   questioning of Oswald about, 207, 227
   as Texas School Book Depository employee,
      59
   and wait for JFK motorcade, 105, 122
   as witness against Oswald, 174, 227
Frazier, William B., 220
Fritz, Will
   "cinched" interview of, 213–14, 242
   and death threats against Oswald, 220

   and guilt of Oswald, 221
   questioning of Oswald by, 182–84, 185–86,
      188–92, 193, 196–97, 207–10, 213–14,
      221–29, 236
   and Ruby's shooting of Oswald, 239
   and transfer of Oswald from city jail to
      county jail, 218–19, 229, 230–32, 233, 234
Frost, Robert: JFK tribute to, 62
Fulbright, J. William, 63
funeral/memorials, JFK
   and burial at Arlington National Cemetery,
      205–6, 250–52
   and Capitol viewing of JFK coffin, 245–47
   and casket for JFK, 166, 170, 175, 176, 180,
      201, 202
   and congressional/Capitol memorial service,
      239, 241–45
   and death of Oswald, 239
   and eulogies for JFK, 203–4, 244–45, 285
   foreign dignitaries at, 248
   and Jackie's last visit with JFK body, 242–43
   and JFK coffin at White House, 201–2,
      242–43
   and LBJ proclamation for national day of
      mourning, 218
   and mass for JFK, 204–5
   media coverage of, 240
   and motorcade from Bethesna Naval
      Hospital to White House, 201–2
   and motorcade to Arlington National
      Cemetery, 206
   plans for, 188, 202, 205–6
   and procession from Capitol to White
      House, 247–48
   and procession from White House to
      Capitol, 243–44
   and procession from White House to St.
      Matthew's, 247, 248–49, 258
   at St. Matthews, 249

Garfield, James, 69
George Washington University: Jackie as
   student at, 262
Georgetown
   Jackie at Harriman home in, 272, 274–75,
      278–79
   Jackie's home in, 278–79, 280
   JFK-Jackie home in, 1–2, 272, 299
Germany
   JFK visit to, 34–37
   Soviet Union and, 35–36
Goldwater, Barry, 283
Government Printing Office, 276
Grassy Knoll (Dallas, Texas), 290

Greer, William "Bill", 102, 150, 201
Grosvenor, Melville Bell, 280

Harriman, W. Averell, 272, 274
Harvard University, Kennedy Library and,
    286–87
Hayden, Carl, 162
"He Had That Special Grace . . ." (Bradlee),
    269–70
Hidell, Alek James
    and Oswald's arrest for Tippit's murder, 182,
    183
    Oswald's denial about use of name, 223
    Oswald's Selective Service cards in name
    of, 209
    and purchase of Oswald's pistol and rifle,
    2, 74
    questioning of Oswald about use of name of,
    184, 209, 223, 224, 229
    See also Oswald, Lee Harvey
Hill, Clint
    bloodstained suit coat of, 161
    and death of JFK, 151–52, 155, 159
    Exceptional Service Award for, 272
    initial reaction to assassination shots by, 134,
    135, 138, 140, 142
    as Jackie's personal agent, 272
    Jackie's relationship with, 272
    as Jackie's Secret Service protection, 134,
    135, 142, 152, 153, 154, 155, 175, 201
    and JFK's Dallas motorcade, 100, 109–10
    and Kennedys in Fort Worth, 92–93
    and Parkland Hospital, 152, 153, 154, 155,
    175
Holmes, Harry D., 222–23, 224
Hoover, J. Edgar, 166–67, 168
Hosty, James, 60, 184–85, 186, 191
House Un-American Activities Committee, 26
Houston, Texas: JFK campaign trip to, 60, 77,
    79–81, 82
Hughes, Sarah T., 176, 178–79
Humphrey, Hubert, 283
Hyannis Port, Massachusetts: White-Jackie
    interview in, 266–69

"I have a rendezvous with death" (Seeger
    poem), 86
"ifs" of JFK assassination, 149–50, 298
In Search of History (T. White), 294–95
Inaugural Addresses of the Presidents of the
    United States from George Washington
    1789 to John F. Kennedy 1961
    (Government Printing Office), 276–77
inauguration, JFK, 2, 28–29, 104

Information Council of the Americas, 44
Ingram, Tommy, 164–66
International Trade Mart. See Trade Mart
"Irish wake" for JFK, 187–88
"iron curtain", 26
Irving, Texas. See Paine, Ruth

Jackson, Bob, 136, 137, 234–35
Jarman, James "Junior", 113–14, 117, 129, 131,
    138, 147, 150, 207
Jenkins, Ron, 148, 155
Johnson, Lady Bird, 70–71, 102, 154, 177, 179,
    192–93, 240, 268, 274
Johnson, Lyndon Baines "LBJ"
    civil rights and, 34, 35
    and conspiracy theories about JFK
    assassination, 296
    and death of JFK, 161–62
    elections of 1960 and, 24, 27, 28, 60, 188, 259
    and Fort Worth trip, 93
    hostile reception in Texas for, 115, 240
    and Houston campaign trip, 80
    and initial reactions to Oswald's shots at
    JFK, 143
    Jackie's relationship with, 258–59, 281–82
    Jackie's views about, 268
    and JFK plans for Texas campaign trip, 60
    JFK relationship with, 192, 259–60
    and JFK's arrival in Dallas, 97
    and JFK's Dallas motorcade, 102, 118
    Kennedy's (Bobby) relationship with, 188,
    259, 281–82, 283, 284
    Oswald's views about, 225
    at Parkland Hospital, 154
    picture of, 89
    return to Washington of, 162, 174–78,
    192–93, 194–95
    and San Antonio campaign trip, 70–71
    Secret Service protection for, 154, 161–62,
    248–49
    in Senate, 167, 259
    Texas ranch of, 60
    and warnings about JFK trip to Dallas, 64
Johnson, Lyndon Baines "LBJ"—as president
    address to joint session of Congress by,
    263–64
    elections of 1964 and, 281–82, 284–85
    elections of 1968 and, 285
    first public statement of, 195
    Jackie's letter to, 257–62
    and Jackie's Oval Office redecoration ideas,
    261–62
    and Jackie's post-funeral stay in White
    House, 271–72

JFK aides relationship with, 188
at JFK funeral/memorials, 248–49, 258,
    275–76
JFK impact on presidency of, 282
and JFK's Presidential Medal of Freedom
    award, 273–74
letters to Caroline and John Jr. from, 195,
    258
national day of mourning for JFK
    proclamation by, 218
presidential limousine for, 294
Reston's views about, 285
swearing in of, 164, 175–80
transition from JFK administration to, 205
Warren Commission and, 264, 284
Johnston, David, 196–97, 200
Justice Department, U.S., 41, 62, 167, 277–78

Kellerman, Roy, 102, 169–70
Kelley, Thomas, 209–10, 224–26
Kennedy, Arabella (daughter), 62
Kennedy, Caroline (daughter)
    and film thanking public for sympathy
        letters, 278
    at funeral/memorial services for JFK, 245,
        249
    Georgetown home of, 279
    and good-byes for Texas trip, 65
    Jackie's talk about JFK's death with, 204
    and JFK body in East Room of White House,
        203
    JFK love for, 270
    JFK playing with, 56
    and Kennedy artifacts at National Archives,
        292–93
    LBJ letter to, 195, 258
    and LBJ transition to White House, 205
    and move out of White House, 272
    and public image of JFK, 55
    and White House school for children, 262
Kennedy, Edward "Ted" (brother), 167–68,
    247, 270
Kennedy, Ethel (sister-in-law), 167, 168
Kennedy, Jacqueline Lee Bouvier "Jackie"
    as American heroine, 279
    arts interests of, 62, 262
    Bartlett article about, 264–65
    and Bobby's assassination, 286
    Bobby's relationship with, 274, 286
    and Christmas of 1963, 276–77
    Communism comment of, 196, 241
    Connally (Nellie) letter from, 294
    courtship and wedding of, 1, 23–24
    death of, 287

elections of 1960 and, 60
elections of 1964 and, 83, 93, 281–82
and film of public thank you for condolence
    letters, 277–78
and flight to Texas, 65–66
in Fort Worth, 90, 92–93
Georgetown homes of, 1–2, 272, 274–75,
    278–79, 280, 299
Hill as personal agent for, 272
Hill/Secret Service protection for, 92–93,
    109–10, 134, 142, 152, 153, 154, 155, 175,
    201
Hill's relationship with, 272
in Houston, 77, 79–81
and "I have a rendezvous with death" poem,
    86
as iconic figure, 280–81
inscription on marble mantle in White
    House by, 272
and JFK as legend/hero, 264–65, 267–69,
    284, 295, 299
JFK first meets, 23, 203
JFK's last night in White House with, 61
JFK's last night with, 83
JFK's last private dinner with, 80
and John Jr.'s birthday party, 217, 254
and Kennedy artifacts at National Archives,
    292
Kennedy Library and, 275, 286–87
LBJ's letter from, 257–62
LBJ's relationship with, 258–59, 281–82
McNamara gift of JFK oil painting for,
    274–75
media and, 264–65, 269, 278–79, 287
as national obsession, 280–81
New York City move of, 280
in 1968, 285, 286, 287
Onassis marriage of, 287
Oswald's (Marina) interest in, 78, 210
Oswald's views about, 210
Oval Office redecoration ideas of, 261–62
and plans for Texas campaign trip, 60, 63
post-funeral stay at White House of, 249–50,
    252, 260–62, 271–74
pre-inauguration parties and, 2
and presidential limousines, 91
and Presidential Medal of Freedom award
    for JFK, 273–74
public image of, 55
and refusal to visit White House, 280, 282
in San Antonio, 70, 71
in seclusion, 217
social activities of, 2, 11, 61, 62
Stevenson and, 63

Kennedy, Jacqueline Lee Bouvier "Jackie" (*cont.*)
  tributes to, 260, 263
  wardrobe of, 64–65, 79
  wedding ring of, 161, 244, 267
  and White House renovation, 56
  and White House school for children, 262
  White House social obligations of, 249–50
  White's interview with, 266–69, 295
Kennedy, Jacqueline Lee Bouvier "Jackie"—and
    Dallas events
  and arrest of Oswald, 196
  and arrival in Dallas, 94–102
  and Dallas motorcade, 108, 112, 119,
    122–23, 138
  and death of JFK, 153, 154, 158, 161
  and initial reactions to shots, 133, 134, 135,
    139, 140, 142
  Jackie's memories/reflections about, 160, 299
  Jackie's silence about, 287
  and JFK's warnings about dangers in Dallas,
    85
  and LBJ's swearing in as president, 177–80
  and media reports about JFK assassination,
    151
  and Oswald's first shot, 125
  at Parkland Hospital, 153–54, 155, 158,
    160–61, 170, 175, 244, 247
  pink suit of, 65, 93, 96–97, 98, 112, 119, 139,
    148, 161, 176, 194, 195, 292–93
  and return of JFK body to Washington, 170,
    174–80, 187–88, 194–95, 202–3, 204, 205,
    217, 248
  and ride to Parkland Hospital, 148, 152–53,
    154, 156
  and roses for Jackie, 71, 77, 97, 98, 99, 160,
    175–76
  as witness in Oswald trial, 212
Kennedy, Jacqueline Lee Bouvier "Jackie"—and
    JFK funeral/memorials
  at Bethesda Naval Hospital, 196, 198
  and casket for JFK, 166, 198
  and eternal flame at JFK grave site, 251–52,
    287
  farewell letters to JFK from, 242–43
  final farewell kiss, 267
  and funeral processions, 247–48
  and "Irish wake" for JFK, 187–88
  and Jackie at funeral/memorial services,
    244–45, 249, 250–52
  and Jackie's visits to Capitol Rotunda,
    245–47
  and JFK body in East Room of White House,
    201, 242–43
  at Lincoln Memorial, 252

and Mansfield eulogy, 244–45
and mass for JFK, 204–5
and plans for JFK funeral, 188, 202, 205–6,
  217
and return to Washington/White House
  from Texas, 170, 174–80, 187–88, 194–95,
  202–3, 204, 205, 217, 248
Kennedy, John F.
  death of, 151–52, 153–54, 155, 158, 161,
    162–63, 238
  eulogies for, 203–4, 244–45, 285
  fatalism of, 85, 86
  health of, 24, 56–57, 293
  indiscretions of, 166
  Jackie first meets, 23, 203
  and Jackie's final farewell kiss, 267
  as legend/hero, 264–65, 267–69, 284, 285,
    294, 295, 299
  luck of, 86
  McNamara gift of oil painting of, 274–75
  meaning of life of, 299
  Oswald's (Marina) views about, 53
  Oswald's views about, 53, 67, 209, 212
  personal and professional background of,
    24, 25
  presidential receptions and dinners of, 61
  PT-109 and, 24, 151
  public image of, 55–56
  ship pictures of, 262
  tributes to, 269–71
  wedding of, 1, 24
Kennedy, John F. Jr. (son)
  birthday party for, 61, 65, 217, 254
  and film thanking public for sympathy
    letters, 278
  at funeral/memorial services for JFK, 245,
    249
  Georgetown home of, 279
  and good-byes for Texas trip, 65, 66
  Jackie's talk about JFK's death with, 204
  and JFK body in East Room of White House,
    203
  JFK love for, 270
  JFK playing with, 56
  LBJ letter to, 195, 258
  and LBJ transition to White House, 205
  and move out of White House, 272, 274
  and photographs of JFK, 275
  Powers photograph with, 277
  and public image of JFK, 55
  salute to JFK of, 249
  and White House school, 262
Kennedy, Joseph "Joe" (father), 25, 28, 66
Kennedy, Patrick Bouvier (son), 57, 60, 62

Kennedy, Robert "Bobby" (brother)
    and Air Force One return to Washington,
        194
    at Arlington National Cemetery, 206, 254,
        256
    assassination of, 285–86, 298
    Capitol Rotunda visit by, 245–46
    Cuba interest of, 41
    and Dallas ad attacking JKF administration,
        85
    and death of JFK, 166–67, 168
    elections of 1964 and, 281–82, 283, 284, 285
    elections of 1968 and, 285
    and funeral procession from White House to
        St. Matthew's Cathedral, 247
    Harvard and, 287
    Hoover's relationship with, 166–67
    and Jackie's film at Justice Department, 277
    Jackie's relationship with, 274, 286
    and JFK artifacts at National Archives, 293
    and JFK autopsy at Bethesda Naval Hospital,
        196
    and JFK body at White House, 201, 242–43
    JFK's relationship with, 270
    and LBJ as vice-presidential selection, 60
    LBJ's relationship with, 188, 259, 281–82,
        283, 284
    and Oswald's arrest, 196
    public service of, 275
    Skelton letter about Dallas trip to, 64
    Walker (Edwin) case and, 12
    and White-Jackie interview, 268
Kennedy, Rose (mother), 192–93
Kennedy Library, 275, 277, 278, 279, 286–87
Khrushchev, Nikita, 32
Kilduff, Malcolm, 158, 161–62, 163
King, Coretta Scott, 285
King, Martin Luther Jr., 34, 35, 285, 298
Korean War, 12, 135

Latin Listening Post (WDSU radio program):
    Oswald as guest on, 41–44, 45
Lawford, Patricia Kennedy, 206, 270
Lawton, Don, 102–3
League of United Latin American Citizens
    (LULAC), 80, 81
Leavelle, James, 231, 233, 234, 235–36
Lee, Robert E., 205
Lee, V.T., 50, 51
Lenin, Vladimir, 25
Life magazine
    and anniversaries of JFK assassination, 284
    "Four Days" article in, 270
    Jackie on cover of, 279

    and JFK assassination as shared event, 291
    and Oswald's funeral, 253
    White's Camelot article in, 266–69, 284
Limited Nuclear Test Ban Treaty (1963), 59
Lincoln, Abraham
    anniversaries of assassination of, 290
    Booth's escape after assassination of, 157
    and Booth's military tribunal, 213
    Booth's motivations for assassinating, 298
    congressional memorial service for, 244
    and Ford's Theatre as museum, 288, 291–92
    funeral/funeral procession for, 188, 217, 243
    Grosvenor story about, 280
    JFK assassination compared with
        assassination of, 69
    JFK compared with, 276
    pistol used in assassination of, 69
    public knowledge about assassination of, 271
    public reaction to assassination of, 262–63
    relics/souvenirs of assassination of, 291–92
    Second Inaugural Address of, 276
    statue of, 245
    and swearing in of vice president, 176
    White House viewing of body of, 202
Lincoln, Evelyn, 102, 196, 205, 293
Lincoln, Mary Todd, 205, 271
Lincoln Memorial
    candlelight tribute to JFK at, 275–76
    Jackie at, 252
Lipscomb (David) High School (Nashville,
        Tennessee): reactions to JFK assassination
        at, 164–66
Little Rock, Arkansas: civil disturbances in, 12
Loewy, Raymond, 65
Look magazine, 270, 280, 284, 291
Love Field (Dallas, Texas). See Dallas, Texas;
    Dallas, Texas—JFK trip to
Lowe, Jacques, 2

MacNeil, Robert, 146
Mafia: and conspiracy theories about JFK
    assassination, 296
The Making of the President: 1960 (T. White), 266
Mansfield, Mike, 244–45
map, Oswald's, 226–27
March on Washington (August 28, 1963), 35
Marine Corps, U.S.
    as honor guards for JFK body at White
        House, 202, 242
    Oswald in, 9, 22, 38–39, 54, 59, 70, 106, 124,
        136, 150
    Oswald's discharge from, 39, 40, 174, 226
    and Oswald's trips to Mexico, 186
    and questioning of Oswald, 189, 226

Marx, Karl, 18, 225
Marxism
    Communism distinguished from, 47, 225
    Oswald's interest in, 37, 48, 225
mass: for JFK, 204–5
McCarthy, Eugene, 285
McCarthy, Joseph, 26
McCormack, John W., 162
McCoy, C.C., 220
McGraw, Preston, 253
McGrory, Mary, 204, 285
McHugh, Godfrey, 187
McIntyre, Bill, 154
McKinley, William, 69
McLean, Randy, 165
McLendon, Gordon, 163
McNamara, Robert, 206, 250, 272, 274–75,
    276–77, 281, 282, 286
media
    and confirmation of JFK's death, 162–63
    coverage of JFK assassination by, 151, 152,
        156, 159, 162–64
    coverage of JFK memorial services/funeral
        by, 240
    and Curry's press conference, 223–24
    Dallas Police relations with, 217, 221, 232
    and death threats against Oswald, 223
    and Fritz's "cinched" interview, 213–14, 242
    initial reactions to JFK assassination by, 146,
        150
    Jackie and, 264–65, 269, 278–79
    Jackie-Onassis marriage and, 287
    JFK's relationship with, 284–85
    LBJ's oath of office and, 176–77
    and Oswald in Dallas Police custody, 182,
        193, 197, 198–200, 213–14
    Oswald's arraignment and, 197
    and Oswald's arrest for Tippit murder, 182
    at Oswald's graveside memorial service, 253
    Oswald's hallway interviews/comments to,
        197, 198–200, 215–16, 233, 236, 242
    and Oswald's pro-Cuban activities, 40–41, 52
    at Parkland Hospital, 154–55
    and report of JFK's death, 152
    and Ruby's shooting of Oswald, 234–35,
        237, 239
    Tippit's murder and, 182
    and transfer of Oswald from city jail to
        county jail, 218–19, 220, 221, 223–24, 231,
        232, 233–35, 242
    Walker (Edwin) shooting and, 17–19
    Zapruder movie and, 217
    See also specific reporter or organization
Medical Museum, U.S. Army, 292

Meredith, James, 12, 34
Mexico
    Oswald's trip to, 58
    questioning of Oswald about, 186, 191
midnight press conference, Oswald's, 198–200
Militant journal, 6, 225
military tribunal
    for Booth (John Wilkes), 213
    rumors about Oswald before a, 212–13
military, U.S.: budget for, 31
missile defense
    elections of 1960 and, 27
    See also Cuban missile crisis
Mohrenschildt, George de, 21, 22
Mohrenschildt, Jeanne, 21
Moorman, Mary, 135
Morgenthau, Robert, 167
mourning buttons, JFK, 255
Moynihan, Daniel Patrick, 204, 246, 285
"Mrs. Kennedy Speaks. Thanks 800,000 Who
    Sent Sympathy" (film), 277–78

Nashville Tennessean, 165
National Aeronautics and Space Administration
    (NASA), 33
National Archives
    JFK assassination relics at, 291, 292–93
    JFK papers and, 287
national day of mourning: LBJ proclamation
    for, 218
National Geographic magazine: special
    assassination issue of, 279–80
National Park Service, 291
National Scholastic Press Association, 165
National Service Corps, 11
Navy, U.S., 40
NBC
    coverage of JFK assassination by, 291
    and Ruby's shooting of Oswald, 234, 239–40
    See also specific journalist
New Frontier, 2, 28, 71
New Orleans
    Oswald in, 21–22, 39–53, 263
    Oswald leaves, 58
    Oswald (Marina) in, 21–22, 39–40, 53, 57
    Oswald on talk radio in, 40–50, 263
    as Oswald's birthplace, 37
New Orleans Police Department, 51
New York City: Jackie's move to, 280
New York Times, 158
Newman family, 135
Newsom, Milton L., 220
Newsweek magazine, Bradlee article about JFK
    in, 269–70

Nixon, Richard M., 24–25, 26–28, 53, 109
Norman, Harold "Hank", 113–14, 117, 129, 130, 131, 147, 150
Novello, Angie, 166–67, 293
nuclear weapons: JFK views about, 58–59

O'Brien, Larry, 180
O'Donnell, Kenneth
  and Dallas motorcade, 108, 118, 122
  and initial reactions to assassination shots, 133, 139–40
  and JFK's comment about Dallas, 96
  and JFK's fatalism, 85
  and LBJ's swearing in as president, 178
  at Parkland Hospital, 154, 155, 158
  and return to Washington from Dallas, 170, 180, 187
Onassis, Aristotle, 287
Oswald: Portrait in Red (album), 263
Oswald, June (daughter), 6, 8, 14, 20, 21, 39, 77, 78, 79, 82, 88, 211, 212, 214, 221
Oswald, Lee Harvey
  "Alek" as Marina's nickname for, 8, 209
  arrests of, 40, 44, 51, 52, 181–82, 196, 207
  burial of, 240, 253
  childhood and youth of, 6, 37–38
  and choice of weapons for assassinations, 4
  death of, 239
  dying declaration by, 236
  education of, 37–38, 50–51
  exhumation of, 253
  fantasies of, 53–54, 67
  financial affairs of, 1, 9–10, 14, 48, 72, 78, 86–87, 140
  fingerprinting of, 199
  jobs of, 9–10, 39, 40, 48, 59
  lawyer for, 192, 196, 200, 208, 215, 216, 221
  magazines devoted to, 261
  Marina's last dinner with, 80
  Marina's marriage to, 39
  Marina's relationship with, 57, 59, 77, 78, 79, 81–82, 149
  marksmanship of, 39, 57–58, 69, 70, 72, 110, 121–22, 128, 136, 150, 189
  paraffin test for, 199
  personal background of, 6, 37, 45
  public image of, 295
  rage of, 6–9
  self-image of, 4, 7, 22, 51, 52, 54
  stealing of tombstone of, 253
  suicide attempt of, 39
  turning points for, 22
  views about JFK of, 53, 67, 209, 212
  wedding ring of, 87, 140
    writings of, 50–51
    See also assassination of JFK; Hidell, Alek James; Oswald, Lee Harvey—shooting of; rifle, Oswald's; specific topic
Oswald, Lee Harvey—shooting of
  Dallas Police reputation and, 288
  and death of Oswald, 239–40
  and dying declaration by Oswald, 236, 239
  and Oswald at Parkland Hospital, 237, 238–39
  public announcement about, 237, 239–40
  public reaction to, 238, 239
  by Ruby, 232–40
  Ruby's justification for, 236–37
  souvenirs/relics of, 292, 294
Oswald, Marguerite (mother), 6, 39, 210–11, 240, 253, 263, 297
Oswald, Marina Prusakova (wife)
  abuse of, 6–7
  and "Alek" as nickname for Oswald, 209
  Buell's information for police about, 174
  Dallas Police questioning of, 187
  and death of Oswald, 236, 240
  FBI visits to, 59–60, 184–85, 209
  fears of, 210
  and "ifs" concerning JFK assassination, 149
  as interested in JFK visit to Dallas, 78–79
  Irving, Texas move of, 57
  Jackie Kennedy as interest of, 78, 210
  and media report about JFK assassination, 150–51, 186–87
  Mohrenschildts as friends of, 21
  in New Orleans, 21–22, 39–40, 53, 57
  and Oswald's fantasies, 53–54
  at Oswald's graveside ceremony, 253
  Oswald's Irving visits with, 59, 75–79, 81–82, 86–88, 140, 227, 289
  Oswald's last dinner with, 80
  Oswald's last meeting with, 210–12
  Oswald's last night with, 81–82
  Oswald's marriage to, 39
  Oswald's note to, 13–15, 17, 20, 141, 212
  and Oswald's post office box, 222
  Oswald's prison visit with, 210–12, 221, 297
  and Oswald's pro-Cuban interests, 40, 53, 54
  Oswald's relationship with, 6–9, 13–14, 20, 57, 59, 77, 78, 79, 81–82, 149, 189
  Oswald's rifle and, 5–6, 20, 52–53, 57, 75 77, 149, 151, 186–87
  and Oswald's (Robert) visit with Oswald, 214
  and Oswald's role in JFK assassination, 212
  and Oswald's sentimental Russian poetry, 82
  and Oswald's wedding ring, 87, 140
  Paine and, 57

Oswald, Marina Prusakova (wife) (*cont.*)
  and photographs of Oswald with his
    weapons, 5–6, 22, 232
  and questioning of Oswald, 184, 189, 208,
    221
  suicide attempt of, 8
  views about JFK of, 53, 209
  Walker assassination attempt and, 16–17, 18,
    19–20, 21, 78, 110, 141, 149, 212
Oswald, Rachel (daughter), 59, 77, 78, 79, 82,
    88, 211, 212, 214, 221
Oswald, Robert (brother), 39, 67, 70, 214–15,
    221, 240, 253, 297
Otten, Al, 93
Oval Office
  Jackie's suggestions for redecoration of,
    261–62
  LBJ transition to, 205

Paine, Ruth
  Dallas police visit home of, 187
  and Frazier's testimony against Oswald, 174
  and "ifs" concerning JFK assassination, 149
  Marina's home with, 57, 59, 75–78, 83, 87,
    88, 140, 187, 189, 208
  and media report about JFK assassination,
    151, 186, 187
  and Oswald's job at Texas School Book
    Depository, 59, 149
  Oswald's rifle at home of, 57–58, 75–79, 83,
    86–88, 140, 149, 151, 174, 186–87, 197,
    208, 289
  and Oswald's visits with Marina in Irving,
    59, 75–79, 81–82, 83, 86–88, 140, 187,
    207, 208, 227, 289
  as suspicious of Oswald, 151
Palm Beach, Florida: attempted assassination
    of JFK in, 86
papers, JFK, 286–87, 293
Pappas, Ike, 234, 235, 237
Parkland Hospital (Dallas)
  cleanup of presidential limousine at, 159–60
  Connally at, 152, 153, 238, 294
  Jackie at, 153–54, 155, 158, 160–61, 170, 175,
    244, 247
  JFK at, 153–56, 158–61, 169–70, 175, 238
  JFK ride to, 143, 148, 152–53
  Oswald at, 237, 238–39, 240
  public announcement of death of JFK at,
    163
  and White-Jackie interview, 267
Peace Corps, 11, 31
Pettit, Tom, 233, 235
Pettus, Patty, 165

pistol, Oswald's
  Dallas police finding of bullets for, 193–94
  and decision not to take pistol to Texas
    School Book Depository, 76–77
  at National Archives, 292
  and Oswald's arrest, 181
  and Oswald's (Robert) visit with Oswald, 214
  at Oswald's rooming house, 75
  photographs of Oswald with, 5–6
  purchase of, 4–5, 206, 208
  questioning of Oswald about, 190, 208
  and Tippit's murder, 171–73, 206
pistols
  as assassination weapons, 69, 85
  and Lincoln's assassination, 69, 291–92
  of Ruby, 294
polygraphs tests: for Oswald, 209
Pony Express (Lipscomb High School paper),
    164–66
Post Office, U.S.: and questioning of Oswald
    about post office box, 222–23, 224–25
Powers, Dave
  defense of Jackie by, 93
  and initial reactions to shooting, 133,
    139–40
  Jackie's Christmas gift to, 277
  as Jackie's Hyannis Port houseguest, 267
  and JFK trip to Dallas, 96, 107–8
  and JFK's Dallas motorcade, 118, 122
  John Jr. photograph with, 277
  at Parkland Hospital, 153, 158, 160
  and return to Washington from Dallas, 180,
    187
presidency: JFK's views about, 62
presidential limousine
  bubble top for, 91, 102, 112–13, 149–50
  cleanup of, 159–60
  as collector's item, 293–94
  as crime scene, 160
  for Dallas motorcade, 67, 77, 91, 98, 99,
    102–3, 107, 112–13, 149–50
  for Fort Worth trip, 91
  and Germany trip, 91
  hardtop for, 159–60
  for Houston motorcade, 77
  and JFK's preference for open car, 91, 107,
    150
  for LBJ, 294
  Lincoln convertible as, 67, 71, 77, 91, 107,
    112–13, 149–50, 195–96, 293–94
  Queen Mary (Cadillac), 71, 102, 142, 175
  return to Washington of, 195–96
  and ride to Parkland Hospital, 156
  for San Antonio motorcade, 71

Secret Service agents as riders on, 102–3, 107, 113

and souvenirs/relics of JFK assassination, 263

Presidential Medal of Freedom

   JFK awarded, 273–74

   JFK makes awards for, 61, 62

Prusakova, Marina. *See* Oswald, Marina Prusakova

Puerto Rican nationalists: shootings by, 69

questioning of Oswald, 182–84, 185–86, 188–92, 193, 196–97, 206–10, 213–14, 221–23, 224–26, 297

Quinn, Mike, 96–97

radio, New Orleans, Oswald on, 40–50, 263

Randall, Linnie Mae Frazier, 87, 149, 227

Rather, Dan, 162–63

Reid, Mrs. Robert A., 138, 145

Reston, James, 284–85

Revolutionary Student Directorate, 44

revolver, Oswald's. *See* pistol, Oswald's

Rickerby, Art, 98

Riedel, Richard, 167–68

rifle, Oswald's

   ammunition for, 83

   Dallas police discover, 206

   as first use of rifle in presidential assassination, 141

   hiding of, 3, 4, 10, 17, 57–58, 75–79, 83, 86–90, 110, 149, 179, 289

   and "ifs" concerning JFK assassination, 149

   importance to Oswald of, 21

   JFK assassination and, 73–77, 105–6, 113, 114, 118, 121–43, 144

   and manhunt for Oswald, 171

   Marina's concern about, 5–6, 20, 52–53, 57

   and Marina's move to Irving, 57–58, 75–79, 83, 86–90, 149

   at National Archives, 292

   and Oswald in New Orleans, 21, 39–40

   Oswald's care/cleaning of, 52–53

   and Oswald's feelings about assassinations, 172

   and Oswald's interviews with media, 216

   and Oswald's marksmanship, 39, 57–58, 69, 70, 72, 110, 121–22, 128, 136, 150

   and Oswald's plans for assassination of JFK, 73–77

   Oswald's practice with, 9, 40, 52–53

   and Oswald's (Robert) visit with Oswald, 214

   at Paine's house in Irving, 57–58, 75–79, 83, 86–88, 140, 149, 151, 174, 186–87, 197, 289

photographs of Oswald and, 5–6, 10, 22, 221–22, 232

purchase of, 4–5, 74, 206, 208, 222–23, 229

questioning of Marina about, 211

questioning of Oswald about, 185–86, 191, 197, 207–8, 222–23, 229

at Texas Book Depository, 89–90, 185, 207, 235

Walker assassination and, 3, 10–11, 16, 17, 18–19, 52–53, 57–58, 110, 128

and when to shoot JFK, 121–24

Roberts, Charles, 99

Roberts, Emory, 103, 154

rooming house, Oswald's Dallas, 75, 78, 157, 171, 184, 189, 190, 192, 208, 226–27, 232, 289

Roosevelt, Eleanor, 12

Roosevelt, Franklin D., 69, 119, 251

Roosevelt, Franklin D. Jr., 267

Roosevelt, Theodore, 69, 119

Rose, Earl, 169–70

Rowland, Arnold, 114–16, 118–19, 126, 132, 149, 150

Rowland, Mrs. Arnold, 119, 149

Rubenstein, Jack. *See* Ruby, Jack

Ruby, Jack, 232–40, 288, 292, 294

Russia. *See* Soviet Union

Salinger, Pierre, 64

San Antonio, Texas: JFK campaign trip to, 60, 70–72, 82

Saturday Evening Post magazine, 270, 280

Saturday Review magazine, 270

Scobey, Lola Sue, 165

Seaport Traders, 4

Secret Service

   autopsy on JFK and, 169–70

   brain of JFK and, 293

   casket for JFK and, 166

   and cleanup of JFK presidential limousine, 159–60

   conspiracy theories about JFK assassination and, 296

   and death of JFK, 152, 161

   flight to Texas and, 65

   and initial reactions to assassination shots, 129, 133, 134, 142–43

   JFK ride to Parkland Hospital and, 152

   JFK trip to Fort Worth and, 90–91, 92–93

   and JFK's Dallas motorcade, 94, 95, 98, 99, 102–3, 107, 108–10, 113, 115–16, 117–18, 121–22

   JFK's relationship with, 107, 170

   and LBJ protection, 154, 161–62, 248–49

Secret Service (cont.)
and manhunt for Oswald, 173–74
Oswald's decision about when to shoot JFK
and, 121–22
Oswald's escape from Texas Book
Depository and, 145
and Oswald's motives for assassinating JFK,
297
Oswald's plans for assassination of JFK and,
73
and Parkland Hospital, 148, 153, 154, 155,
159, 169–70, 175
presidential limousines and, 71, 142, 159–60,
195–96, 293–94
presidential motorcades and, 71, 150
and questioning of Oswald, 183, 191, 192,
206, 209–10, 225–26, 229, 297
and return of presidential limousine to
Washington, 195–96
and return to Washington from Dallas, 175,
176
weather concerns of, 90–91, 102
See also specific agent
Seeger, Alan, 86
Seigenthaler, John, 165–66
Selective Service cards, Oswald's, 209, 229
Senate, U.S., LBJ-JFK relationship and, 259–60
Seward, William, 213
Shaw, Mark, 2
Shaw, Maud, 204
Shelley, Bill, 183
Shepard, Taz, 202
Sheriff's Office, Dallas: and death threats
against Oswald, 219, 220
Sidey, Hugh, 160
Skelton, Byron, 63–64
Slatter, Bill, 44–50
Smith, Jean Kennedy, 206
Smith, Merriman, 137, 146, 152, 154–55,
158–59, 163, 179
socialism: Oswald's interest in, 37
Sorrels, Forrest, 148, 191, 192, 229, 232
Southeast Asia, Communism in, 31
souvenirs/relics
of JFK assassination, 261, 262, 263, 291–94
of Lincoln assassination, 262–63, 291–92
Soviet Union
Bay of Pigs invasion and, 30
Communism in, 25–26
Cuban missile crisis and, 31–32
Cuban relations with, 30, 42, 45
elections of 1960 and, 28
Germany and, 35–36
and "ifs" concerning JFK assassination, 149

JFK's views about, 24, 58–59
media questioning of Oswald about, 199
Nixon's views about, 24
nuclear weapons and, 58–59
Oswald as puppet of, 49
and Oswald's attempt to visit Cuba, 58
Oswald's interest in, 37, 174, 225, 297
and Oswald's motivation for assassinating
JFK, 297
Oswald's trip to, 10, 39, 40, 45, 46, 48, 49, 52,
67, 174, 226, 295
Oswald's views about, 45, 297
questioning of Oswald about, 186, 190, 192,
225, 226
space program and, 32–33
U.S. competition with, 32
in World War II, 26
space programs, 32–33, 72
Spalding, Chuck, 267
Sputnik space program, 32
St. Matthew's Cathedral, 247, 248, 258
Stalin, Joseph, 25
Stanton, Frank, 270–71
State Department, U.S.: Cuban relations and, 49
Steensland, Bill, 165
Steichen, Edward, 273
Stevenson, Adlai, 63, 115, 116, 240
Stoughton, Cecil, 176, 178–79, 180
Stuckey, William, 41–44, 45, 47, 48, 50
Supreme Court, U.S., 33, 61, 62, 177, 273

Talbert, C.E., 220–21
taxes: as concern in early years of JFK
presidency, 33
technology: and conspiracy theories about JFK
assassination, 296–97
television
elections of 1960 and, 27
and JFK assassination as shared event, 291
Texas
LBJ's hostile reception in, 240
as victim of Oswald's actions, 240–42
Texas, JFK trip to
and flight to Texas, 65–66
Jackie's wardrobe for trip to, 64–65
JFK plans for campaign trip to, 60, 62, 63–64
warnings about JFK's trip to, 63–64, 240–41
See also specific city
Texas School Book Depository (Book
Depository)
discovery of Oswald's rifle at, 206
and JFK motorcade route, 68, 70, 79, 95,
111–19, 121–28
and manhunt for Oswald, 174

as museum, 288–90
Oswald as employee at, 59, 68–69, 70, 79, 112, 144, 149, 183, 184, 189–91, 226–27, 228
and Oswald's comments to the media, 216
and Oswald's decision to assassinate JFK, 68–69, 70
Oswald's escape from, 144–45, 146, 147–48, 157, 225, 289
Oswald's map and, 226–27
and Oswald's plans for assassination of JFK, 72–77
Oswald's rifle at, 89–90, 185, 207, 235
and Oswald's shooting of JFK, 123–41
and Oswald's wait for the motorcade, 105–6, 110–20, 121–23
and Paine's and Marina's concerns about role in assassination, 151
post-assassination activities at, 156
questioning of Oswald about, 189–91, 207–8, 227–29
roll call of employees at, 158, 173
and souvenirs/relics of JFK assassination, 263
visitors to, 289–90
Texas Theatre
Oswald's arrest at, 181–82, 183, 185, 207, 237, 289
and questioning of Oswald, 207
Thomas, Albert, 77, 81
Thomas, George, 79
Time magazine, 160
Tippit, J.D.
arraignment of Oswald for murder of, 196–97, 200
burial of, 252–53
and conspiracy theories about JFK assassination, 296
Oswald's actions after murder of, 180–82
Oswald's arrest for murder, 181–82
and Oswald's comments to the media, 215
Oswald's denial about murdering, 207, 222
pistol involved in murder of, 171–73, 206
and plans for Oswald's trial, 213
questioning of Oswald for murder of, 182–86, 190
reasons for murder of, 173–74
and shooting of Tippit, 171–73, 289
Tobias, Mahlon, 9
Trade Mart (Dallas)
JFK planned lunch and speech at, 67, 68, 82–83, 94, 95, 98, 103–4, 105, 107, 122, 156, 158
Oswald's demonstrations at, 40

transfer, Oswald's: from City Hall jail to County jail, 218–19, 220–21, 223–24, 229, 230–35, 242
Treasury, U.S. Department of the, 272
trial, Oswald's
and Oswald's death, 242
Wade's comments about plans for, 212–13
tributes to JFK
Kennedy Library funding and, 279–80
at Lincoln Memorial, 275–76
Truly, Roy, 144, 145, 185
Truman, Harry S., 12, 69

United Nations, 63
United Press International (UPI), 146, 151, 159, 163, 253
United States
and American exceptionalism, 103–4
Oswald's views about, 67
University of Alabama: desegregation at, 34
University of Mississippi: desegregation at, 12, 34

van der Rohe, Mies, 273
Vietnam, 31, 85, 285, 298
Vogue magazine: Jackie and, 262

Wade, Henry, 183, 212–13, 215, 221
Walker, Bob, 94–95
Walker, Edwin A.
civil rights movement and, 12
Communism and, 11–12
and JFK relief of Walker from command, 12
media coverage of shooting of, 17–19
Oswald at rally with, 59
Oswald's attempted assassination of, 13, 15–16, 17, 39, 52–53, 57–58, 70, 78, 110, 128, 129, 141, 146, 149, 157, 212, 242
Oswald's fixation on, 11–12
Oswald's justification for assassination of, 12
Oswald's plans for assassination of, 3–4, 10–12, 19
Oswald's reactions to shooting, 17–20, 21, 22, 110
professional background of, 12
souvenir of shooting of, 19–20
and turning points for Oswald, 22
Wall Street Journal, 93
Wallace, George, 34
War Is Hell (movie), 181
Warren Commission, 264, 283, 284, 292, 295–97
Warren, Earl, 241–42, 264, 284

Washington, D.C.
and anniversaries of Lincoln assassination,
290
Kennedys and Johnsons return to, 169–70,
174–78, 187–88, 192–93, 194–95
return of presidential limousine to, 195–96
Washington, D.C, *See also specific location*
Washington Star
Bartlett article about Jackie in, 264–65
Jackie as reporter for, 287
WDSU-TV/radio (New Orleans), Oswald on,
40–50, 263
West, B.C., 274
Western Cartridge Company, 9
White, Byron, 62, 286
White House
funeral procession from Capitol to, 247–48
funeral procession to St. Matthew's from,
247, 248–49, 258
Jackie's inscription on marble mantle in, 272
Jackie's post-funeral stay at, 249–50, 252,
260–62, 271–74
Jackie's refusal to visit, 280, 282
Jackie's return from Texas to, 202–3
and Jackie's suggestions for redecorating the
Oval Office, 261–62
Jackie's thank-you notes to staff at, 272
JFK body in East Room of, 201, 202, 203,
219, 242–43

JFK-Jackie's last night in, 61
JFK motorcade from Bethesda Naval
Hospital to, 201–2
JFK receptions and dinners at, 61
LBJ invitations to Jackie to return to,
282
LBJ move into, 205
Lincoln family stay in, 205, 271
Marine honor guard for JFK body at, 202,
242
renovation of, 56
school for children at, 262
White, Theodore, 266–69, 294–95
Whitman, Walt, 290
Wicker, Tom, 158
Wilder, Thornton, 273
Williams, Bonnie Ray, 113–14, 117, 127–28,
129, 130, 131, 147, 150
Wilson, Edmund, 273
Worker journal, 6, 11, 12
World War II, 12, 24, 26, 35, 58, 135, 151, 166,
181
Worrell, James, 126–27, 130
Wyeth, Andrew, 273

Youngblood, Rufus, 143, 154, 162, 174

Zapruder, Abraham, 119–20, 122, 124, 126,
130, 133, 135, 143, 217, 290

ABOUT THE AUTHOR

Photo by Lisa Nipp

James L. Swanson is the author of the *New York Times* bestsellers *Manhunt: The 12-Day Chase for Lincoln's Killer* and its sequel, *Bloody Crimes: The Funeral of Abraham Lincoln and the Chase for Jefferson Davis*. *Manhunt* won an Edgar Award for the best nonfiction crime book of the year. Swanson's other books include the bestselling classic *Chasing Lincoln's Killer*, an adaptation of *Manhunt* for young readers, and *Bloody Times*, the young adult version of *Bloody Crimes*. His pictorial book *Lincoln's Assassins: Their Trial and Execution*, is an acclaimed photo history of the crime, the pursuit of the conspirators, and their fates. He was awarded a Historic Deerfield Fellowship in Early American History, and he serves on the advisory council of the Ford's Theatre Society. He has degrees in history and in law from the University of Chicago and UCLA, and he has held a number of government and think-tank posts in Washington, D.C., including at the United States Department of Justice. Follow him on Twitter @JamesLSwanson.

KENNEDY EXTRA
PRESIDENT DEAD
Shot by Assassin
American
ghlights of Kennedy's Career